# Recent Advances in Riemannian and Lorentzian Geometries

CONTEMPORARY MATHEMATICS

337

# Recent Advances in Riemannian and Lorentzian Geometries

AMS Special Session
Recent Advances in Riemannian and Lorentzian Geometries
January 15–18, 2003
Baltimore, Maryland

Krishan L. Duggal
Ramesh Sharma
Editors

American Mathematical Society
Providence, Rhode Island

2000 *Mathematics Subject Classification.* Primary 53C25, 53C30, 53C40, 53C50, 53C55, 83C25, 83C75; Secondary 22E25, 35L30, 53D10.

---

**Library of Congress Cataloging-in-Publication Data**

Recent advances in Riemannian and Lorentzian geometries / Krishan L. Duggal, Ramesh Sharma, editors.

p. cm. — (Contemporary mathematics, ISSN 0271-4132 ; 337)

Includes bibliographical references.

ISBN 0-8218-3379-0 (alk. paper)

1. Geometry, Riemannian—Congresses. 2. Geometry, Differential—Congresses. I. Duggal, Krishan L., 1929– II. Sharma, Ramesh, 1953– III. Contemporary mathematics (American Mathematical Society) ; v. 337.

QA641.R43 2003
516.3′73—dc22 2003062907

---



∞ The paper used in this book is acid-free and falls within the guidelines
established to ensure permanence and durability.
Visit the AMS home page at `http://www.ams.org/`

10 9 8 7 6 5 4 3 2 1 08 07 06 05 04 03

# Contents

# Preface

This volume covers the proceedings of a special session "Recent advances in Riemannian and Lorentzian geometries" of the annual meeting of American Mathematical Society, held at Baltimore, January 15–18, 2003. The speakers presented their research on Riemannian, Lorentzian, and pseudo-Riemannian manifolds.

The topics covered include classification of curvature-related operators, curvature-homogeneous Einstein 4-manifolds, linear stability/instability, singularity and hyperbolic operators of spacetimes, spectral geometry, cut loci of nilpotent Lie groups, conformal geometry of almost Hermitian manifolds and also submanifolds of complex and contact submanifolds.

This special session presented a great setting for differential geometers to interact among themselves and to expose the interplay/exchange between Riemannian and Lorentzian geometries. All the topics published in this volume were formally refereed.

We are grateful to the American Mathematical Society for the support in publishing this volume. Also thanks to all the participants and the contributors for their interest and work in publication of these proceedings.

Contemporary Mathematics
Volume **337**, 2003

# On Some Complex Manifolds with Torus Symmetry

Kinetsu Abe, Dimitar Grantcharov, and Gueo Grantcharov

## 1. Introduction

One of the first examples of complex non-Kähler manifolds are the products $S^{2n+1} \times S^{2m+1}$ of odd-dimensional spheres equipped with the Calabi-Eckmann complex structure. A standard approach to this construction is to consider $S^{2n+1} \times S^{2m+1}$ as the total space of a principal torus bundle over $\mathbb{C}P^n \times \mathbb{C}P^m$. Wang [**12**] generalized the construction and characterized all compact complex homogeneous spaces as the total spaces of torus bundles over Kählerian C-spaces (or generalized flag manifolds). Other generalizations of the Calabi-Eckmann construction were given by Abe [**1, 2**] and Blair-Ludden-Yano [**5**]. They used the tools of contact geometry to construct the so called bicontact manifolds in the latter two papers.

In the present note we consider the total spaces of principal torus bundles over Kähler manifolds. We start with the observation that if the fiber is even-dimensional and the characteristic classes are of type (1,1), then the total space admits a complex structure - a fact which is implicit in many papers. The aim of the paper is to consider two partial cases of this construction in more details. The construction of the complex structures makes use of connection 1-forms, a situation which resembles some properties of the contact geometry.

After the preliminaries which contain the general results, in Section 3 we consider a generalization of the bicontact structures. Analogously to the results of [**2**], we show that such structures satisfy a local Darboux Lemma and their complex subspaces admit fibrations. Another case of the torus construction is considered in Section 4 - the examples of manifolds constructed by Lopez di Medrano and Verjovsky [**9**] and Meersseman [**10**]. From [**11**], it is known that under an additional restriction the manifolds arise from the torus bundle construction above. Our observation is that they also appear as the zero set of a well-known moment map in symplectic geometry. We hope that this will be useful in the further investigation of their

1991 *Mathematics Subject Classification.* Primary 32L05, Secondary 53C15, 53D20.
*Key words and phrases.* toric bundles, moment map.
The first author is partially supported by NSF grant CCR-0226504.
The third author is supported by NSF grant DMS-0209306 and EDGE.

properties. Finally, in Section 5 we present an explicit example of the total space $M$ of a bundle over the complex surface $X = \mathbb{C}P^2\#2\overline{\mathbb{C}P^2}$. $M$ was also considered in [**11**], where it was implicitly identified with a $S^1$-quotient of $\#5S^3 \times S^4$. We identify $M$, using an old result by Goldsein and Linninger [**8**]. In particular, we obtain an explicit example of a complex non-Kähler six-manifold with vanishing first Chern class. The geometry of such manifolds became of interest recently because of its relation to string theory [**3, 4**].

## 2. Preliminaries

In this section we make some general observations about the even-dimensional torus bundles over complex manifolds. The next proposition is probably well-known, but for the sake of completeness we provide the proof. We assume that the Chern class of the bundle is identified with a k-tuple of classes in $H^2(X, \mathbb{R})$ and that a class is of type (1,1) if it admits a representative 2-form which is of type (1,1) with respect to the complex structure on $X$.

PROPOSITION 2.1. *Suppose that $M$ is the total space of an even-dimensional principal torus bundle over a complex manifold $X$ with characteristic classes of type (1,1). Then $M$ admits a complex structure. Moreover, any free holomorphic action of a compact complex torus with orbits which are complex submanifolds arises in this form.*

PROOF. (Sketch) Since we have a complex structure on the base and connection forms with curvature of type (1,1), we can construct an almost complex structure on the total space in the standard way:

i) On the horizontal spaces $\mathcal{H}$ we have the horizontal lift of the base complex structure.

ii) The vertical spaces $\mathcal{V}$ are tangent spaces to even-dimensional tori hence carry a natural complex structure.

Finally to check integrability we use the horizontal lifts of base (1,0) vector fields $X^h, Y^h$ and vertical (1,0) vector fields $V, W$. Then one can check that: $[V, W] \in \mathcal{V}^{(1,0)}$, $[V^h, X^h] = 0$ and $[X^h, Y^h] = \omega(X, Y) = 0$, where $\omega$ is the curvature form (vertical valued after some identifications). It is of type (1,1) and vanishes on any two (1,0) vectors.

For the converse statement we use a general result about locally free group actions that provide the structure of a principal torus bundle of $M$. Since the actions are holomorphic, the quotients are equipped with a complex structure. Finally, the characteristic classes of any holomorphic bundle are of type (1,1). □

Another feature of the above spaces is that they are in general non-Kähler. More precisely we have:

THEOREM 2.1. *Suppose that the base of the fibration $M \to X$ is Kähler and the characteristic classes of the fibration are linearly independent. Then the complex structure on $M$ is non-Kähler.*

PROOF. We first note that by taking a subbundle we may assume that the fiber is of dimension two. Then we see, using the spectral sequence arguments from Lemma 5.1 (see section 5), that the Betti numbers satisfy $b_2(M) = b_2(X) - 2$ for a general base $X$. Moreover, any class in $H^2(M, \mathbb{R})$ is a pull-back of a class in $H^2(X, \mathbb{R})$. Therefore, there is no class $\omega \in H^2(M, \mathbb{R})$ with $\omega^n \neq 0$. This implies that $M$ does not admit Kähler structure. □

Next we consider the Riemannian metrics on $M$. The following result was first proved by Berard-Bergery [**6**]:

THEOREM 2.2. *The total space of a torus bundle over a manifold with a positive Ricci curvature admits a metric with positive Ricci curvature provided that it has a finite fundamental group.*

## 3. Contact-type Structures and Torus Actions

Here we consider a special type of torus actions which generalizes the notion of bicontact structure of [**2, 5**].

Let $X_1, X_2, ..., X_{2k}$ be commuting non-vanishing vector fields on a Hermitian manifold $(M, J, g)$, preserving the Hermitian structure and satisfying $X_{2i} = JX_{2i-1}$ for $i = 1, ..., k$. Consider the forms $u_i$ defined via $u_i(X_j) = \delta_i^j$,and each $u_i$ vanishes on the orthogonal complement of $span\{X_1, X_2, ..., X_{2k}\}$ which is also a complex subspace.

DEFINITION 3.1. *We say that the action of a 2k-torus on $(M, J, g)$ is holomorphic totally non-degenerate if there are forms $u_i$ as above that satisfy the following conditions:*

*i) At least one of the forms $u_i$ has exterior derivative a (1,1) form (hence all have).*

*ii) There are integers $n_i$ not all 0 with $n_1 + n_2 + .. + n_{2k} = dim_c M - k$ such that the induced 1-forms satisfy $(du_1)^{n_1+1} = 0, (du_2)^{n_2+1} = 0, ..., (du_{2k})^{n_{2k}+1} = 0$, and that $u_1 \wedge (du_1)^{n_1} \wedge u_2 \wedge (du_2)^{n_2} .. \wedge u_{2k} \wedge (du_{2k})^{n_{2k}} \neq 0$ everywhere.*

One can easily check that $u_{2i} = Ju_{2i-1}$ and that the orbits are holomorphic submanifolds. When $k = 1$, we have the bicontact manifolds of [**2, 5**].

As in the bicontact case, we have:

THEOREM 3.1. *(Darboux Lemma) There are local coordinates $x_i^j$ on $M$ such that $u_i = dx_i^0 - \Sigma x_i^{2j-1} dx_i^{2j}$. In case when the base space X is a manifold, it is locally a product of Kähler manifolds.*

PROOF. Since the proof is identical to the one in [**5**], we only sketch the ideas. Define the subspaces $T_i$ of the tangent space $T_pM$ as

$$T_i = \cap_{\alpha \neq i} Ker(u_\alpha) \cap_{\alpha \neq i} Ker(du_\alpha) \in TM.$$

Then $\sum_i T_i = TM$ and $dimT_i|_p = n_i$. We consider $T_i$'s as subbundles and we easily see that they are involutive and that the forms $u_i$ are contact on the corresponding (transversal at each point) integral submanifolds. Applying the standard Darboux lemma completes the proof. □

In analogy to the bicontact case, we have the following Theorem:

THEOREM 3.2. *Every complex subvariety of $M$ is fibred by $l$-dimensional $l \geq 1$ subvarietes of the fibres of $\pi : M \to X$ .*

PROOF. Again we only sketch the proof which is analogous to the one given in [**2**]. If $N$ is a subvariety with the (necessary dense) set of regular points $R(N)$, then we have to prove that $T_xR(N) \cap T_x(\pi^{-1}(\pi(x)) \neq 0$. This is clear if $dim_cN \geq n-k$, where $n = dim_cM$. If $dim_cN \leq n-k$ and $N_c$ is not contained in a fiber, then there is an open set of points where $N_c$ is transversal to the fibers. From Definition 3.1(ii) follows that $d(u_1 + ... + u_{2k})^{n-k} \neq 0$ when evaluated on complex planes transversal to the vertical subspace. Hence $\int_N d(u_1 + ... + u_{2k})^{dim_cN} \neq 0$. On the other hand, it vanishes by Stokes' theorem. □

## 4. LM-V Manifolds and Moment Maps

Here we consider the manifolds constructed by Lopez de Medrano and Verjovsky [**9**], and generalized by Meersseman [**10**]. We give a geometric description for the case when they are torus bundles over complex manifolds. We also identify them with the zero sets of appropriate moment maps.

We start with the description of the LM-V manifolds. Fix integers $m$ and $n$ with $n > 2m$. Take $(\Lambda_1, ..., \Lambda_n)$ to be a set of $n$ vectors in $\mathbb{C}^m$ satisfying the following conditions:

i) The Siegel condition: $0 \in CH(\Lambda_1, ..., \Lambda_n)$, where $CH$ denotes the convex hull.

ii) The weak hyperbolicity condition: $0 \notin CH(\Lambda_{i_1}, ..., \Lambda_{i_{2m}})$ for any 2m-tuple $1 \leq i_1 \leq ... \leq i_{2m} \leq n$.

Such an n-tuple of vectors is said to form an *"admissible configuration"*. Then using a remark by Haefliger about the leaf space of a transversally holomorphic foliation, Lopez de Medrano and Verjorvsky [**9**] and Meersseman [**9**] constructed a complex structure on the following manifold:

$$M = \{[z] \in \mathbb{C}P^{n-1} | \Sigma_{i=1}^n \Lambda_i |z_i|^2 = 0\}.$$

Clearly $M$ is not a complex submanifold of $\mathbb{C}P^n$. The manifold has many interesting properties - see [**9, 10**]. Here we consider a partial case by imposing further the following restriction on $(\Lambda_1, ..., \Lambda_n)$

DEFINITION 4.1. *An admissible configuration fulfills condition (K) if for the real space of solutions of the system:*

$$\Sigma_{i=1}^n x_i \Lambda_i = 0, \Sigma_{i=1}^n x_i = 0, \tag{1}$$

*there exists a basis with integer coordinates.*

It is known that for any LM-V manifold there exists a natural fibration $G$ defined by commuting holomorphic vector fields. In the case that the given LM-V manifold satisfies condition (K), Meersseman and Verjovsky proved the following:

THEOREM 4.1. [**11**] *Let* $(\Lambda_1, ..., \Lambda_n)$ *form an admissible configuration satisfying condition (K). Then*

*i) The leaves of the fibration* $G$ *are compact complex tori and the quotient space is a toric projective variety with only orbifold singularities* $X(\Delta)$.

*ii) The projection* $M \to X(\Delta)$ *is a holomorphic Seifert bundle.*

Moreover, one can see that $X(\Delta)$ is a manifold if the configuration $\Lambda$ with condition (K) satisfies an additional assumption that the basis of (1) can be completed to a basis of $\mathbb{Z}^n$. In [**11**] the manifold $M$ has also been identified as a presymplectic (or Poisson) reduction of $\mathbb{C}^n$. Now we want to describe the fibration $M \to X(\Delta)$ in terms of "genuine" moment map of a torus action on the complex projective space $\mathbb{C}P^n$. This will also provide another proof of the above theorem for the smooth case and will lead to additional insight into the properties of $M$.

Consider the standard action of the real torus $T^n = \{(e^{2\pi i x_1}, e^{2\pi i x_2}, ..., e^{2\pi i x_n}) \mid x_i \in \mathbb{R}\}$ on $\mathbb{C}P^{n-1}$, which in homogeneous coordinates has the form:

$$[z_1, ...z_n] \to [e^{2\pi i x_1} z_1, ..., e^{2\pi i x_n} z_n].$$

The moment map of this action with respect to the Fubini-Study form is:

$$\mu([\mathbf{z}]) = (1/\Sigma_{i=1}^n |z_i|^2)(|z_1|^2, ..., |z_n|^2).$$

A more general form of the above action is the weighted $T^{2m}$-action on $\mathbb{C}P^{n-1}$ induced by $(\Lambda_1, ..., \Lambda_n)$:

$$[z_1, ...z_n] \to [e^{2\pi i <\alpha_1, \mathbf{x}>} z_1, e^{2\pi i <\beta_1, \mathbf{x}>} z_1, ..., e^{2\pi i <\alpha_1, \mathbf{x}>} z_n, e^{2\pi i <\beta_1, \mathbf{x}>} z_n],$$

where $\mathbf{x} = (x_1, ..., x_{2m}) \in \mathbb{R}^{2m}$, $\Lambda_j = \alpha_j + i\beta_j$. $< \cdot, \cdot >$ is the the standard bilinear form of $\mathbb{R}^{2m}$. The moment map for this action with respect to the Fubini-Study form is:

$$\mu_\Lambda([\mathbf{z}]) = (\Sigma_{i=1}^n \alpha_i |z_i|^2 / \Sigma_{i=1}^n |z_i|^2, \Sigma_{i=1}^n \beta_i |z_i|^2 / \Sigma_{i=1}^n |z_i|^2).$$

Then the zero level set of $\mu_\Lambda$ is exactly

$$Z = \{[\mathbf{z}] \mid \Sigma_{i=1}^n \Lambda_i |z_i|^2 = 0\}.$$

Thus we have proved:

THEOREM 4.2. *The LM-V manifold* $M$ *is diffeomorphic to* $Z = \mu_\Lambda^{-1}(0)$.

It is known that $Z \to X(\Delta)$ is a principal torus bundle in the smooth case and a Seifert fibration in the orbifold case [**7**]. From Theorem 2.1 we obtain a complex structure on $Z$.

A corollary for the Riemannian geometry of such bundles is:

COROLLARY 4.1. *The LM-V manifolds satisfying condition (K) with base Fano manifold admit a metric with positive Ricci curvature.*

## 5. Examples in Dimension Six

In this section $X = X(\Delta) = \mathbb{CP}^2 \# 2\overline{\mathbb{CP}^2}$. We fix the basis $\{\omega_1, \omega_2, \omega_3\}$ of $H^2(X, \mathbb{Z})$ dual to $\{\mathcal{O}(1), E_1, E_2\}$, where $E_1$ and $E_2$ are the exceptional divisors of $X$. The structure of $H^*(X, \mathbb{Z})$ is determined by the relations $\omega_1^2 = -\omega_2^2 = -\omega_3^2$ and $\omega_1 \wedge \omega_2 = \omega_2 \wedge \omega_3 = \omega_1 \wedge \omega_3 = 0$. In example 5.3 in [**11**], Verjovsky and Meerssemman show that the toric manifold $X$ is a base of a specific $T^2$-fiber bundle $N$ arising from their construction. We start with a more general case:

LEMMA 5.1. *Let $N$ with $\pi_1(N) = 0$ be the total space of a principal $T^2$-bundle over $X$ such that the Chern classes of the bundle are part of a set of generators of $H^2(X, \mathbb{Z})$. Then $H^2(N, \mathbb{Z}) = \mathbb{Z}$ and the cohomology ring of $N$ has no torsion.*

PROOF. There exist connection forms $(\theta_1, \theta_2)$ on $N$ with curvatures $(d\theta_1, d\theta_2)$. Since the curvatures represent the first Chern classes of the bundle, the condition in the lemma implies that we may assume $d\theta_1 = \omega_1, d\theta_2 = \omega_2$ . Then $\theta_1|_{T_x^2}, \theta_2|_{T_x^2}, \theta_1 \wedge \theta_2|_{T_x^2}$ generate the $\mathbb{Z}$-module $H^*(T_x^2, \mathbb{Z})$ when restricted to any fiber $T_x^2, x \in X$. This and the standard Lerray's theorem imply that the $E_2$-term of the Lerray spectral sequence for the bundle

$$T^2 \to N \to X$$

is:

| | | | | |
|---|---|---|---|---|
| $\mathbb{Z} =< \theta_1 \wedge \theta_2 >$ | 0 | $\mathbb{Z}^3$ | 0 | $\mathbb{Z}$ |
| $\mathbb{Z}^2 =< \theta_1, \theta_2 >$ | 0 | $\mathbb{Z}^2 \otimes \mathbb{Z}^3$ | 0 | $\mathbb{Z}^2$ |
| $\mathbb{Z}$ | 0 | $\mathbb{Z}^3 =< \omega_1, \omega_2, \omega_3 >$ | 0 | $\mathbb{Z} =< \omega_1^2 >$ |

Here $< e_1, ..., e_k >$ denotes the $\mathbb{Z}$-module generated by $e_1, ..., e_k$. For $d_2$ we have:

$$d_2 : E_2^{0,1} \to E_2^{2,0}, \theta_1, \theta_2, \to \omega_1, \omega_2$$

so it is an injection in this term. Similarly,

$$d_2 : E_2^{0,2} \to E_2^{2,1}; \theta_1 \wedge \theta_2 \to \omega_1 \wedge \theta_2 + \omega_2 \wedge \theta_1$$

is an injection. Consequently,

$$d_2 : E_2^{2,2} \to E_2^{4,1}; \theta_1 \wedge \theta_2 \wedge \omega \to \omega_1 \wedge \theta_2 \wedge \omega + \omega_2 \wedge \theta_1 \wedge \omega$$

for any $\omega \in H^2(X, \mathbb{R})$. In particular, it is a surjection in this term. Finally

$$d_2 : E_2^{2,1} \to E^{4,0}; \theta_i \wedge \omega_j \to \omega_i \wedge \omega_j$$

is surjective. Hence the $E_3$-term is:

| | | | | |
|---|---|---|---|---|
| 0 | 0 | $\mathbb{Z}$ | 0 | $\mathbb{Z}$ |
| 0 | 0 | $\mathbb{Z}^4$ | 0 | 0 |
| $\mathbb{Z}$ | 0 | $\mathbb{Z}$ | 0 | 0 |

In particular the spectral sequence collapses at the $E_3$-term and we complete the proof. ☐

Let's fix $N$ from Lemma 5.1. As a corollary of the lemma, the Betti numbers of $N$ are $b_0(N) = b_2(N) = b_4(N) = b_6(N) = 1$ and $b_3(N) = 4$, so $b_2(N) = b_2(X) - 2$ as claimed in Theorem 2.1. We need the following result about the classification of simply connected 6-manifolds with a free $S^1$-action:

THEOREM 5.1. [**8**] *Let $M$ be a compact smooth simply-connected 6-manifold with torsion-free cohomology and $b_2(M) = 1$. Then $M$ is diffeomorphic to one of the following:*

*i)$S^2 \times S^4 \# 2S^3 \times S^3$ if $M$ is spin and*

*ii) $S^2 \overline{\times} S^4 \# 2S^3 \times S^3$ if $M$ is not spin. Here $S^2 \overline{\times} S^4$ denotes the non-trivial $S^4$-bundle over $S^2$.*

From Lemma 5.1 and the Theorem 5.1, we can identify $N$ as soon as we know if it is spin or not. Thus, we obtain our main result:

THEOREM 5.2. *The spaces $S^2 \times S^4 \# 2S^3 \times S^3$ and $S^2 \overline{\times} S^4 \# 2S^3 \times S^3$ admit complex structures and Ricci positive metrics. The moduli space of the complex structures on each of them is disconnected. Moreover, the first space admits also a complex structure with vanishing first Chern class.*

PROOF. In order to see when $N$ is spin, we use the standard splitting $T^{(1,0)}(N) = \pi^* T^{(1,0)}(X) \oplus \mathbb{C}^2$, where $\mathbb{C}^2$ is the trivial tangent bundle to the fiber. Then $c_1(N) = \pi^*(c_1(X))$, and this class is proportional to the generator of $H^2(N, \mathbb{Z})$. In our case, $c_1(X) = 3\omega_1 + \omega_2 + \omega_3$. Changing the basis of $H^2(X, \mathbb{Z})$, we obtain both spin and non-spin manifolds $N$. More precisely, $\omega_1' := k\omega_1 + \omega_3$, $\omega_2' := \omega_2$, and $\omega_3' = (k-1)\omega_1 + \omega_3$ form a basis of $H^2(X, \mathbb{Z})$ for every $k \in \mathbb{Z}$. Then we can choose $\lambda := \pi^*(\omega_3')$ as a generator of $H^2(N, \mathbb{Z})$ and easily calculate:

$$\pi^*(c_1(X)) = \pi^*(3\omega_1 + \omega_2 + \omega_3) = \pi^*((4-k)\omega_1' + \omega_2' + (k-3)\omega_3') = (k-3)\lambda$$

Therefore, $N$ is spin manifold for an odd $k$ and non-spin for an even $k$. We can now identify $N$ using Theorem 5.1. In particular, we also obtain that for different $k$'s we have complex structures in different components of the moduli space, since they have different first Chern classes. The positivity of the Ricci tensor follows from Theorem 2.2. □

REMARK 5.1. *The existence of complex structures on the spaces in Theorem 5.2 is also obtained in* [**11**]. *However, the existence of different components of the moduli space as well as the existence of a complex structure with vanishing first Chern class has not been known. We also note that the previously known explicit examples of manifolds with a positive Ricci curvature on connected sums are the manifolds $\# kS^m \times S^m$.*

## References

[1] K.Abe, *A generalisation of Hopf fibration*, Tohoku Math. J. **30** (1978), 177–210.

[2] K.Abe, *On a class of Hermitian manifolds*, Invent. Math. **51** (1979), 103–121.

[3] K. Becker, M. Becker, K. Dasgupta, P. Green, *Compactifications of Heterotic Theory on non-Kähler complex manifolds*, hep-th/0301161.

[4] K. Becker, K. Dasgupta, *Heterotic strings with torsion*, J. High Energy Phys. **11** (2002) 006, 30 pp.

[5] D. E. Blair, G.Ludden, K.Yano, *Geometry of complex manifolds similar to Calabi-Eckman manifolds*, J. Diff. Geom. **9** (1974), 263–274.

[6] L. Berard-Bergery, *Certains fibrés à courbure de Ricci positive*, C. R. Acad. Sci. Paris **286** (1978), 929–931.

[7] V. Gulliemin, *Moment maps and combinatorial invariants of Hamiltonian $T^n$-spaces*, Progress in Math. 122, Birkhäuser, Boston (1994).

[8] R. Goldstein, L. Linninger, *Six manifolds with free $S^1$ action*, Lect. Notes Math. **298** (1972), 316–323.

[9] S.Lopez di Medrano, A.Verjovsky, *A new family of complex, compact, non symplectic manifolds*, Bol Soc. Brasil. Mat. **28** (1997), 243–267.

[10] L.Meersseman, *A new geometric construction of compact complex manifolds in any dimension*, Math. Ann. **317** (2000), 79–115.

[11] L.Meersseman, A.Verjovsky, *Holomorphic principal bundles over projective toric varietes*, preprint 2002.

[12] H.C.Wang, *Closed manifolds with homogeneous complex structure*, Amer. J. Math. **76** (1955), 1–32.

[13] K. Yano, M. Kon, *Structures on manifolds*, World Scientific (1984).

Department of Mathematics, University of Connecticut, Storrs, CT 06269, USA

*E-mail address*: abe@math.uconn.edu

Department of Mathematics, University of California, Riverside, CA 92521, USA

*E-mail address*: grandim@math.ucr.edu

Department of Mathematics, Florida International University, Miami, FL 33199, USA

*E-mail address*: grantchg@fiu.edu

Contemporary Mathematics
Volume **337**, 2003

# Curvature singularities and abstract boundary singularity theorems for space-time

Michael J. S. L. Ashley and Susan M. Scott

Abstract. The abstract boundary construction of Scott and Szekeres is a general and flexible way to define singularities in General Relativity. The abstract boundary construction also proves of great utility when applied to questions about more general boundary features of space-time. Within this construction an essential singularity is a non-regular boundary point which is accessible by a curve of interest (e.g. a geodesic) within finite (affine) parameter distance and is not removable. Ashley and Scott proved the first theorem linking abstract boundary essential singularities with the notion of causal geodesic incompleteness for strongly causal, maximally extended space-times. The relationship between this result and the classical singularity theorems of Penrose and Hawking has enabled us to obtain abstract boundary singularity theorems. This paper describes essential singularity results for maximally extended space-times and presents our recent efforts to establish a relationship between the strong curvature singularities of Tipler and Królak and abstract boundary essential singularities.

## 1. Introduction

There are many convincing reasons for using the abstract boundary construction ($a$-boundary) of Scott and Szekeres [**22**] when considering the issue of singularities for space-time as opposed to using the $g$, $c$ or $b$-boundary constructions of Geroch [**10**], Geroch, Kronheimer and Penrose [**11**], and Schmidt [**21**], respectively. The $g$, $c$ and $b$-boundaries are plagued with problems of a topological nature, and in certain circumstances produce non-intuitive results [**3, 23**]. An overview of the literature dealing with this issue has been given by Ashley [**1**, Chapter 2]. The common theme presented there is that the $a$-boundary provides a superior alternate structure in which to consider space-time singularities and other boundary features of space-time. Recently it has become apparent that the $a$-boundary formalism might allow the proof of curvature singularity theorems for space-time. This is a result of direct application of the abstract boundary classification scheme to

2000 *Mathematics Subject Classification.* Primary: 53C50, 83C75; Secondary: 53C23, 57R15.

*Key words and phrases.* singularity, abstract boundary, space-time.

This research was supported by an Australian Postgraduate Award (MJSLA) and in part by a Fulbright Postgraduate Award (MJSLA).

Theorem 3 of Hawking [**13**, p. 271][1]. The recent result of Ashley and Scott [**1**, Theorem 4.12] applies directly to the maximally extended space-times covered by Hawking's theorem and the link this forms with curvature appears exploitable to produce a curvature singularity theorem.

In this paper we briefly describe the present state of research into singularity theorems for space-time and our recent attempts to include curvature in this framework. We then formulate what we consider to be the absolute minimum considerations for the proof of a curvature singularity theorem applicable to physically realistic situations. Finally, we review how much of this structure is already in place and state which additional results are required to complete any such theorem.

## 2. Singularity Theorems for Space-time

The earliest attempts to produce theorems about the existence of singularities in space-time were negative in nature. It was believed that singularities in the then known cosmological and stellar model space-times were due to their highly symmetrical nature. The view that singularities would not be a feature of general solutions was put forward by Lifshitz and Khalatnikov [**18**] where it was proposed that Cauchy data giving rise to singularities comprised a set of measure zero in the set of all possible Cauchy data. Those authors later retracted this view in [**2**].

The first positive theorems about the existence of singularities were pioneered by Penrose [**20**] and Hawking [**12**]. Instead of dealing with the properties of solutions of the Einstein Field Equations directly, the Penrose-Hawking singularity theorems applied the techniques of differential topology and used only the most fundamental topological properties of the space-time and the behaviour predicted by General Relativity. In these singularity theorems timelike, null or causal geodesic incompleteness is deduced from a structure which is roughly composed of three parts:

(1) **Energy condition** — usually the strong energy or null convergence condition,
(2) **Causality condition** — the strong causality or chronology condition,
(3) **Closure or Curve Trapping condition** — these sometimes take the form of a topological condition (e.g. the presence of a compact spacelike hypersurface) or a trapping condition on the causal geodesics defined using a differential geometric construct such as the null second fundamental form or expansion scalar.

The energy condition guarantees that gravity is attractive for all matter and energy in the space-time. The causality condition prevents the escape of causal geodesics from the process of being focussed by curvature through causal violations. Finally, the trapping condition sets up the elements necessary for gravitational collapse or for some sort of cosmological or primordial singularity.

The primary reason for clarifying the status of these 'singularity theorems' is that, as mentioned above, there are two inequivalent philosophies for the definition of singular behaviour for space-time. Examples of space-time exist (see [**19**]) where geodesic incompleteness occurs without the presence of any curvature pathology. These situations typically involve *imprisoned incompleteness* [**13**, § 8.5, p. 289]

[1]It may be beneficial to the reader to compare this proof to that given in [**12**].

or *quasi-regularity* [**7, 8, 9**]. So in fact the term 'singularity theorem' is a misnomer since these theorems are actually geodesic incompleteness theorems. It is commonly believed that real singularities must be a type of curvature singularity and present astrophysical evidence supports the formation of singularities only in space-times derived from realistic collapse scenarios. In the case of primordial singularities, these might be formed from reasonable boundary assumptions (such as those arising from the Weyl curvature hypothesis of Penrose). Nevertheless, the Penrose-Hawking geodesic incompleteness theorems have had enormous success in most physical situations. There is, however, strong physical motivation for the specific prediction of 'unbounded curvature'.

## 3. Problems in Proving Curvature Singularity Theorems

Much of the difficulty in proving a curvature result has been due to

(1) the challenge in ascribing boundary points to space-time — the *boundary construction problem*, and
(2) the variety of ways in which curvature may not be smooth in a given space-time — the *curvature pathology classification problem.*

Hence the most significant hurdles to date have not been the proof of a curvature singularity theorem itself but instead in the precise formulation of the theorem which one wants to prove.

The abstract boundary is a promising way to overcome (1) for the following reasons. As discussed previously, one encounters problems of a topological nature when attempting to generate boundary points for a space-time purely from the pseudo-Riemannian manifold which is used to represent it. Within this philosophy of boundary construction, it appears that there is no single construction which has all the requisite properties and remains well-behaved. In essence, the expectation that there exists a single boundary construction, which is physically significant, topologically well-behaved, and easy to construct for any space-time is overly optimistic. The $a$-boundary, however, uses a completely different philosophy based only on the invariant properties of boundary points in the countless many possible representations of the boundary. Thus, it avoids the problems described in [**1**] by not enforcing a particular boundary construction upon a space-time, but instead allowing us to derive invariant results from whichever representation of the boundary is best suited for a specific problem. In addition, we obtain a precise definition for an essential singularity, which is independent of its envelopment, and for which the previous objections described in [**1**] do not apply. We refer the reader to Definitions 37-42 of [**22**] which define $a$-boundary essential singularities. The reader should also note that we have chosen the family $\mathcal{C}$ of curves satisfying the bounded parameter property to be the family of affinely parametrised causal geodesics and we employ these curves to define the essential singularities which are central to the following analysis.

Although we have a solution to the boundary construction problem (1), we are still left with the problems provided by (2). The literature embodies a large number of different definitions of a curvature singularity. If we consider singularities, not in the sense of curve incompleteness, but as a 'place' resisting metric extension, then we find many specific definitions. Clarke [**4**] and Ellis and Schmidt [**8**] developed schemes aimed at classifying this notion of singularities. Their work generalised the fundamental definitions of *scalar polynomial curvature singularities* and *parallelly* *propagated curvature singularities* discussed in Hawking and Ellis [**13**] to include not only divergent behaviour but also non-smoothness in the derivatives of the respective curvature based objects. In addition, Ellis and Schmidt [**8**] defined these singularities as singular boundary points of the $b$-completion of the space-time. The previous definitions of Hawking and Ellis [**13**] only discussed curves *corresponding* to a curvature singularity and not the boundary points themselves, since no particular boundary construction was employed. Finally Clarke, Ellis and Schmidt defined an additional singularity type — the *quasi-regular singularity* — for which no type of curvature pathology occurs at all.

In order to gain some feeling for the issue in the abstract boundary context we redefine below and broaden Clarke, Ellis and Schmidt's definitions in terms of $a$-boundary concepts.

DEFINITION 3.1 ($C^k$ Curvature Singularity). Let $q \in \partial_\phi \mathcal{M} \subset \widehat{\mathcal{M}}$ be a $C^k$ essential singularity, $k \geqslant 0$. The boundary point $q$ is termed a $C^k$ *curvature singularity*[2] if there exists a curve $\gamma : [0, a) \to \mathcal{M}$, $a \in \mathbb{R}^+$ in $\mathcal{C}$ which has $q$ as a limit point such that some component $R_{abcd}$ of the Riemann curvature tensor, when evaluated in a parallelly propagated orthonormal tetrad $\{E_a(s)\}$ along $\gamma(s)$, satisfies the condition

$$\limsup_{s \to a} |R_{abcd;e_1 \ldots e_k}(\gamma(s))| = \infty .$$

In this case $\gamma$ is said to *correspond* to the $C^k$ curvature singularity.

DEFINITION 3.2 ($C^k$ Quasi-regular Singularity). Let $q \in \partial_\phi \mathcal{M} \subset \widehat{\mathcal{M}}$ be a $C^k$ essential singularity, $k \geqslant 0$. The boundary point $q$ is termed a $C^k$ *quasi-regular singularity* if it is not a $C^k$ curvature singularity.

DEFINITION 3.3 ($C^k$ Scalar Curvature Singularity). Let $q \in \partial_\phi \mathcal{M} \subset \widehat{\mathcal{M}}$ be a $C^k$ curvature singularity. The boundary point $q$ is termed a $C^k$ *scalar curvature singularity* if there exists a curve $\gamma : [0, a) \to \mathcal{M}$, $a \in \mathbb{R}^+$ in $\mathcal{C}$ which has $q$ as a limit point and a scalar polynomial function $P(x)$ on $\mathcal{M}$ composed only of $g_{ab}$, $\eta_{abcd}$ (volume form) and $R_{abcd}$ such that

$$\limsup_{s \to a} |P(\gamma(s))| = \infty .$$

As above, $\gamma$ is said to *correspond* to the $C^k$ scalar curvature singularity.

DEFINITION 3.4 ($C^k$ Non-scalar Curvature Singularity). Let $q \in \partial_\phi \mathcal{M} \subset \widehat{\mathcal{M}}$ be a $C^k$ curvature singularity. The boundary point $q$ is termed a $C^k$ *non-scalar curvature singularity* if it is not a $C^k$ scalar curvature singularity.

Using the four definitions above we might then wish to define an abstract boundary point $[p] \in \mathcal{B}(\mathcal{M})$ as being one of the above types of singularity — e.g. a curvature singularity — if it possesses a boundary point representative which is itself that type of singularity. In order to show that this concept is well-defined, the next step in the $a$-boundary approach is then to prove the invariance of this classification under re-envelopment so that the property passes to all boundary point members of the equivalence class.

[2] This concept is traditionally referred to in the literature as a *parallelly propagated curvature singularity*. Similarly, the concept in Definition 3.3 is referred to as a *scalar polynomial curvature singularity*.

PROPOSITION 3.5. *The boundary point* $q \in \partial_\phi \mathcal{M} \subset \widehat{\mathcal{M}}$ *and the boundary point* $p \in \partial_{\phi'} \mathcal{M} \subset \widehat{\mathcal{M}}'$ *are assumed to be equivalent i.e.* $q \sim p$. *If the boundary point* $q$ *is a* $C^k$ *curvature singularity,* $C^k$ *quasi-regular singularity,* $C^k$ *scalar curvature singularity or* $C^k$ *non-scalar curvature singularity, then the boundary point* $p$ *is a singularity of the same type.*

PROOF. If the boundary point $q$ is any of these four types of singularity, then it is, by definition, a $C^k$ essential singularity. It has been shown in Theorem 46 of [**22**] that the property of being an essential singularity passes to the abstract boundary $\mathcal{B}(\mathcal{M})$. Thus the boundary point $p$ is also a $C^k$ essential singularity.

Now suppose that $q$ is a $C^k$ curvature singularity. By Definition 3.1, there exists a curve $\gamma : [0, a) \to \mathcal{M}$, $a \in \mathbb{R}^+$ in $\mathcal{C}$ which has $q$ as a limit point such that some component $R_{abcd}$ of the Riemann curvature tensor, when evaluated in a parallelly propagated orthonormal tetrad $\{E_a(s)\}$ along $\gamma(s)$, satisfies the condition

$$\limsup_{s \to a} |R_{abcd;e_1 \ldots e_k}(\gamma(s))| = \infty .$$

Since $p$ covers $q$, then by Theorem 17 of [**22**], $p$ is also a limit point of the curve $\gamma$ and thereby satisfies the requirements to be a $C^k$ curvature singularity.

If $q$ is a $C^k$ quasi-regular singularity, then by Definition 3.2, it is not a $C^k$ curvature singularity. By the previous argument $p$ cannot then be a $C^k$ curvature singularity and is therefore a $C^k$ quasi-regular singularity.

Now suppose that $q$ is a $C^k$ scalar curvature singularity. This means that $q$ is a $C^k$ curvature singularity and thus $p$ is also a $C^k$ curvature singularity. By Definition 3.3, there exists a curve $\gamma : [0, a) \to \mathcal{M}$, $a \in \mathbb{R}^+$ in $\mathcal{C}$ which has $q$ as a limit point and a scalar polynomial function $P(x)$ on $\mathcal{M}$ composed only of $g_{ab}$, $\eta_{abcd}$ (volume form) and $R_{abcd}$ such that

$$\limsup_{s \to a} |P(\gamma(s))| = \infty .$$

Since $p$ covers $q$, then by Theorem 17 of [**22**], $p$ is also a limit point of the curve $\gamma$ and thereby satisfies the requirements to be a $C^k$ scalar curvature singularity.

If $q$ is a $C^k$ non-scalar curvature singularity, then by Definition 3.4, it is a $C^k$ curvature singularity which is not a $C^k$ scalar curvature singularity. The boundary point $p$ is also a $C^k$ curvature singularity. By the previous argument $p$ cannot be a $C^k$ scalar curvature singularity and is therefore a $C^k$ non-scalar curvature singularity.

It has thus been shown that the categories of curvature singularity, quasi-regular singularity, scalar curvature singularity and non-scalar curvature singularity all pass to the abstract boundary $\mathcal{B}(\mathcal{M})$. □

Note that $C^k$ scalar curvature singularities are clearly $C^k$ curvature singularities but the converse is not necessarily true — i.e. a $C^k$ curvature singularity can be a $C^k$ non-scalar curvature singularity. There exist space-times where the curvature components in a parallelly propagated tetrad are ill-behaved but all curvature scalar polynomials are zero [**8**]. One should also note the essential differences between the definitions presented above and our goal of describing physical singularities due to unbounded curvature only. The definitions above include non-smooth behaviour other than simply the divergence of curvature components or a curvature scalar polynomial. There may be oscillatory behaviour in the curvature components or a curvature scalar polynomial and/or the derivatives of the curvature components may diverge or themselves exhibit oscillatory behaviour.

It is conceivable that fast changes in the curvature derivatives alone will not lead to the destruction of physical objects[3]. So we must consider the question of exactly what type of curvature pathology leads to the destruction of any object nearing a singularity. Thus, although the previous concepts of singularity given in Definitions 3.1—3.4 are both useful and physically significant, the idea of *strong curvature* described in Ellis and Schmidt [**8**, p. 944] deals with this notion specifically. Strong curvature has other features which make it an attractive definition of a curvature singularity to use for the proof of curvature singularity theorems. We will now introduce the concept of strong curvature and its definition in terms of Jacobi fields.

## 4. Jacobi Fields and Strong Curvature

DEFINITION 4.1 (Jacobi Field). If $\gamma : [0, a) \to \mathcal{M}$, $a \in \mathbb{R}^+ \cup \{\infty\}$ is a geodesic with affine parameter $s$, then the smooth vector field $J : [0, a) \to T\mathcal{M}$ along $\gamma$ is a *Jacobi field* if it satisfies the *Jacobi equation*

$$J'' + R(J, \gamma')\gamma' = 0,$$

where $J'' = \dfrac{\mathrm{D}^2 J}{\mathrm{d}s^2} = \nabla_{\gamma'}(\nabla_{\gamma'} J)$ and $R(X, Y)Z = \nabla_X \nabla_Y Z - \nabla_Y \nabla_X Z - \nabla_{[X,Y]} Z$ is the curvature operator.

Intuitively, a Jacobi field along $\gamma$ represents the relative displacement between $\gamma$ and nearby geodesics. Now let $\gamma$ be a timelike geodesic with unit tangent vector $\gamma'$ and let $J$ be a Jacobi field along $\gamma$ which is orthogonal to $\gamma'$. Then $\gamma$ represents the path of a freely falling massive point particle and $J$ the displacement vector between it and a second massive point particle following a nearby timelike geodesic. Thus $J' = \nabla_{\gamma'} J$ represents the relative velocity and $J''$ the relative (or tidal) acceleration of the second particle as measured by the first. This allows us to physically interpret the Jacobi equation as linking curvature to the relative acceleration between neighbouring particles moving on geodesics.

We will now restrict our attention to Jacobi fields with particular properties. Let $\gamma : [0, a) \to \mathcal{M}$, $a \in \mathbb{R}^+ \cup \{\infty\}$ be a timelike (respectively null) geodesic with affine parameter $s$. We define $J_b(\gamma)$ for $b \in [0, a)$ to be the set of smooth vector fields $Z : [b, a) \to T\mathcal{M}$ along $\gamma$ such that

(1) $Z(s) \in T_{\gamma(s)}\mathcal{M}$,
(2) $Z(b) = 0$,
(3) $\dfrac{\mathrm{D}^2 Z}{\mathrm{d}s^2} = \nabla_{\gamma'}(\nabla_{\gamma'} Z) = R(\gamma', Z)\gamma'$
(4) $g\left( \left.\dfrac{\mathrm{D}Z}{\mathrm{d}s}\right|_b, \gamma'(b) \right) = 0$

Thus $J_b(\gamma)$ is comprised of Jacobi fields which vanish at $\gamma(b)$ and which lie in the space orthogonal to $\gamma'$ at each point $\gamma(s)$.

For a timelike geodesic $\gamma$, it is possible to choose a set of three linearly independent Jacobi fields $\{Z_1, Z_2, Z_3\}$, $Z_\alpha \in J_b(\gamma)$ $(\alpha = 1, 2, 3)$ along $\gamma$, where for each $\alpha$, $Z_\alpha(s)$ is a spacelike vector. A spacelike 3-volume element $V(s) =$

---

[3]The reader may wish to consult Ellis and Schmidt [**8**] for a physically motivated discussion of weak curvature singularities.

$Z_1(s) \wedge Z_2(s) \wedge Z_3(s)$ can then be defined at each point $\gamma(s)$. In the case of a null geodesic $\lambda$, a set of two linearly independent Jacobi fields $\{\widehat{Z}_1, \widehat{Z}_2\}$, $\widehat{Z}_\alpha \in J_b(\lambda)$ $(\alpha = 1, 2)$ may be chosen along $\gamma$, where for each $\alpha$, $\widehat{Z}_\alpha(s)$ is a spacelike vector. A spacelike 2-volume element $\widehat{V}(s) = \widehat{Z}_1(s) \wedge \widehat{Z}_2(s)$ can then be defined at each point $\gamma(s)$. We now present the strong curvature definitions of Tipler [**24**] and Królak [**16**].

DEFINITION 4.2 (Strong Curvature Singularity - Tipler [**24**]). Let $\gamma : [0, a) \to \mathcal{M}$, $a \in \mathbb{R}^+$ be a timelike (respectively null) geodesic with affine parameter $s$. A *Tipler strong curvature condition* is said to be satisfied along $\gamma$ if for all $b \in [0, a)$ and all linearly independent Jacobi fields $Z_1, Z_2, Z_3 \in J_b$ (respectively, $\widehat{Z}_1, \widehat{Z}_2 \in J_b$ if $\gamma$ is a null geodesic)

$$\liminf_{s \to a} V(s) = 0 \text{ (or, respectively, } \liminf_{s \to a} \widehat{V}(s) = 0).$$

DEFINITION 4.3 (Strong Curvature Singularity - Królak [**16**]). Let $\gamma : [0, a) \to \mathcal{M}$, $a \in \mathbb{R}^+$ be a timelike (respectively null) geodesic with affine parameter $s$. The *Królak strong curvature condition* is said to be satisfied along $\gamma$ if for all $b \in [0, a)$ and all linearly independent Jacobi fields $Z_1, Z_2, Z_3 \in J_b$ (respectively, $\widehat{Z}_1, \widehat{Z}_2 \in J_b$ if $\gamma$ is a null geodesic) there exists a $c \in [b, a)$ such that

$$\left.\frac{\mathrm{d}V}{\mathrm{d}s}\right|_c < 0 \qquad \left(\text{respectively, } \left.\frac{\mathrm{d}\widehat{V}}{\mathrm{d}s}\right|_c < 0\right).$$

The concept of a strong curvature singularity was first introduced by Ellis and Schmidt [**8**]. They loosely considered it a type of singularity at which no object could arrive intact due to the presence of unbounded tidal forces. Tipler [**24**] was the first to formalise this definition to the one above using only geometrically defined objects but with the same physical content as the earlier definition. Essentially, Tipler's definition requires that any object have its volume crushed to zero as the singularity is approached. The Królak [**16**] definition is weaker than the Tipler version and arose out of investigations into the cosmic censorship conjecture [**15, 14, 17**]. In this literature the Królak definition is sometimes referred to as the *limiting focusing condition.*

Our primary reason for considering the use of strong curvature is the existence of the following results due to Clarke and Królak [**6**][4].

PROPOSITION 4.4. *For both the timelike and the null cases, if the Tipler strong curvature condition is satisfied, then for some component* $R^i{}_{0j0}$ *of the Riemann tensor in a parallelly propagated frame the integral*

$$I^i{}_j(v) = \int_0^v dv' \int_0^{v'} dv'' |R^i{}_{0j0}(v'')|$$

*does not converge as* $v \to a$.

PROPOSITION 4.5. *If* $\lambda(v)$ *is a null geodesic and the Tipler strong curvature condition is satisfied then, with respect to a parallelly propagated frame, either the*

[4]We recommend that the reader refer to Clarke [**5**] instead of Clarke and Królak [**6**] for the proofs of these theorems. The original paper is full of misleading typographic errors which have been corrected in the proofs contained in Clarke [**5**].

*integral*

$$K(v) = \int_0^v dv' \int_0^{v'} dv'' R_{00}(v'')$$

*or the integral*

$$L^m{}_n(v) = \int_0^v dv' \int_0^{v'} dv'' \left( \int_0^{v''} dv''' |C^m{}_{0n0}(v''')| \right)^2$$

*for some $m, n$ does not converge as $v \to a$.*

Similar results dealing with the Królak definition were also proven by Clarke and Królak [**6**] but with one integral less in each case.

These results are significant since they pertain to parallelly propagated components of the Riemann, Ricci and Weyl curvature tensors and verify that strong curvature singularities not only imply the divergence of various parallelly propagated components of the Riemann, Ricci and Weyl tensors but also the divergence of their integrals, as defined above in Proposition 4.4 and Proposition 4.5, along geodesics approaching them. In addition, Clarke and Królak [**6**] proved some sufficiency conditions on integrals of the Ricci and Weyl curvature tensor components which ensure the existence of Tipler and Królak-type strong curvature singularities.

In summary, strong curvature has three salient properties for application to the curvature singularity problem:

(1) Strong curvature is a physically intuitive and mathematically precise way to define the presence of infinite tidal forces.
(2) Clarke and Królak established the equivalence between the various notions of strong curvature and divergences in the parallelly propagated components of the Riemann, Ricci and Weyl tensors and in integrals of these curvature components along the geodesic.
(3) The form of the strong curvature condition is similar to the trapping condition of Theorem 4.7 (given below) by Hawking [**13**, Theorem 3, p. 271].

It is this final point which we will proceed to examine in more detail. Previous arguments have made it clear that the production of a curvature singularity theorem will require the connection of three key concepts:

(1) Essential Singularities
(2) Curve Incompleteness
(3) Unbounded Curvature

Concept (1) is important since a singularity predicted by any resulting theorem should not be removable simply by re-embedding the space-time. The recent Ashley and Scott result [**1**, Theorem 4.12] stated below as Theorem 4.6 provides the necessary, and previously unknown, link between (1) and (2). Concept (2) is also significant since, intuitively, realistic singularities all cause the incompleteness of physically relevant curve families.

THEOREM 4.6. *Let $(\mathcal{M}, g)$ be a strongly causal, $C^l$ maximally extended, $C^k$ space-time $(1 \leqslant l \leqslant k)$ and $\mathcal{C}$ be the family of affinely parametrised causal geodesics in $(\mathcal{M}, g)$. Then the abstract boundary $\mathcal{B}(\mathcal{M})$ of $\mathcal{M}$ contains a $C^l$ essential singularity iff there is an incomplete causal geodesic in $(\mathcal{M}, g)$.*

It has proven difficult to establish theorems on divergent curvature directly from the types of conditions typically employed in the geodesic incompleteness theorems.

A link between (2) and (3), however, is provided using strong curvature and the following theorem by Hawking [**13**, Theorem 3, p. 271], [**12**].

THEOREM 4.7. *If the space-time $(\mathcal{M}, g)$ satisfies the following conditions:*

(1) $R_{ab}K^aK^b \geqslant 0$ *for all non-spacelike vectors* $K$*;*
(2) *the strong causality condition holds on* $(\mathcal{M}, g)$*;*
(3) *there is some past-directed unit timelike vector* $W$ *at a point* $p$ *and a positive constant* $b$ *such that if* $V$ *is the unit tangent vector to the past-directed timelike geodesics through* $p$*, then on each such geodesic the expansion* $\theta \equiv V^a{}_{;a}$ *of these geodesics becomes less than* $-3c/b$ *within a distance* $b/c$ *from* $p$*, where* $c \equiv -W^aV_a$*,*

*then there is a past incomplete causal geodesic through* $p$*.*

The result which follows comes directly from Theorem 4.7 and the proof of Theorem 4.6. Within the context of the singularity theorems, it provides the desired link between curve incompleteness and the existence of essential singularities. It says that if there is an incomplete causal geodesic $\gamma$ in $\mathcal{M}$ predicted by the Hawking theorem, then $\gamma$ corresponds to an $a$-boundary essential singularity of $(\mathcal{M}, g)$. In accordance with the conditions imposed for the Penrose-Hawking singularity theorems as well as Theorem 4.6, it will be assumed that the space-time $(\mathcal{M}, g)$ is maximally extended.

COROLLARY 4.8. *Suppose that* $(\mathcal{M}, g)$ *is a maximally extended space-time which satisfies the conditions of the Hawking theorem. Let* $\gamma$ *be a past incomplete causal geodesic through* $p \in \mathcal{M}$ *predicted by that theorem. Then there exists a* $C^\infty$ *embedding* $\phi : \mathcal{M} \to \widehat{\mathcal{M}}$ *(where* $\widehat{\mathcal{M}}$ *is a 4-dimensional manifold) and a boundary point* $q \in \partial_\phi \mathcal{M} \subset \widehat{\mathcal{M}}$ *such that* $q$ *is a limit point of* $\gamma$ *and is an a-boundary essential singularity.*

This result is important since it provides the connection between the focusing condition on the expansion of the timelike geodesic congruence through $p$ (condition 3 of Theorem 4.7) and the existence of an incomplete causal geodesic which corresponds to an essential singularity. All that would now be required to obtain a curvature singularity theorem is to link this condition with our chosen concept of a strong curvature singularity. Since it is possible to determine the expansion $\theta$ of a congruence of geodesics in terms of the volume element $V(s)$ derived from the Jacobi fields, namely,

$$\theta = \frac{1}{V(s)} \frac{\mathrm{d}V(s)}{\mathrm{d}s} ,$$

we are able to rewrite the strong curvature definition of Królak in terms of the expansion. This is a profitable exercise since the expansion scalar obeys the Raychaudhuri equation. Specifically, for the case of $\hat{\theta}$ being the expansion scalar for a congruence of null geodesics, the Raychaudhuri equation takes the form [**13**, p. 88]

$$\frac{\mathrm{d}\hat{\theta}}{\mathrm{d}s} = -R_{ab}K^aK^b - 2\hat{\sigma}^2 - \frac{1}{2}\hat{\theta}^2 ,$$

where $\hat{\sigma}(s)$ is the shear of the congruence and $K^a(s)$ is the tangent vector of the null geodesic along which $\hat{\theta}$ is being determined. If we define a length scale $\hat{x}^2 = \hat{V}(s)$

where $\hat{V}(s)$ is the spacelike 2-volume element then

$$\hat{\theta} = \frac{1}{\hat{V}(s)} \frac{\mathrm{d}\hat{V}(s)}{\mathrm{d}s} = \frac{2}{\hat{x}} \frac{\mathrm{d}\hat{x}}{\mathrm{d}s} ,$$

and the Raychaudhuri equation becomes

$$\frac{\mathrm{d}^2 \hat{x}}{\mathrm{d}s^2} = -\frac{1}{2}[R_{ab}K^a K^b + 2\hat{\sigma}^2]\hat{x} . \tag{1}$$

Pursuing the details of this propagation equation for the two definitions of the strong curvature condition we obtain the following lemma and definition.

LEMMA 4.9 (Tipler Strong Curvature Condition [**5**, p. 164]). *Let $\lambda : [0, a) \to \mathcal{M}$, $a \in \mathbb{R}^+$ be a null geodesic with affine parameter $s$. The Tipler strong curvature condition implies that for all $b \in [0, a)$ and all solutions $\hat{x}(s)$ of Equation* (1) *with initial conditions $\hat{x}(b) = 0$, $\liminf_{s \to a} |\hat{x}(s)| = 0$.*

DEFINITION 4.10 (Królak Strong Curvature Condition [**14**]). Let $\lambda : [0, a) \to \mathcal{M}$, $a \in \mathbb{R}^+$ be a future-inextendible null geodesic with affine parameter $s$. It will be said that $\lambda$ terminates in a *strong curvature singularity* in the future if, for all points $p$ on $\lambda$ and for all initial value conditions for the Raychaudhuri equation and the shear equation at $p$, there exists a point $q$ on $\lambda$ to the future of $p$ such that $\hat{\theta} = [2/\hat{x}(s)][\mathrm{d}\hat{x}(s)/\mathrm{d}s]$ is negative.

It is the version of Definition 4.10 relating to a past-inextendible null geodesic which looks especially promising for application to Corollary 4.8. The results in the Clarke and Królak paper [**6**] would then link this to the presence of parallelly propagated curvature components which are unbounded.

## 5. Future Directions in Proving Curvature Singularity Theorems

This paper has outlined a program within which curvature singularity theorems might be proven. In order to complete this program, a specific definition of strong curvature must be chosen. Physical intuition would lead one to choose the Tipler condition, however it may be easier in practice to use the Królak condition. Later extensions which include the Tipler condition should then be relatively straightforward.

In conclusion, it appears that this program of research, incorporating the use of the abstract boundary construction, offers some exciting opportunities for proving curvature singularity theorems. Given the importance of the long-standing investigation into singularities, geodesic incompleteness and curvature, such a result would form the critical step in determining the physical content of singularity production in General Relativity.

## References


1. Michael John Siew Lueng Ashley, *Singularity theorems and the abstract boundary construction*, Ph.D. thesis, The Australian National University, 2002.
2. V. A. Belinskii, I. M. Khalatnikov, and E. M. Lifshitz, *Oscillatory approach to a singular point in relativistic cosmology*, Adv. in Phys. **19** (1970), 523–573.
3. B. Bosshard, *On b-boundaries of special space-time models*, Gen. Rel. Grav. **10** (1979), 963–966.
4. C. J. S. Clarke, *The classification of singularities*, Gen. Rel. Grav. **6** (1975), 35–40.
5. ———, *The analysis of space-time singularities*, Cambridge University Press, Cambridge, 1993.

6. C. J. S. Clarke and A. Królak, *Conditions for the occurrence of strong curvature singularities*, J. Geom. Phys. **2** (1985), 127–143.
7. C. J. S. Clarke and B. G. Schmidt, *Singularities: The state of the art*, Gen. Rel. Grav. **8** (1977), 129–137.
8. G. F. R. Ellis and B. G. Schmidt, *Singular space-times*, Gen. Rel. Grav. **8** (1977), 915–953.
9. ———, *Classification of singular space-times*, Gen. Rel. Grav. **10** (1979), 989–997.
10. R. Geroch, *Local characterization of singularities in general relativity*, J. Math. Phys. **9** (1968), 450–465.
11. R. Geroch, E. H. Kronheimer, and R. Penrose, *Ideal points in space-time*, Proc. R. Soc. Lond. **A327** (1972), 545–567.
12. S. W. Hawking, *The occurrence of singularities in cosmology. III Causality and singularities*, Proc. R. Soc. Lond. **A300** (1967), 187–201.
13. S. W. Hawking and G. F. R. Ellis, *The large scale structure of space-time*, Cambridge University Press, 1973.
14. A. Królak, *Strong curvature singularities and causal simplicity*, J. Math. Phys. **33** (1986), 701–704.
15. ———, *Towards the proof of the cosmic censorship hypothesis*, Class. Quant. Grav. **3** (1986), 267–280.
16. ———, *Towards the proof of the cosmic censorship hypothesis in cosmological space-times*, J. Math. Phys. **28** (1987), 138–141.
17. A. Królak and W. Rudnicki, *Singularities, trapped sets, and cosmic censorship in asymptotically flat space-times*, Inter. J. Theor. Phys. **32** (1993), 137–142.
18. E. M. Lifshitz and I. M. Khalatnikov, *Investigations in relativistic cosmology*, Adv. in Phys. (Phil. Mag. Suppl.) **12** (1963), 185–249.
19. C. W. Misner, *Taub-NUT spaces as a counterexample to almost anything*, Relativity Theory and Astrophysics I: Relativity and Cosmology (J. Ehlers, ed.), Lectures in Applied Mathematics, vol. 8, American Mathematical Society, 1967, pp. 160–169.
20. R. Penrose, *Gravitational collapse and space-time singularities*, Phys. Rev. Lett. **14** (1965), 57–59.
21. B. G. Schmidt, *A new definition of singular points in general relativity*, Gen. Rel. Grav. **1** (1971), 269–280.
22. S. M. Scott and P. Szekeres, *The abstract boundary–a new approach to singularities of manifolds*, J. Geom. Phys. **13** (1994), 223–253.
23. L. Szabados, *Causal boundary for strongly causal space-times*, Class. Quant. Grav. **5** (1988), 121–134.
24. F. J. Tipler, *Singularities in conformally flat spacetimes*, Phys. Lett. **64A** (1977), 8–10.

Department of Physics, Faculty of Science, The Australian National University, Canberra ACT 0200, AUSTRALIA

*Current address*: Center for Gravitational Wave Physics, Department of Physics, Eberly College of Science, The Pennsylvania State University, 104 Davey Laboratory - PMB 005, University Park, PA 16802-6300

*E-mail address*: `ashley@gravity.psu.edu`

Department of Physics, Faculty of Science, The Australian National University, Canberra ACT 0200, AUSTRALIA

*E-mail address*: `Susan.Scott@anu.edu.au`

Contemporary Mathematics
Volume **337**, 2003

# Curvature-Homogeneous Indefinite Einstein Metrics in Dimension Four: the Diagonalizable Case

Andrzej Derdzinski

## §0. Introduction

A pseudo-Riemannian manifold $(M,g)$ is called *curvature-homogeneous* if the algebraic type of its metric/curvature pair $(g,R)$ is the same at all points, i.e., if for any $x,y\in M$ some isomorphism $T_xM\to T_yM$ sends $g(x),R(x)$ to $g(y),R(y)$.

Every locally homogeneous pseudo-Riemannian manifold is, obviously, curvature-homogeneous. The converse proposition fails; counterexamples with positive-definite metrics were first found by Takagi [**17**] and, on compact manifolds, by Ferus, Karcher and Münzner [**11**]; see also [**2**]. Analogous examples with indefinite metrics have been known even longer ([**3**] – [**7**], [**13**]).

The present paper provides a classification, up to local isometries, of all those curvature-homogeneous pseudo-Riemannian four-manifolds $(M,g)$ which are Einstein and have, at some (or every) point $x$, a *complex-diagonalizable* curvature operator $R(x):[T_xM]^{\wedge 2}\to[T_xM]^{\wedge 2}$. (The last condition means that the complex-linear extension of $R(x)$ to the complexification of the bivector space $[T_xM]^{\wedge 2}$ is diagonalizable.) It turns out that all such manifolds are locally homogeneous and, in fact, either locally symmetric, or locally isometric to a Lie group with a left-invariant indefinite metric of a specific type; see Theorems 5.1, 6.1 and 7.1. In those theorems we assume constancy of eigenvalues of the curvature operator, which sounds weaker than curvature-homogeneity, but, in the complex-diagonalizable case, is actually equivalent to it; cf. [**10**], p. 701 and Remark 6.19 on p. 472.

The metric $g$ can have any signature. Using a sign change, we may assume that $g$ is *Riemannian*, *neutral* or *Lorentzian*, that is, has one of the sign patterns

$$(1)\qquad ++++\,,\qquad --++\,,\qquad -+++\,.$$

Two known families of curvature-homogeneous Einstein four-manifolds, one Lorentzian [**3**] and one neutral ([**10**], p. 705), give rise to infinite-dimensional spaces of local-isometry types. By contrast, for the manifolds classified here, the analogous space is clearly finite-dimensional (see above). Also, our diagonalizability assumption always holds for *Riemannian* manifolds, as the curvature operator is

2000 *Mathematics Subject Classification.* 53B30.
*Key words and phrases.* Einstein metric, curvature-homogeneity, Lorentz metric, neutral metric.

self-adjoint, and for Riemannian metrics our theorem becomes the result of [**10**], mentioned below. In the Lorentzian case, the complex-diagonalizability condition means that the curvature is of the Petrov type I at each point, cf. [**10**], p. 659.

Some types of curvature-homogeneous Einstein four-manifolds have already been classified. This includes locally symmetric spaces ([**8**], [**9**]; cf. [**10**], pp. 662–663); Brans's classification [**3**] of Lorentzian Einstein metrics representing the Petrov type III at every point (a condition that implies curvature-homogeneity); as well as the Riemannian case ([**10**], Corollary 7.2, p. 476), in which the metrics in question are all locally symmetric (see also §7).

The text is organized as follows. In sections 2 – 4 we introduce our "model spaces", using a construction basically due to Petrov [**15**]. The classification result is stated in sections 5 – 7 and then proved in sections 8 – 13.

## §1. Preliminaries

Our conventions about the curvature tensor $R = R^\nabla$ of any connection $\nabla$ in a real/complex vector bundle $\mathcal{E}$ over a manifold $M$, its Ricci tensor Ric when $\mathcal{E}$ is the tangent bundle $TM$, and the scalar curvature $\mathrm{s}$ in the case where $\nabla$ is the Levi-Civita connection of a given pseudo-Riemannian metric $g$ on $M$, are

$$(2)\qquad \begin{aligned} &\text{i)}\quad R(u,v)\psi \,=\, \nabla_v\nabla_u\psi \,-\, \nabla_u\nabla_v\psi \,+\, \nabla_{[u,v]}\psi\,,\\ &\text{ii)}\quad \mathrm{Ric}\,(u,w) \,=\, \mathrm{Trace}\,[v\mapsto R(u,v)w]\,, \qquad \mathrm{s} \,=\, \mathrm{Trace}_g\mathrm{Ric}\,,\end{aligned}$$

for any (local) $C^2$ sections $u, v, w$ of $TM$ and $\psi$ of $\mathcal{E}$.

A pseudo-Riemannian manifold $(M,g)$ with $\dim M = n$ is said to be an *Einstein manifold* [**1**] if $n \ge 3$ and $\mathrm{Ric} = \mathrm{s}g/n$, while, if $n \ge 4$, formulae $\sigma = \mathrm{Ric} - (2n-2)^{-1}\mathrm{s}g$ and $W = R - (n-2)^{-1}g\wedge\sigma$ define the *Schouten tensor* $\sigma$ and *Weyl tensor* $W$ of $(M,g)$. Here $\wedge$ is the exterior product of 1-forms valued in 1-forms, obtained using the valuewise multiplication which is also provided by $\wedge$, so that the result is a 2-form valued in 2-forms.

For $(M,g)$ as above, we denote $[TM]^{\wedge 2}$ the vector bundle of bivectors over $M$, with the fibres $[T_xM]^{\wedge 2}$, $x \in M$. There exists a unique pseudo-Riemannian fibre metric $\langle\,,\rangle$ in $[TM]^{\wedge 2}$ such that $\langle v\wedge u, v'\wedge u'\rangle = g(v,v')\,g(u,u') - g(v,u')\,g(u,v')$ for any $x \in M$ and $v, u, v', u' \in T_xM$. Both $R, W$ are four-times covariant tensor fields on $M$ sharing the (skew)symmetry properties of the curvature tensor, which allows us to treat them as morphisms acting on bivectors and self-adjoint relative to $\langle\,,\rangle$ at each point; in this way, $R$ gives rise to the *curvature operator*

$$(3)\qquad R : [TM]^{\wedge 2} \to [TM]^{\wedge 2} \quad \text{with} \quad \langle R(u\wedge v), w\wedge w'\rangle \,=\, g(R(u,v)w, w')$$

for $x \in M$ and $u, v, w, w' \in T_xM$. When $(M,g)$ is four-dimensional and oriented, another important morphism $[TM]^{\wedge 2} \to [TM]^{\wedge 2}$ is the *Hodge star* $*$, given by $*(e_1\wedge e_2) = \varepsilon_3\varepsilon_4\, e_3\wedge e_4$ for any $x \in M$ and any positive-oriented orthonormal basis $e_1, \ldots, e_4$ of $T_xM$, where $\varepsilon_a = g(e_a, e_a) \in \{1, -1\}$ (no summation). This well-known description of $*$ (cf. [**10**], formula (37.13) on p. 639) is clearly equivalent to its more common definition $\alpha\wedge\beta = \langle *\alpha, \beta\rangle\,\mathrm{vol}$ for any bivectors $\alpha, \beta$, where vol is the *volume four-vector*, equal to $e_1\wedge\ldots\wedge e_4$ for any $e_1, \ldots, e_4$ as above.

Let $(M,g)$ be an oriented pseudo-Riemannian 4-manifold. Then $[W, *] = 0$, that is, the morphisms $W, * : [TM]^{\wedge 2} \to [TM]^{\wedge 2}$ commute (cf. [**16**]), while our formula for $*$ gives $*^2 = \mathrm{Id}$ if $g$ is Riemannian $(++++)$ or neutral $(--++)$, and $*^2 = -\,\mathrm{Id}$ when $g$ is Lorentzian $(-+++)$. In the Lorentzian case, this turns $[TM]^{\wedge 2}$ into a *complex vector bundle* of fibre dimension 3, in which $*$ is the multiplication by $i$, and, as $[W, *] = 0$, the Weyl tensor $W$ is a *complex-linear bundle morphism* $[TM]^{\wedge 2} \to [TM]^{\wedge 2}$. In the Riemannian and neutral cases, the self-adjoint involution $* : [TM]^{\wedge 2} \to [TM]^{\wedge 2}$ gives rise to the orthogonal decomposition $[TM]^{\wedge 2} = \Lambda^+ M \oplus \Lambda^- M$, where $\Lambda^{\pm} M$, the $(\pm 1)$-eigenspace bundles of $*$, are real vector bundles of fibre dimension 3, called the bundles of *self-dual* and *anti-self-dual* bivectors in $(M, g)$. As $[W, *] = 0$, both $\Lambda^{\pm} M$ are $W$-invariant, which leads to the restrictions $W^{\pm} : \Lambda^{\pm} M \to \Lambda^{\pm} M$ of $W$, called the *self-dual* and *anti-self-dual* Weyl tensors of $(M, g)$. See [**16**] and [**10**], pp. 637–651.

REMARK 1.1. For a pseudo-Riemannian *Einstein* manifold $(M, g)$ of dimension $n \geq 4$, the difference $R - W : [TM]^{\wedge 2} \to [TM]^{\wedge 2}$ of the morphisms $R, W$ clearly equals the constant $\mathrm{s}/[n(n-1)]$ times $\mathrm{Id} = (g \wedge g)/2$. If $n = 4$ and $M$ is oriented, relation $[W, *] = 0$ thus gives $[R, *] = 0$, i.e., in the Lorentzian case the curvature operator $R : [TM]^{\wedge 2} \to [TM]^{\wedge 2}$ is complex-linear, while in the Riemannian and neutral cases both $\Lambda^{\pm} M$ are $R$-invariant; we will call the restriction $R^+ : \Lambda^+ M \to \Lambda^+ M$ of $R$ the *self-dual curvature operator* of $(M, g)$.

REMARK 1.2. Let $x$ be a point in an oriented pseudo-Riemannian 4-manifold $(M, g)$ having one of the sign patterns (1), and let $u \in T_x M$ be a vector such that $g(u, u) \neq 0$ and the subspace $u \wedge u^{\perp}$ of $[T_x M]^{\wedge 2}$ formed by all $u \wedge v$ with $v \in u^{\perp}$ is invariant under the Weyl tensor $W(x) : [T_x M]^{\wedge 2} \to [T_x M]^{\wedge 2}$. Then

(a) In the Riemannian and neutral cases, the restriction $u \wedge u^{\perp} \to \Lambda_x^+ M$ of the orthogonal projection $[T_x M]^{\wedge 2} \to \Lambda_x^+ M$ is a linear isomorphism under which $W(x) : u \wedge u^{\perp} \to u \wedge u^{\perp}$ corresponds to $W^+(x) : \Lambda_x^+ M \to \Lambda_x^+ M$.

(b) In the Lorentzian case, the real subspace $u \wedge u^{\perp}$ spans $[T_x M]^{\wedge 2}$ as a complex vector space, and the Weyl tensor $W(x) : [T_x M]^{\wedge 2} \to [T_x M]^{\wedge 2}$ is the unique complex-linear extension of $W(x) : u \wedge u^{\perp} \to u \wedge u^{\perp}$.

In fact, our formula for $*$ applied to $e_1, \ldots, e_4$ with $u = r e_1$ for some $r > 0$ shows that $\mathcal{H} = u \wedge u^{\perp}$ and its $*$-image $*\mathcal{H}$ together span $[T_x M]^{\wedge 2}$, and so, for dimensional reasons, $\mathcal{H} \cap *\mathcal{H} = \{0\}$. This gives (b). Now let $g$ be Riemannian or neutral. As $\mathcal{H} \cap *\mathcal{H} = \{0\}$, the space $\mathcal{H}$ contains no nonzero (anti)self-dual bivectors. The projection $[T_x M]^{\wedge 2} \to \Lambda_x^+ M$, which has the kernel $\Lambda_x^- M$, is therefore injective on $\mathcal{H}$, i.e., constitutes an isomorphism $\mathcal{H} \to \Lambda_x^+ M$. Finally, since $\Lambda^{\pm} M$ are $W$-invariant, the projection commutes with $W(x)$, and (a) follows.

## §2. One particular family of metrics

The construction described here goes back to Petrov; see [**15**], p. 185.

Let $\mathcal{X}$ be a real vector space of any dimension $n \geq 3$ with a codimension-one subspace $V \subset \mathcal{X}$ and an element $u \in \mathcal{X} \smallsetminus V$, and let $\langle \, , \rangle$ be a nondegenerate symmetric bilinear form in $V$. If a linear operator $F : V \to V$ is *self-adjoint* relative to $\langle \, , \rangle$, that is, $\langle Fv, v' \rangle = \langle v, Fv' \rangle$ for all $v, v' \in V$, then, choosing any $\delta \in \{1, -1\}$, we define a Lie-algebra multiplication $[\, , ]$ in $\mathcal{X}$ and a nondegenerate symmetric bilinear form $g$ in $\mathcal{X}$ by

$$(4) \quad \begin{array}{ll} \text{i)} & [u, v] = Fv, \qquad [v, v'] = 0 \qquad \text{whenever} \quad v, v' \in V, \\ \text{ii)} & g(u, u) = \delta, \quad g(u, v) = 0, \quad g(v, v') = \langle v, v' \rangle \quad \text{for all} \quad v, v' \in V. \end{array}$$

Let there also be given an $n$-dimensional real manifold $M$ such that $\mathcal{X}$, rather than being just an abstract Lie algebra, is a *simply transitive Lie algebra of vector fields* on $M$, as defined below in the appendix. An explicit description of such $M$ is given in the last paragraph of this section; another option is to choose $M$ to be the underlying manifold of a Lie group $G$ whose Lie algebra of left-invariant vector fields is isomorphic to $\mathcal{X}$. Formula (4.ii) now defines a pseudo-Riemannian metric $g$ on $M$ such that $g(u,v)$ is constant whenever $u,v\in\mathcal{X}$, or, in Lie-group terms, $g$ is invariant under left translations in $G$. If $\nabla, R$ and Ric denote the Levi-Civita connection, curvature tensor and Ricci tensor of this metric $g$, and $v,v',w\in V$ are treated, along with $u$, as vector fields on $M$, then

$$(5)\quad \begin{aligned} &\text{a)}\ \ \nabla_u u = \nabla_u v = 0\,, \quad \nabla_v u = -Fv\,, \quad \nabla_v w = \delta\langle Fv,w\rangle u\,,\\ &\text{b)}\ \ R(u,v)u = -F^2v\,, \quad R(v,w)u = 0\,, \quad R(u,w)v = \delta\langle F^2w,v\rangle u\,,\\ &\text{c)}\ \ R(v,v')w = \delta\langle Fv',w\rangle Fv - \delta\langle Fv,w\rangle Fv', \quad \text{d)}\ \ R(u\wedge v) = -\delta u\wedge F^2v, \end{aligned}$$

with $R(u\wedge v)$ as in (3). Namely, by (4) with $\langle Fv,v'\rangle = \langle v,Fv'\rangle$, the connection $\nabla$ in $TM$ *defined* by (5.a) is torsionfree and $\nabla g = 0$, so that $\nabla$ must be the Levi-Civita connection of $g$. Next, (5.b, c) follow from (2.i), (5.a) and (4.i), while (5.d) is clear since, by (3) and (5.b, c), $R(u\wedge v) + u\wedge F^2v$ is orthogonal to $u\wedge w$ and $v'\wedge w$ for $v',w\in V$ (cf. §1), and hence to all bivectors at every point. Also,

$$(6)\quad \mathrm{Ric}\,(u,u) = -\,\mathrm{Trace}\,F^2, \quad \mathrm{Ric}\,(u,v) = 0, \quad \mathrm{Ric}\,(v,w) = -\delta\langle Fv,w\rangle\,\mathrm{Trace}\,F,$$

for $v,w\in V$. In fact, by (2.ii) and (4.ii), $\mathrm{Ric}\,(u,u)$, $\mathrm{Ric}\,(v,w) - \delta g(R(u,v)u,w)$ and $\mathrm{Ric}\,(u,v)$ are the traces of the operators $V\to V$ given by $v\mapsto R(u,v)u$, $v'\mapsto R(v,v')w$, and $w\mapsto \mathrm{pr}[R(u,w)v]$, where $\mathrm{pr}:\mathcal{X}\to V$ is the orthogonal projection, so that (6) is immediate from (5.b, c).

An $n$-dimensional manifold $M$, admitting a simply transitive Lie algebra $\mathcal{X}$ of vector fields (cf. the appendix) with a vector subspace $V\subset\mathcal{X}$ and a linear operator $F:V\to V$ such that $\dim V = n-1$ and the Lie bracket in $\mathcal{X}$ satisfies (4.i) for some $u\in\mathcal{X}\smallsetminus V$, can be constructed as follows. We fix $V$ and $F:V\to V$, then set $M = V\times(0,\infty)$ and let $\mathcal{X} = V + \mathbf{R}u$ be the space of vector fields on $M$ spanned by $u$ and $V$, where $u$ is the linear vector field with $u(x,t) = (-Fx,t)$ for $(x,t)\in M$, and each $v\in V$ is identified with the constant vector field $(v,0)$. Now (4.i) follows since $[v,w] = d_vw - d_wv$ for vector fields $v,w$ on any open subset $U$ of a finite-dimensional vector space $\mathcal{X}$, treated as functions $U\to\mathcal{X}$.

## §3. The Einstein case

Let $\mathcal{X}, n, V, u, F, \langle\,,\rangle, \delta, M$ have the properties listed in §2, and let $g$ be the pseudo-Riemannian metric with (4.ii) on the $n$-dimensional manifold $M$. Then $g$ is Einstein if and only if one of the following conditions holds:

(i) $F$ equals some real scalar $\lambda$ times the identity.
(ii) $F\neq 0$, while $\mathrm{Trace}\,F = 0$ and $F^2 = 0$.
(iii) $F^2\neq 0$ and $\mathrm{Trace}\,F = \mathrm{Trace}\,F^2 = 0$.

To see this, note that each of (i) – (iii) implies, by (6), that $g$ is Einstein. Conversely, let $g$ be Einstein; then either $\mathrm{Trace}\,F\neq 0$ (and hence (6) for $v,w$ yields (i)), or $\mathrm{Trace}\,F = 0$ and so, by (6), $\mathrm{Trace}\,F^2 = 0$, which in turn gives (i) (when $F = 0$), or (ii) (when $F\neq 0$ and $F^2 = 0$), or, finally, (iii) (when $F^2\neq 0$).

LEMMA 3.1. *For $\mathcal{X}, n, V, u, F, \langle\,,\rangle, \delta, M$ and $g$ as above, let $g$ be an Einstein metric, so that we have* (i), (ii) *or* (iii). *Then*

*In case* (i), *$g$ has the constant sectional curvature $-\delta\lambda^2$.*

*In case* (ii), *under the additional assumption that* $n = 4$, *the metric* $g$ *is flat. In case* (iii), $g$ *is Ricci-flat but* not *locally symmetric.*

In fact, the assertion about (i) follows from (5.b, c). Next, if $n = 4$, (ii) gives $F(V) \subset \operatorname{Ker} F \neq V$ and $\dim[F(V)] + \dim[\operatorname{Ker} F] = \dim V = 3$, i.e., $\dim[F(V)] = 1$ and $\dim[\operatorname{Ker} F] = 2$. Thus, the right-hand sides in (5.b, c) both vanish: the former since $F^2 = 0$, the latter in view of skew-symmetry in $Fv, Fv' \in F(V)$ with $\dim[F(V)] = 1$. This proves our claim about (ii).

Finally, in case (iii), Ric $= 0$ by (6), while $(\nabla_w R)(u,v)v' = \nabla_w[R(u,v)v'] - R(\nabla_w u, v)v' - R(u, \nabla_w v)v' - R(u,v)\nabla_w v'$ for $v, v', w \in V$, and so $\delta(\nabla_w R)(u,v)v' = -\langle F^2 v, v'\rangle Fw + \langle Fv, v'\rangle F^2 w - \langle F^2 w, v'\rangle Fv + \langle Fw, v'\rangle F^2 v$ by (5). If $R$ were parallel, setting $w = v$ and applying $g(\,\cdot\,, v'')$ with any $v'' \in V$ (see §1) we would get $Fv \wedge F^2 v = 0$ for all $v \in V$. Every $Fv \in F(V) \smallsetminus \{0\}$ thus would be an eigenvector of $F$, making $F^2$ a multiple of $F$, contrary to (iii) (cf. Remark 3.2). $\square$

REMARK 3.2. If $\operatorname{Trace} F = 0$ for an operator $F : V \to V$ in a finite-dimensional vector space $V$ and $F^2$ equals a nonzero scalar times $F$, then $F = 0$. In fact, let $rF^2 = 2F$ and $r \neq 0$. Then $A^2 = \mathrm{Id}$ for the operator $A = \mathrm{Id} - rF$ in $V$, and so $A = \pm \mathrm{Id}$ on some subspaces $V_\pm$ with $V = V_+ \oplus V_-$. Hence $\operatorname{Trace} A = n_+ - n_-$, where $n_\pm = \dim V_\pm$, while $\operatorname{Trace} A = \dim V = n_+ + n_-$ as $A = \mathrm{Id} - rF$ and $\operatorname{Trace} F = 0$. Thus, $n_- = 0$, i.e., $V = V_+$, so that $A = \mathrm{Id}$ and $F = 0$.

## §4. The curvature operator

Given a fixed sign $\pm$, formulae

(a) $V = \mathbf{C} \times \mathbf{R}$ and $\langle (z,t), (z',t')\rangle = \operatorname{Im} zz' \pm tt'$ for $(z,t), (z',t') \in V$,

(b) $F(z,t) = (pqz, pt)$, with $q = e^{2\pi i/3}$ and any fixed $p \in \mathbf{R} \smallsetminus \{0\}$,

define a real vector space $V$ with $\dim V = 3$, a nondegenerate symmetric bilinear form $\langle\,,\rangle$ in $V$ with the sign pattern $- \pm +$, and a self-adjoint operator $F : V \to V$ satisfying (iii) in §3. (See also the last paragraph of this section.)

In fact, $\langle F(z,t), (z',t')\rangle = \langle (pqz, pt), (z',t')\rangle = p(\operatorname{Im} qzz' \pm tt')$ is symmetric in $(z,t), (z',t')$, while (iii) holds for $F$ since $F^2(z,t) = (p^2q^2z, p^2t)$, $q = (\sqrt{3}\,i - 1)/2$ and $q^2 = q^{-1} = \bar{q}$.

REMARK 4.1. Let $B : V \to V$ be a linear operator in an $n$-dimensional real vector space $V$. As in §0, we call $B$ *complex-diagonalizable* if its complex-linear extension $B : V^{\mathbf{C}} \to V^{\mathbf{C}}$ to the complexification of $V$ is diagonalizable. Clearly, (a) if $B$ is diagonalizable, it is complex-diagonalizable; (b) $B$ is complex-diagonalizable whenever its characteristic polynomial has $n$ distinct complex roots; (c) if $V$ is the underlying real space of a complex vector space in which $B$ acts complex-linearly, then complex-diagonalizability of $B$ is equivalent to diagonalizability of $B$ as a complex-linear operator.

EXAMPLE 4.2. Let a four-manifold $M$ and an indefinite metric $g$ on $M$ be chosen as in §2 using $n = 4$, some $\delta \in \{1, -1\}$, and $V, \langle\,,\rangle, F$ defined in (a), (b) above for any fixed sign $\pm$ and $p \in \mathbf{R} \smallsetminus \{0\}$. According to Lemma 3.1, $g$ is Ricci-flat but not locally symmetric. By (4.ii), the sign pattern of $g$ is $- \pm + +$ (when $\delta = 1$) or $- - \pm +$ (when $\delta = -1$). We consider two cases:

(i) $\delta = 1$ and the sign $\pm$ is $+$. Thus, $g$ is a Ricci-flat Lorentzian metric.

(ii) $\delta = \mp 1$, so that $g$ is a neutral $(- - + +)$ Ricci-flat metric.

In both cases, $(M, g)$ is locally homogeneous and, by (4.ii), locally isometric to a Lie group with a left-invariant metric. (See Corollary A.3 in the appendix.)

Also, the curvature operator $R$ of $(M,g)$ is complex-diagonalizable at every point. Namely, by (5.d), $R$ leaves invariant the subbundle $\mathcal{H}$ of $[TM]^{\wedge 2}$ spanned by all $u \wedge v$ with $v \in V$. Also, again by (5.d), $R : \mathcal{H} \to \mathcal{H}$ is, at every point, algebraically equivalent to $-\delta F^2 : V \to V$. On the other hand, $F^2$ is complex-diagonalizable by Remark 4.1(b), since $F^2/p^2$ has the characteristic roots $1, q, \overline{q}$, and our assertion follows, in case (i), from Remark 1.2(b) for any fixed orientation of $M$, combined with Remark 4.1(c), and, in case (ii), from Remark 1.2(a) applied to both orientations of $M$.

REMARK 4.3. As we just saw, for $(M,g)$ obtained in Example 4.2, the curvature operator $R$ (case (i)), or its self-dual restriction $R^+$, for either orientation (case (ii)), has the complex eigenvalues $\lambda, \lambda e^{2\pi i/3}, \lambda e^{4\pi i/3}$ with $\lambda \in \mathbf{R} \smallsetminus \{0\}$. Also, for every locally symmetric pseudo-Riemannian Einstein 4-manifold with a complex-diagonalizable curvature operator, $R$ (or, $R^+$) has a multiple eigenvalue ([**10**], pp. 662–663). Finally, according to sections 5 – 7 below, Example 4.2 describes, locally, all possible 4-dimensional curvature-homogeneous pseudo-Riemannian Einstein manifolds with the sign patterns (1), which are not locally symmetric. Thus, the algebraic types of curvature operators realized by curvature-homogeneous pseudo-Riemannian Einstein 4-manifolds are quite special, in analogy with the result of [**14**] for curvature-homogeneous *Riemannian* manifolds of dimension 4.

The claim made in the three lines following (a), (b) above remains valid if one replaces (b) with $F(z,t) = (\pm it, \operatorname{Re} z)$. This leads, as in Example 4.2, to another locally homogeneous Ricci-flat pseudo-Riemannian 4-manifold $(M,g)$, except that, for analogous reasons, its curvature operator is *not* complex-diagonalizable.

## §5. A classification theorem for the Lorentzian case

In the following theorem, proved in §13, the diagonalizability assumption about the curvature operator amounts to its complex-diagonalizability; see Remark 4.1(c).

THEOREM 5.1. *Let $(M,g)$ be an oriented four-dimensional Lorentzian Einstein manifold whose curvature operator, treated as a complex-linear vector bundle morphism $R : [TM]^{\wedge 2} \to [TM]^{\wedge 2}$, is diagonalizable at every point and has complex eigenvalues that form constant functions $M \to \mathbf{C}$. Then $(M,g)$ is locally homogeneous, and one of the following three cases occurs:*

(a) *$(M,g)$ is a space of constant curvature.*

(b) *$(M,g)$ is locally isometric to the Riemannian product of two pseudo-Riemannian surfaces having the same constant Gaussian curvature.*

(c) *$(M,g)$ is locally isometric to Petrov's Ricci-flat manifold of* Example 4.2(i). *Furthermore, $(M,g)$ is locally symmetric in cases* (a) – (b), *but not in* (c), *and in case* (c) *it is locally isometric to a Lie group with a left-invariant metric.*

## §6. A classification theorem in the neutral case

The next theorem will be proved at the end of §13. For the definitions of $R^+$ and complex-diagonalizability, see Remarks 1.1 and 4.1.

THEOREM 6.1. *Let the self-dual curvature operator $R^+ : \Lambda^+ M \to \Lambda^+ M$ of an oriented four-dimensional Einstein manifold $(M,g)$ of the metric signature $--++$ be complex-diagonalizable at every point, with complex eigenvalues forming* *constant functions* $M \to \mathbf{C}$. *If* $\nabla R^+ \neq 0$ *somewhere in* $M$, *then* $(M, g)$ *is locally homogeneous, namely, locally isometric to a Lie group with a left-invariant metric.*

*More precisely,* $(M, g)$ *then is locally isometric to one of Petrov's Ricci-flat manifolds, described in* Example 4.2(ii).

Theorem 6.1 sounds much stronger than its Riemannian analogue, i.e., Theorem 7.2 in §7: an assumption about $R^+$ yields a complete local description of the metric in the former result, but only an assertion about $R^+$ in the latter. However, if the clause "$\nabla R^+ \neq 0$ somewhere" were to be included among the hypotheses of Theorem 7.2, as it is in Theorem 6.1, the conclusion of Theorem 7.2 would become an equally strong nonexistence statement.

## §7. The Riemannian case

For Riemannian metrics, our assertion amounts to the following theorem, in which the assumption of complex-diagonalizability is redundant, as the curvature operator is self-adjoint at every point; cf. Remark 4.1(a). See also [**12**].

THEOREM 7.1 ([**10**], Corollary 7.2 on p. 476). *If the curvature operator of a four-dimensional Riemannian Einstein manifold* $(M, g)$, *acting on bivectors, has the same eigenvalues at every point* $x \in M$, *then* $(M, g)$ *is locally symmetric.*

This is immediate from the next result, proved in §13 (and, originally, in [**10**]):

THEOREM 7.2 ([**10**], p. 476, Theorem 7.1). *If* $(M, g)$ *is an oriented Riemannian Einstein four-manifold and its self-dual curvature operator* $R^+ : \Lambda^+M \to \Lambda^+M$ *has the same eigenvalues at every point, then* $R^+$ *is parallel.*

## §8. Further basics

Unless stated otherwise, all tensor fields are of class $C^\infty$. For 1-forms $\xi, \eta$, vector fields $u, v, w$ and a pseudo-Riemannian metric $g$ on any manifold, $\xi \wedge \eta$, $d\xi$ and the Lie derivative $\mathcal{L}_w g$ are given by $(\xi \wedge \eta)(u, v) = \xi(u)\eta(v) - \xi(v)\eta(u)$, $(d\xi)(u, v) = d_u[\xi(v)] - d_v[\xi(u)] - \xi([u, v])$ and, with $[\,,\,]$ denoting the Lie bracket,

$$(\mathcal{L}_w g)(u, v) = d_w[g(u, v)] - g([w, u], v) - g(u, [w, v]) . \tag{7}$$

On a pseudo-Riemannian manifold $(M, g)$ we use the same symbol, such as $u$, for a vector field and the corresponding 1-form $g(u, \cdot\,)$. Similarly, a vector-bundle morphism $\alpha : TM \to TM$ is treated as a twice-contravariant tensor field, and as a twice-covariant one with $\alpha(u, v) = g(\alpha u, v)$ for vector fields $u, v$. In particular, a bivector field $\alpha$ (such as $v \wedge u$) is also regarded as a differential 2-form, or a morphism $\alpha : TM \to TM$ with $\alpha^* = -\alpha$ (i.e., skew-adjoint at each point). Specifically, for bivector fields $\alpha, \alpha'$ and $C^1$ vector fields $u, v, w$,

$$\begin{array}{ll} \text{a)} & v \wedge u = v \otimes u - u \otimes v\,, \qquad dw = P - P^*, \quad \text{where } P = \nabla w\,, \\ \text{b)} & (v \otimes u)w = \langle v, w\rangle u\,, \qquad (v \wedge u)w = \langle v, w\rangle u - \langle u, w\rangle v\,, \\ \text{c)} & \langle \alpha, v \wedge u\rangle = g(\alpha v, u)\,, \qquad \langle \alpha, \alpha'\rangle = -\operatorname{Trace}(\alpha \circ \alpha')/2\,, \\ \text{d)} & \alpha \circ (v \wedge u) = v \otimes (\alpha u) - u \otimes (\alpha v), \quad \text{e) } \operatorname{Trace}[\alpha \circ (v \wedge u)] = -2g(\alpha v, u)\,, \\ \text{f)} & 2P^*w = d\langle w, w\rangle\,, \qquad \text{where } P = \nabla w \text{ and } \langle w, w\rangle = g(w, w)\,, \end{array} \tag{8}$$

with $\langle v, w\rangle = g(v, w)$, $\langle u, w\rangle = g(u, w)$ in b). (Cf. §1.) Here d) follows from b), and implies e), as $\operatorname{Trace}(v \otimes u) = g(v, u)$, while $P = \nabla w : TM \to TM$ in a), f) acts by $Pv = \nabla_v w$, and so $2\langle v, P^*w\rangle = 2\langle Pv, w\rangle = d_v\langle w, w\rangle$, which gives f).

REMARK 8.1. Let $(M,g)$ be a pseudo-Riemannian Einstein manifold. Then $\operatorname{div} W = 0$. If, in addition, $M$ is oriented, $\dim M = 4$, and the sign pattern of $g$ is $++++$ or $--++$, then also $\operatorname{div} W^+ = \operatorname{div} W^- = 0$.

In fact, these are well-known consequences of the second Bianchi identity (cf. [**10**], pp. 460, 468). Here $\operatorname{div}\alpha$, for any covariant tensor field $\alpha$, is the $g$-contraction of $\nabla\alpha$ involving the first argument of $\alpha$ and the differentiation argument.

By a *complex vector field* on a real manifold $M$ we mean a section $w$ of its complexified tangent bundle. Sections of the ordinary ("real") tangent bundle of $M$ may be referred to as *real vector fields* on $M$. Thus, $w = u + iv$ with real vector fields $u = \operatorname{Re} w$, $v = \operatorname{Im} w$. Complex *bivector* fields are defined similarly.

All real-multilinear operations involving real vector/bivector fields will, without further comment, be extended to complex vector (or, bivector) fields $v, w$ (or, $\alpha, \alpha'$), so as to become complex-linear in each argument. This includes the Lie bracket $[v,w]$, the covariant derivative $\nabla_v w$ relative to any connection in the tangent bundle, the inner product $g(v,w)$, the Lie derivative $\mathcal{L}_w g$ for any given pseudoRiemannian metric $g$, the composite $\alpha \circ \alpha'$, as well as $\langle \alpha, \alpha' \rangle$, $*\alpha$ and $\alpha v$ (cf. (8.c)). Note that $g(v,w)$ and $\langle \alpha, \alpha' \rangle$ are *complex-bilinear* (not sesquilinear!) in $v, w$ or $\alpha, \alpha'$, while a $C^\infty$ complex vector field $w$ is a Killing field, i.e., $\mathcal{L}_w g = 0$, if and only if its real and imaginary parts both are real Killing fields.

Although the bivector bundle $[TM]^{\wedge 2}$ of an oriented Lorentzian four-manifold $(M,g)$ is a complex vector bundle with the multiplication by $i$ provided by $*$ (see §1), it is also convenient to use the complexification $([TM]^{\wedge 2})^{\mathbf{C}}$ of its underlying real vector bundle. Then $([TM]^{\wedge 2})^{\mathbf{C}} = \Lambda^+ M \oplus \Lambda^- M$, where $\Lambda^\pm M$ are, this time, the complex vector bundles of fibre dimension 3, obtained as the $(\pm i)$-eigenspace bundles of $*$. This is clear since $*^2 = -\operatorname{Id}$, cf. §1, and the complex-conjugation antiautomorphism $([TM]^{\wedge 2})^{\mathbf{C}} \to ([TM]^{\wedge 2})^{\mathbf{C}}$ obviously sends $\Lambda^+ M$ onto $\Lambda^- M$.

## §9. A unified treatment of all three cases

Throughout this section $(M,g)$ stands for a fixed oriented pseudo-Riemannian four-manifold with a metric $g$ of one of the sign patterns (1), while $\mathcal{E}$ is a complex vector bundle of of fibre dimension 3 over $M$, and $W^{(+)}$ is a complex-linear bundle morphism $\mathcal{E} \to \mathcal{E}$. Our choices of $\mathcal{E}$ and $W^{(+)}$ are quite specific. Namely, when $g$ is Riemannian or neutral, $\mathcal{E} = [\Lambda^+ M]^{\mathbf{C}}$ is the complexification of the subbundle $\Lambda^+ M$ of $[TM]^{\wedge 2}$ (§1) and $W^{(+)}$ is the unique $\mathbf{C}$-linear extension of $W^+ : \Lambda^+ M \to \Lambda^+ M$ to $[\Lambda^+ M]^{\mathbf{C}}$, while, if $g$ is Lorentzian, $\mathcal{E} = \Lambda^+ M \subset ([TM]^{\wedge 2})^{\mathbf{C}}$ (see end of §8) and $W^{(+)}$ is the restriction to $\mathcal{E}$ of the $\mathbf{C}$-linear extension of $W : [TM]^{\wedge 2} \to [TM]^{\wedge 2}$ to $([TM]^{\wedge 2})^{\mathbf{C}}$. (The latter extension leaves $\mathcal{E}$ invariant, since $[W, *] = 0$, cf. §1.)

We will use the symbol $\nabla$ for the connection in $\mathcal{E}$ induced by the Levi-Civita connection of $g$, and let $h$ stand for the complex-bilinear fibre metric in $\mathcal{E}$ which, in the Riemannian/neutral (or, Lorentzian) case is the unique complex-bilinear extension of $\langle\,,\rangle$ (see §1) from $\Lambda^+ M$ to $[\Lambda^+ M]^{\mathbf{C}}$ (or, respectively, the restriction to $\mathcal{E} = \Lambda^+ M$ of the complex-bilinear extension of $\langle\,,\rangle$ from $[TM]^{\wedge 2}$ to $([TM]^{\wedge 2})^{\mathbf{C}}$). Note that, in all cases, $\nabla h = 0$ and $\Lambda^+ M$ is a $\nabla$-parallel subbundle of $[TM]^{\wedge 2}$ or $([TM]^{\wedge 2})^{\mathbf{C}}$, since the Levi-Civita connection of $g$ makes both $g$ and $*$ parallel.

In this and the next two sections, the indices $j, k, l$ always vary in the range $\{1, 2, 3\}$ and repeated indices are summed over, unless explicitly stated otherwise.

Given $M, g, \mathcal{E}, W^{(+)}, \nabla, h$ as above, let us now fix any $C^\infty$ local sections $\alpha_j$ of $\mathcal{E}$ which trivialize $\mathcal{E}$ on an open set $U \subset M$. This gives rise to complex-valued functions $h_{jk}$ and 1-forms $\xi_j^k$ with

$$(9)\quad \begin{aligned} &\text{a) } \nabla\alpha_j = \xi_j^l \otimes \alpha_l\,, \text{ i.e., } \nabla_v \alpha_j = \xi_j^l(v)\alpha_l \text{ for every tangent vector field } v,\\ &\text{b) } dh_{jk} = \xi_{jk} + \xi_{kj}\,, \quad \text{where } \xi_{jk} = \xi_j^l h_{lk} \text{ and } h_{jk} = h(\alpha_j, \alpha_k). \end{aligned}$$

Thus, $h_{jk}$ are the component functions of the fibre metric $h$ and $\xi_j^k$ are the connection forms of $\nabla$, relative to the $\alpha_j$, while (9.b) states that $\nabla h = 0$. For div as in Remark 8.1 and all tangent vectors $v$ we have, with summation over $k$,

$$(10)\quad \begin{aligned} &\text{i) } [\nabla_v W]\alpha_j = \theta_j^k(v)\alpha_k\,, &&\text{ii) } [\operatorname{div} W]\alpha_j = \alpha_k\,\theta_j^k\,, \quad \text{where}\\ &\text{iii) } W\alpha_j = W_j^k\alpha_k \quad \text{and} &&\text{iv) } \theta_j^l = dW_j^l + W_j^k\xi_k^l - W_k^l\xi_j^k\,. \end{aligned}$$

In fact, $\nabla_v$ applied to iii) gives i) (by (9.a)), and contracting i) we get ii) . (Here $W$ might be replaced by $W^{(+)}$, as $\nabla W^{\pm}$ are the $\Lambda^{\pm}M$ components of $\nabla W$.)

Remark 9.1. If $g$ is Riemannian or neutral, $W^{(+)}$ is the $\mathbf{C}$-linear extension of $W^+$ to $[\Lambda^+M]^{\mathbf{C}}$. Thus, in the Riemannian case, the eigenvalues of $W^{(+)}$ at every point are all real, as $W : [TM]^{\wedge 2} \to [TM]^{\wedge 2}$ is self-adjoint.

If $g$ is Lorentzian, $W^{(+)}$ is, at each point, algebraically equivalent to $W$ acting in $[TM]^{\wedge 2}$, since $\alpha \mapsto \alpha - i[*\alpha]$ is an isomorphism $[TM]^{\wedge 2} \to \Lambda^+M$ of complex vector bundles, sending $W$ onto $W^{(+)}$ (as $[W,*] = 0$, cf. §1).

## §10. Calculations in a local orthonormal frame

As in §9, the indices $j,k,l$ vary in the set $\{1,2,3\}$. The *Ricci symbol* $\varepsilon_{jkl}$ will always stand for the signum of the permutation $(j,k,l)$ of $(1,2,3)$, if $j \neq k \neq l \neq j$, while $\varepsilon_{jkl} = 0$ if $j = k$ or $k = l$ or $l = j$. From now on we assume (cf. Remark 10.3 below) that, for our $\alpha_j$ and some $\varepsilon_j \in \mathbf{R}$,

$$(11)\quad \begin{aligned} &\text{i) } \varepsilon_j\alpha_j \circ \alpha_j = -\mathrm{Id} \text{ and } \alpha_j \circ \alpha_k = \varepsilon_l\alpha_l = -\alpha_k \circ \alpha_j \text{ if } \varepsilon_{jkl} = 1,\\ &\text{ii) } \varepsilon_1\varepsilon_2\varepsilon_3 = 1 \quad \text{and} \quad \varepsilon_j \in \{1,-1\}\,,\ j = 1,2,3. \end{aligned}$$

(No summing over $j,l$.) For a complex vector field $w$, the complex vector fields

$$(12)\qquad v_j = \alpha_j w\,, \qquad j = 1,2,3\,,$$

satisfy, in view of (11) and skew-adjointness of the $\alpha_j$, the relations

$$(13)\quad \begin{aligned} &\text{a) } \langle v_j, v_k\rangle = \varepsilon_j\langle w,w\rangle\,\delta_{jk} \text{ (no summation), } \langle w, v_j\rangle = 0,\ j,k = 1,2,3,\\ &\text{b) } \alpha_j v_k = -\alpha_k v_j = \varepsilon_l v_l,\ \alpha_j v_j = -\varepsilon_j w \text{ (no summing) if } \varepsilon_{jkl} = 1, \end{aligned}$$

with $\langle\,,\rangle$ standing for $g(\ ,\ )$. From (11), (9.b) and (8.c), $h_{jk} = 2\varepsilon_j\delta_{jk}$ (no summing), and so, again by (9.b), the $\xi_{jk}$ are skew-symmetric in $j,k$. Therefore, as $\varepsilon_k\varepsilon_l = \varepsilon_j$ when $\{j,k,l\} = \{1,2,3\}$ (by (11.ii)), we have, from (9.a), (11.ii),

$$(14)\quad \begin{aligned} &\text{i) } \xi_j^j = 0 \text{ (no summing) and } \xi_j^k = \varepsilon_j\xi_l\,,\ \xi_j^l = -\varepsilon_j\xi_k \text{ if } \varepsilon_{jkl} = 1\,,\\ &\text{ii) } \varepsilon_j\nabla\alpha_j = \xi_l \otimes \alpha_k - \xi_k \otimes \alpha_l \text{ (no summing) whenever } \varepsilon_{jkl} = 1\,, \end{aligned}$$

with the 1-forms $\xi_j$ defined by $\xi_j = \varepsilon_j\xi_{kl}$ if $\varepsilon_{jkl} = 1$. Next, we define complex-valued functions $\lambda_j, \mu_j$, $j = 1,2,3$, by

$$(15)\quad \lambda_j = W_j^j\,, \quad \text{and} \quad \mu_j = \varepsilon_l W_k^l \quad \text{if } \{j,k,l\} = \{1,2,3\} \quad \text{(no summing).}$$

We always have $\operatorname{Trace} W^{(+)} = 0$ ([**10**], p. 650); thus, for any function s on $U$,

$$(16)\qquad \begin{aligned} &\text{a) } \lambda_1 + \lambda_2 + \lambda_3 = 0\,,\\ &\text{b) } L_1 + L_2 + L_3 = 0 \text{ if } L_j = (\lambda_k - \lambda_l)(\lambda_j + \mathrm{s}/12) \text{ whenever } \varepsilon_{jkl} = 1\,, \end{aligned}$$

REMARK 10.1. By (10.iv), (14.i) and (15), $\theta_j^j = d\lambda_j + 2\mu_k\xi_k - 2\mu_l\xi_l$, $\theta_j^k = \varepsilon_k d\mu_l + \varepsilon_j(\lambda_j - \lambda_k)\xi_l + \varepsilon_j\varepsilon_k\mu_j\xi_k - \mu_k\xi_j$, $\theta_j^l = \varepsilon_l d\mu_k + \varepsilon_j(\lambda_l - \lambda_j)\xi_k - \varepsilon_j\varepsilon_l\mu_j\xi_l - \mu_l\xi_j$ (no summing), if $\varepsilon_{jkl} = 1$. Hence, from (10.ii) and (11.i), $[\operatorname{div} W]\alpha_j = \alpha_j[d\lambda_j + w_k - w_l]$ whenever $\varepsilon_{jkl} = 1$ (no summing), for the complex vector fields $w_j$ given by $w_j = 2\mu_j\xi_j + \alpha_j[d\mu_j + \varepsilon_k\varepsilon_l(\lambda_k - \lambda_l)\xi_j + \varepsilon_k\mu_k\xi_l - \varepsilon_l\mu_l\xi_k]$ if $\varepsilon_{jkl} = 1$ (no summing).

Consequently, if $\operatorname{div} W^{(+)} = 0$ and the $\lambda_j$ are all constant, then there exists a complex vector field $w$ with $w_1 = w_2 = w_3 = w$. This is clear if one applies $\alpha_j$ to the above formula for $[\operatorname{div} W]\alpha_j$ and uses (11.i).

REMARK 10.2. For $M, g, \mathcal{E}, W^{(+)}, U, \alpha_j, \varepsilon_j, \xi_j$ as above, with (11),

(i) $dv_j = \varepsilon_j\xi_l \wedge v_k - \varepsilon_j\xi_k \wedge v_l + P_j$ if $\varepsilon_{jkl} = 1$, for $v_j, P_j$ given by (12) and

$$(17)\qquad P_j = \alpha_j \circ P + P^* \circ \alpha_j \qquad \text{with} \quad P = \nabla w\,,$$

where $w$ is any given complex $C^\infty$ vector field defined on $U$.

(ii) If $W_j^k$ in (10.iii) are constant, the following two conditions are equivalent:

a) $\nabla W^{(+)} = 0$ everywhere in $U$.

b) $\mu_1\xi_1 = \mu_2\xi_2 = \mu_3\xi_3$ and $(\lambda_k - \lambda_l)\xi_j + \varepsilon_l\mu_k\xi_l - \varepsilon_k\mu_l\xi_k = 0$ for $\lambda_j, \mu_j$ given by (15) and any $j, k, l$ with $\{j, k, l\} = \{1, 2, 3\}$.

(iii) If $g$ is Einstein and $\varepsilon_{jkl} = 1$, then $d\xi_j + \varepsilon_j\xi_k \wedge \xi_l = -(W + \mathrm{s}/12)\alpha_j$.

In fact, $\nabla v_j = \varepsilon_j\xi_l \otimes v_k - \varepsilon_j\xi_k \otimes v_l + \alpha_j \circ P$ if $\varepsilon_{jkl} = 1$, by (14.ii), and so (8.a) yields (i). Next, (ii) is clear from (10.i), (15) and the formulae for $\theta_j^j, \theta_j^k, \theta_j^l$ in Remark 10.1. Finally, let $g$ be Einstein. For fixed $j, k, l$ with $\varepsilon_{jkl} = 1$, (2.i), (14.ii) and the formulae for $\xi \wedge \eta$, $d\xi$ in §8 give $-\varepsilon_k\varepsilon_l\omega/2 = d\xi_j + \varepsilon_j\xi_k \wedge \xi_l$, where $\omega$ is the complex-valued 2-form with $\omega(u, v) = h(R^\nabla(u, v)\alpha_k, \alpha_l)$ for any vector fields $u, v$, with $R^\nabla$ denoting the curvature of our connection $\nabla$ in $\mathcal{E}$ (§9). However, $R^\nabla(u, v)\alpha_k = [R(u, v), \alpha_k]$, where $[\,,\,]$ also stands for the commutator of bundle morphisms $[TM]^{\mathbf{C}} \to [TM]^{\mathbf{C}}$, and $R(u, v) : [TM]^{\mathbf{C}} \to [TM]^{\mathbf{C}}$ is defined as in (2.i); this is easily seen using (2.i) and the Leibniz-rule equality $\nabla_u\alpha = [\nabla_u, \alpha]$ for such morphisms $\alpha$ (with the commutator applied, this time, to operators acting on vector *fields*). Hence, by Lemma 5.3 on p. 460 of [**10**], $\omega = R[\alpha_k, \alpha_l]$ with $[\alpha_k, \alpha_l] = \alpha_k \circ \alpha_l - \alpha_l \circ \alpha_k$, i.e., as $[\alpha_k, \alpha_l] = 2\varepsilon_j\alpha_j$ (by (11.i)) and $R = W + \mathrm{s}/12$ (Remark 1.1), we have $\omega = 2\varepsilon_j(W + \mathrm{s}/12)\alpha_j$, and (iii) follows.

REMARK 10.3. Let $M, g, \mathcal{E}, W^{(+)}$ be as in §9. If $W^{(+)}(x) : \mathcal{E}_x \to \mathcal{E}_x$ is diagonalizable for every $x \in M$ and the set of its eigenvalues does not depend on $x$, then a suitable connected neighborhood $U$ of any given point of $M$ admits $C^\infty$ local trivializing sections $\alpha_j$ of $\mathcal{E}$, $j = 1, 2, 3$, satisfying conditions (11) along with (14.ii) for suitable $\varepsilon_j, \xi_j$, and such that the corresponding complex-valued functions $\lambda_j, \mu_j$ in (15) are all constant, with $\mu_j = 0$. Thus, $W\alpha_j = \lambda_j\alpha_j$ (no summing), for $j = 1, 2, 3$, i.e., the $\lambda_j$ then are the (constant) eigenvalues of $W^{(+)}$.

Namely, by Lemma 6.15(ii),(iii) of [**10**], p. 468, $W\alpha_j = \lambda_j\alpha_j$ for some $C^\infty$ sections $\alpha_j$ trivializing $\mathcal{E}$ on such a set $U$, and constants $\lambda_j$. As $W^{(+)}$ is self-adjoint relative to $h$, cf. §1, while $h$ is nondegenerate, the $\alpha_j$ may be chosen so that $h_{jk} = 2\varepsilon_j\delta_{jk}$ (no summing) with $\varepsilon_j \in \{1, -1\}$. Next, for sections $\alpha, \beta$

of $\mathcal{E}$, the anticommutator $\{\alpha,\beta\} = \alpha\circ\beta + \beta\circ\alpha$ equals $-h(\alpha,\beta)$ times Id, and the commutator $[\alpha,\beta] = \alpha\circ\beta - \beta\circ\alpha$ is a section of $\mathcal{E}$. (For $\{\alpha,\beta\}$ one can verify this, in the Lorentzian case, using a basis of $\Lambda_x^+ M$, $x\in M$, obtained by replacing each $\alpha$ by $\alpha - i[*\alpha]$ in a basis of $[TM]^{\wedge 2}$ of the form (37.28) in [**10**], p. 642; about the Riemannian and neutral cases, and for $[\alpha,\beta]$, see [**10**], formulae (37.31), (37.29) on pp. 642, 643.) Thus, by (8.c), $\varepsilon_j\alpha_j\circ\alpha_j = -\mathrm{Id}$, $j = 1,2,3$, and $\alpha_j\circ\alpha_k = \delta_l\varepsilon_l\alpha_l = -\alpha_k\circ\alpha_j$ whenever $\varepsilon_{jkl} = 1$, with some $\delta_j\in\{1,-1\}$, $j = 1,2,3$. (In view of (8.c), $\alpha_j\circ\alpha_k = [\alpha_j,\alpha_k]/2$ is $h$-orthogonal to $\alpha_j,\alpha_k$.) Now, as $(\alpha_j\circ\alpha_k)\circ\alpha_l = \alpha_j\circ(\alpha_k\circ\alpha_l)$, we get $\delta_l = \delta_j$. Consequently, $\delta_1 = \delta_2 = \delta_3$. Applying an odd permutation to the $\alpha_j$ and/or replacing them by $-\alpha_j$, if necessary, we now obtain (11).

## §11. The main structure theorem

The following result is a crucial step in our classification argument. We establish it using a refined version of the proof of Theorem 7.1 in [**10**] (pp. 477–479).

THEOREM 11.1. *Suppose that $(M,g)$ is an oriented pseudo-Riemannian Einstein four-manifold with one of the sign patterns (1), such that $W^{(+)}:\mathcal{E}\to\mathcal{E}$, defined as in §9, is diagonalizable at every point and has constant eigenvalues.*

(i) *If $g$ is positive definite, the self-dual Weyl tensor $W^+$ is parallel.*

(ii) *If $g$ is Lorentzian $(-+++)$ or neutral $(--++)$ and $W^{(+)}$ is not parallel, then any given point of $M$ has a neighborhood $U$ with $C^\infty$ complex vector fields $w, v_1, v_2, v_3$ which are linearly independent at every point of $U$, commute with every real Killing field defined on any open subset of $U$, and satisfy the inner-product and Lie-bracket relations*

$$(18)\qquad \begin{aligned} &g(w,w) = g(v_j,v_j) = \gamma \ \text{(no summing)},\ \ g(w,v_j) = g(v_j,v_k) = 0 \ \text{if } j\neq k,\\ &[w,v_j] = \rho_j v_j \quad \text{(no summing)}, \qquad [v_j,v_k] = 0 \quad \text{for all } a,b\in\{1,2,3\}, \end{aligned}$$

*for some $\gamma\in\mathbf{C}\smallsetminus\{0\}$, where $\rho_j\in\mathbf{C}$ are the three cubic roots of $\gamma^2$, and both $g$, $[\,,\,]$ act complex-bilinearly on complex vector fields.*

PROOF. Given $x\in M$, let us choose $\mathcal{E}, W^{(+)}, U, \alpha_j, \varepsilon_j, \xi_j, \lambda_j$ as in Remark 10.3, with $x\in U$. Since $(M,g)$ is Einstein, $\operatorname{div} W^{(+)} = 0$ (Remark 8.1) and so, by Remark 10.1, $w_1 = w_2 = w_3 = w$ for some complex vector field $w$, where the $w_j$ are as in Remark 10.1 with $\mu_1 = \mu_2 = \mu_3 = 0$. Now

(a) $v_j = (\lambda_l - \lambda_k)\xi_j$ if $\varepsilon_{jkl} = 1$, for the complex vector fields $v_j = \alpha_j w$,

(b) $d\xi_j + \varepsilon_j\xi_k\wedge\xi_l = -(\lambda_j + \mathrm{s}/12)\alpha_j$ whenever $\varepsilon_{jkl} = 1$,

s being the scalar curvature; in fact, (a) follows if one applies $\alpha_j$ to the formula for $w_j$ in Remark 10.1 (with $\mu_j = 0$ and $w_j = w$), using (11.i, iii), while (b) is obvious from Remark 10.2(iii) with $W\alpha_j = \lambda_j\alpha_j$. We now define a constant $\phi$ by

$$(19)\qquad \phi = (\lambda_j - \lambda_k)(\lambda_k - \lambda_l)(\lambda_l - \lambda_j) \qquad \text{whenever} \quad \varepsilon_{jkl} = 1\,.$$

Throughout this proof we will write $\langle\,,\rangle$ instead of $g(\;,\;)$. For $P_j$ given by (17),

$$(20)\qquad \begin{aligned} &\text{i)}\ \ (\lambda_j - \lambda_k)(\lambda_j - \lambda_l)P_j + 2\varepsilon_j(\lambda_k - \lambda_l)v_k\wedge v_l = -(\lambda_j + \mathrm{s}/12)\,\phi\,\alpha_j,\\ &\text{ii)}\ \ (\lambda_j - \lambda_k)(\lambda_j - \lambda_l)\operatorname{div} w + 2(\lambda_k - \lambda_l)\langle w,w\rangle = -(2\lambda_j + \mathrm{s}/6)\,\phi\,, \end{aligned}$$

if $\varepsilon_{jkl} = 1$. In fact, since the $\lambda_j$ are constant, multiplying (b) above by $\phi$ and using (a) we obtain $(\lambda_j - \lambda_k)(\lambda_j - \lambda_l)\,dv_j + \varepsilon_j(\lambda_k - \lambda_l)v_k\wedge v_l = -(\lambda_j + \mathrm{s}/12)\phi\alpha_j$, if $\varepsilon_{jkl} = 1$. In view of Remark 10.2(i), (a) above, and (19), this is nothing else than

(20.i). Also, as $\operatorname{div} w = \operatorname{Trace}_{\mathbf{C}} \nabla w$, taking the complex trace of the composites of both sides of (20.i) with $\alpha_j$, we obtain (20.ii) from (8.e), (13) and (11.i), since, by (17), $2 \operatorname{Trace} P = -\varepsilon_j \operatorname{Trace}(\alpha_j \circ P_j)$ (no summation). Next, for $\phi$ as in (19),

$$(21) \qquad \phi = 0 \quad \text{if and only if} \quad \nabla W^{(+)} = 0 \quad \text{identically.}$$

Namely, if $\phi = 0$, by (19), (a) above, (12) and (11.i), $w = 0$ and $(\lambda_l - \lambda_k)\xi_j = 0$ whenever $\varepsilon_{jkl} = 1$, and so (as $\mu_j = 0$, $j = 1, 2, 3$), Remark 10.2(ii) yields $\nabla W^{(+)} = 0$. Conversely, let $\nabla W^{(+)} = 0$. Remark 10.2(ii) with $\mu_j = 0$ now gives $(\lambda_l - \lambda_k)\xi_j = 0$ whenever $\varepsilon_{jkl} = 1$. Hence $\phi = 0$, for if we had $\phi \neq 0$, the last relation and (19) would imply $\xi_j = 0$, $j = 1, 2, 3$, i.e., from (b) above, $\lambda_1 = \lambda_2 = \lambda_3 = -\mathrm{s}/12$, and, by (19), $\phi$ would be zero anyway.

Since our assertion is immediate when $\nabla W^{(+)} = 0$, we now assume that

$$(22) \qquad \phi \neq 0, \qquad \text{i.e.,} \quad \lambda_1 \neq \lambda_2 \neq \lambda_3 \neq \lambda_1 .$$

(Cf. (19), (21).) We may treat (20.ii) as a system of three linear equations with two unknowns: $\operatorname{div} w$ and $\langle w, w\rangle$. This system's matrix has the $2 \times 2$ subdeterminants equal, by (16.a), to $\pm 6\lambda_j(\lambda_k - \lambda_l)^2$, $\varepsilon_{jkl} = 1$. They cannot be all zero, or else (16.a) would give $0 = \lambda_j(\lambda_k - \lambda_l) = -(\lambda_k + \lambda_l)(\lambda_k - \lambda_l) = \lambda_l^2 - \lambda_k^2$, if $\varepsilon_{jkl} = 1$, i.e., any two of the $\lambda_j$ would coincide up to a sign, proving that, with the $\lambda_j$ suitably rearranged, $\lambda_2 = \lambda_3 = \pm\lambda_1$, contrary to (22). Hence the system (20.ii) has rank two, and can be solved for $\operatorname{div} w$ and $\langle w, w\rangle$ using determinants. Thus, $\langle w, w\rangle$ is constant since so are the coefficients of (20.ii) (cf. (19)); in addition, $\langle w, w\rangle \neq 0$. Namely, if $\langle w, w\rangle$ were zero, we would have $\operatorname{div} w = (\lambda_k - \lambda_l)(2\lambda_j + \mathrm{s}/6)$, $\varepsilon_{jkl} = 1$, by (20.ii), (19) and (22); summed over $j = 1, 2, 3$, this would yield $\operatorname{div} w = 0$ (cf. (16.b)); hence $(\lambda_k - \lambda_l)(2\lambda_j + \mathrm{s}/6) = 0$, $\varepsilon_{jkl} = 1$, which, in view of (22), would imply that $2\lambda_j = -\mathrm{s}/6$ for $j = 1, 2, 3$, contrary to (22). Next,

$$(23) \qquad \begin{array}{ll} \text{i)} \ \nabla_{v_j} w = \lambda_j(\lambda_k - \lambda_l)\, v_j & \text{whenever} \quad \varepsilon_{jkl} = 1 , \\ \text{ii)} \ \operatorname{div} w = 0 , & \text{iii)} \ \mathrm{s} = 0 , \ \text{ i.e., } (M, g) \text{ is Ricci-flat.} \end{array}$$

In fact, both sides of (20.i) may be treated as bundle morphisms $[TM]^{\mathbf{C}} \to [TM]^{\mathbf{C}}$, and hence applied to the complex vector field $v_j = \alpha_j w$, giving, by (17), (13) and (8.b), $(\lambda_j - \lambda_k)(\lambda_j - \lambda_l)\alpha_j P v_j = \varepsilon_j(\lambda_j + \mathrm{s}/12)\,\phi w$, whenever $\varepsilon_{jkl} = 1$. (Note that, from (8.b) and (13.a), $(v_k \wedge v_l)v_j = 0$, while, as $\langle w, w\rangle$ is constant, (13.b) and (8.f) yield $P^*\alpha_j v_j = 0$.) Now, applying $\alpha_j$ to both sides of the last equality, we obtain $(\lambda_j - \lambda_k)(\lambda_j - \lambda_l) P v_j = -(\lambda_j + \mathrm{s}/12)\phi v_j$ from (11.i) and (12). Thus, since $Pv = \nabla_v w$ for all vectors $v$, (19) and (22) imply that $\nabla_{v_j} w = (\lambda_k - \lambda_l)(\lambda_j + \mathrm{s}/12)\, v_j$ whenever $\varepsilon_{jkl} = 1$. Also, $w, v_1, v_2, v_3$ form an orthogonal trivialization of $[TU]^{\mathbf{C}}$ (by (13.a) with $\langle w, w\rangle \neq 0$). Evaluating $\operatorname{div} w$ in that trivialization, we get, from (13.a), $\langle w, w\rangle \operatorname{div} w = \langle w, w\rangle \operatorname{Trace}_{\mathbf{C}} \nabla w = \sum_{j=1}^{3} \varepsilon_j \langle v_j, \nabla_{v_j} w\rangle$, since $\langle w, \nabla_w w\rangle = 0$ as $\langle w, w\rangle$ is constant. Therefore, (23.ii) is immediate from the above formula for $\nabla_{v_j} w$ and (13.a), (16.b). Finally, we have $(\lambda_k - \lambda_l)\langle w, w\rangle = -(\lambda_j + \mathrm{s}/12)\,\phi$, $\varepsilon_{jkl} = 1$, from (20.ii) and (23.ii). Summed over $j$ this gives $\mathrm{s}\phi = 0$ (by (16.a)), and so (22) yields (23.iii), while (23.iii) and our formula for $\nabla_{v_j} w$ imply (23.i).

Next, (20.ii) and (23.ii, iii) give $\phi\lambda_j = (\lambda_l - \lambda_k)\langle w, w\rangle$, $\varepsilon_{jkl} = 1$, i.e., by (19) and (22), $\langle w, w\rangle = \lambda_j(\lambda_j - \lambda_k)(\lambda_j - \lambda_l)$ if $\varepsilon_{jkl} = 1$. However, this means that $\langle w, w\rangle = 2\lambda_j^3 + \lambda_1\lambda_2\lambda_3$, $j = 1, 2, 3$. (Note that, whenever $\varepsilon_{jkl} = 1$, (16.a) yields $(\lambda_j - \lambda_k)(\lambda_j - \lambda_l) = \lambda_j^2 - \lambda_j(\lambda_k + \lambda_l) + \lambda_k\lambda_l = 2\lambda_j^2 + \lambda_k\lambda_l$.) Thus, $\lambda_j^3 = \mu$ for some complex number $\mu$, not depending on $j \in \{1, 2, 3\}$ and, by (22), the $\lambda_j$ are the three cubic roots of $\mu$, so that $\lambda_1\lambda_2\lambda_3 = \mu$. As $\langle w, w\rangle = 2\lambda_j^3 + \lambda_1\lambda_2\lambda_3$, we have $\lambda_j^3 = -\gamma$, $j = 1, 2, 3$, for the constant $\gamma = -\langle w, w\rangle/3 \in \mathbf{C} \smallsetminus \{0\}$.

Hence, by (22), the $\lambda_j$ cannot be all real. Therefore, according to Remark 9.1, (22) implies that $g$ is not Riemannian, i.e., $\phi = 0$ in the Riemannian case, which, in view of (21), proves assertion (i).

We may now assume that $g$ is Lorentzian or neutral and $\nabla W^{(+)} \neq 0$. Thus,

$$(24)\quad \begin{array}{l} \text{i) } \lambda_k = z\lambda_j,\ \lambda_l = \bar{z}\lambda_j \text{ if } \varepsilon_{jkl} = 1, \quad \text{ii) } \lambda_k - \lambda_l = \pm i\sqrt{3}\,\lambda_j,\ \ \varepsilon_{jkl} = 1, \\ \text{iii) } [w, v_j] = \lambda_j(\lambda_l - \lambda_k)\, v_j \text{ if } \varepsilon_{jkl} = 1, \quad \text{iv) } [v_j, v_k] = 0 \text{ for all } j,\ k, \end{array}$$

where $z = e^{\pm 2\pi i/3}$ for a suitable sign $\pm$ and $[\ ,\ ]$ is the Lie bracket. In fact, i) follows since $\lambda_j^3 = -\gamma \neq 0$, while ii) is obvious from i) as $z - \bar{z} = \pm i\sqrt{3}$. Next, applying (20.i) to $w$ and using (17), (12), (8.b), (13.a), (23.ii), (19) and (22), we obtain the formula $\alpha_j \nabla_w w = -(\nabla w)^* v_j + \lambda_j(\lambda_k - \lambda_l)\, v_j$, if $\varepsilon_{jkl} = 1$. By (23.i), $\langle (\nabla w)^* v_j, v_k\rangle = \langle v_j, (\nabla w) v_k\rangle = 0$ whenever $k \neq j$ (cf. (13.a)). Since $w, v_1, v_2, v_3$ form a complex orthogonal basis at every point, our formula for $\alpha_j \nabla_w w$ thus shows that, at each point, $\alpha_j \nabla_w w$ is a combination of $w$ and $v_j$, i.e., by (12), (13.b) and (11.i), $\nabla_w w = \psi_j w + \chi_j v_j$ for some functions $\psi_j$, $\chi_j$. As this is true for all $j \in \{1, 2, 3\}$ and $w$ does not depend on $j$, we have $\chi_j = 0$, while $\psi_j = 0$ since $\langle w, w\rangle$ is constant. Consequently, $\nabla_w w = 0$. Furthermore, $\nabla_w \alpha_j = 0$ for all $j$ in view of (14.ii) and the relation $\langle \xi_j, w\rangle = 0$ (immediate from (a) above, (13.a) and (22)), and so (12) with $\nabla_w w = 0$ gives $\nabla_w v_j = 0$, $j = 1, 2, 3$. This, combined with (23.i) and the fact that $\nabla$ is torsionfree, proves (24.iii). Next, since the $v_j$ are mutually orthogonal by (13.a), and every $\xi_j$ is a multiple of $v_j$ in view of (a) with (22), we have, by (14.ii), $\nabla_{v_j} \alpha_k = \varepsilon_k \langle \xi_j, v_j\rangle \alpha_l$ (no summation) whenever $\varepsilon_{jkl} = 1$. Hence, by (a) and (13.a), $\nabla_{v_j} \alpha_k = \varepsilon_j \varepsilon_k (\lambda_l - \lambda_k)^{-1} \langle w, w\rangle \alpha_l$, i.e., from (11.ii) and (12), $[\nabla_{v_j} \alpha_k] w = \varepsilon_l (\lambda_l - \lambda_k)^{-1} \langle w, w\rangle v_l$, $\varepsilon_{jkl} = 1$. On the other hand, by (23.i) and (13.b), $\alpha_k(\nabla_{v_j} w) = -\varepsilon_l \lambda_j (\lambda_k - \lambda_l) v_l$. From (12) and our expressions for $[\nabla_{v_j} \alpha_k] w$ and $\alpha_k(\nabla_{v_j} w) = 0$ we now obtain $\nabla_{v_j} v_k = \nabla_{v_j}(\alpha_k w) = [\nabla_{v_j} \alpha_k] w + \alpha_k(\nabla_{v_j} w) = 0$ if $\varepsilon_{jkl} = 1$, as (24.ii) with $\lambda_j^3 = -\gamma = \langle w, w\rangle/3$ gives $(\lambda_l - \lambda_k)^{-1}\langle w, w\rangle = \pm i\sqrt{3}\,\lambda_j^2$. Similarly, $\nabla_{v_j} v_l = 0$ if $\varepsilon_{jkl} = 1$. Thus, $\nabla_{v_j} v_k = 0$ when $j \neq k$, proving (24.iv).

Since, by (13.a), $\langle w, w\rangle = \langle v_j, v_j\rangle = -3\gamma$ (no summing) for $j = 1, 2, 3$, the new complex vector fields $\tilde{w} = \pm iw/\sqrt{3}$ and $\tilde{v}_j = iv_j/\sqrt{3}$, with the same sign $\pm$ as in (24.ii), have $\langle \tilde{w}, \tilde{w}\rangle = \langle \tilde{v}_j, \tilde{v}_j\rangle = \gamma$ and are pairwise orthogonal by (13.a), while, from (24.ii) – (24.iv), $[\tilde{v}_j, \tilde{v}_k] = 0$, and $[\tilde{w}, \tilde{v}_j] = \lambda_j^2 \tilde{v}_j$ (no summation). As $\lambda_j^3 = -\gamma$, replacing $w$, $v_j$ with $\tilde{w}$, $\tilde{v}_j$ and setting $\rho_j = \lambda_j^2$, we now obtain (ii).

Finally, $w, v_j$ and $\tilde{w}, \tilde{v}_j$ commute with all Killing fields since, up to permutations and sign changes, they are invariant under all isometries between connected open subsets of $M$. Namely, by (22), relations $W\alpha_j = \lambda_j \alpha_j$, (11.i), (14.ii) and (a) above determine the $\alpha_j, \xi_j, v_j$ and $w$ uniquely up to permutations and sign changes. This completes the proof. □

## §12. Complex Lie algebras and real manifolds

Given a real/complex vector space $\mathcal{Z}$ of sections of a real/complex vector bundle $\mathcal{E}$ over a manifold $M$, we will say that $\mathcal{Z}$ *trivializes* $\mathcal{E}$ if it consists of $C^\infty$ sections of $\mathcal{E}$ and, for every $x \in M$, the evaluation operator $\psi \mapsto \psi(x)$ is an isomorphism $\mathcal{Z} \to \mathcal{E}_x$. This amounts to requiring that $\dim \mathcal{Z}$ coincide with the fibre dimension of $\mathcal{E}$ and each $v \in \mathcal{Z}$ be either identically zero, or nonzero at every point of $M$. Equivalently, a basis of $\mathcal{Z}$ then is a $C^\infty$ trivialization of $\mathcal{E}$.

For instance, a simply transitive Lie algebra of vector fields on a manifold $M$ (see the appendix) is nothing else than a real vector space of vector fields on $M$, trivializing its (real) tangent bundle, and closed under the Lie bracket.

Let the real/complexified tangent bundle of a manifold $M$ be trivialized by a real/complex vector space $\mathcal{Z}$ of real/complex vector fields on $M$ (cf. end of §8). We will say that a real/complex vector field $w$ defined of any open subset $U$ of $M$ *commutes with* $\mathcal{Z}$, and write $[w,\mathcal{Z}] = \{0\}$, if $w$ is of class $C^\infty$ and $[w,v] = 0$ for every $v \in \mathcal{Z}$. In view of the Jacobi identity, real/complex vector fields $\mathcal{Z}$ defined on a given open set $U$ and commuting with $\mathcal{Z}$ form a Lie algebra.

LEMMA 12.1. *Let $\mathcal{Z}$ be a real/complex Lie algebra of real/complex vector fields on a real manifold $M$, trivializing its real/complexified tangent bundle. Then, the real/complexified tangent bundle of any sufficiently small connected neighborhood $U$ of any given point $x$ of $M$ is trivialized by the Lie algebra $\mathcal{Y}$ of all real/complex vector fields defined on $U$ and commuting with $\mathcal{Z}$.*

In fact, let D be the unique connection in the real/complexified tangent bundle $\mathcal{T}$ with $\mathrm{D}_v w = [v,w]$ for all $v \in \mathcal{Z}$ and all $C^1$ sections $w$ of $\mathcal{T}$. Thus, D is flat: by (2.i), $R^{\mathrm{D}}(v,w)u = [w,[v,u]] - [v,[w,u]] + [[v,w],u]$ whenever $v,w,u \in \mathcal{Z}$, which is zero by the Jacobi identity. (As $\mathcal{Z}$ is a Lie algebra, $[v,w] \in \mathcal{Z}$, and so $\mathrm{D}_{[v,w]}u = [[v,w],u]$.) Now $\mathcal{Y}$ consists of all D-parallel sections of $\mathcal{T}$ on $U$. $\square$

For instance, the real Lie algebra $\mathcal{X}$ of left-invariant vector fields on a Lie group $G$ trivializes its real tangent bundle. A real vector field on an open connected subset $U$ of $G$ commutes with $\mathcal{X}$ if and only if it is the restriction to $U$ of a right-invariant vector field on $G$. In fact, right-invariant fields $w$ all commute with $\mathcal{X}$, since the flow of $w$ (or, of any $v \in \mathcal{X}$) consist of left (or, right) translations, while left and right translations commute due to associativity. The converse follows since both Lie algebras are of dimension $\dim G$ (Lemma 12.1).

REMARK 12.2. If a real/complex vector space $\mathcal{Z}$ of real/complex vector fields on a manifold $M$ trivializes its real/complex tangent bundle, then any real/complex vector field $w$ on $M$ with $[w,\mathcal{Z}] = \{0\}$ is a real/complex Killing field on $(M,g)$ for any pseudo-Riemannian metric $g$ on $M$ such that $g(u,v)$ is constant whenever $u,v \in \mathcal{Z}$. In fact, (7) then gives $(\mathcal{L}_w g)(u,v) = 0$ for all $u,v \in \mathcal{Z}$.

Let a complex Lie algebra $\mathcal{Z}$ of complex vector fields on a manifold $M$ trivialize its complexified tangent bundle $[TM]^{\mathbf{C}}$. We say that $\mathcal{Z}$ *admits a real form* if $\mathrm{Re}\, w \in \mathcal{Z}$ for every $w \in \mathcal{Z}$. This is obviously equivalent to the existence of a real Lie algebra $\mathcal{X}$ of real vector fields on $M$, trivializing its ordinary tangent bundle $TM$, and such that $\mathcal{Z} = \mathcal{X} + i\mathcal{X}$, i.e., $\mathcal{Z}$ is the complexification of $\mathcal{X}$ (or, $\mathcal{X}$ is a *real form* of $\mathcal{Z}$). Clearly, $\mathcal{X}$ then is uniquely determined by $\mathcal{Z}$, as $\mathcal{X} = \{\mathrm{Re}\, w : w \in \mathcal{Z}\} = \{w \in \mathcal{Z} : \mathrm{Im}\, w = 0\}$. Thus, $\mathcal{Z}$ admits a real form if and only if the real vector fields which are elements of $\mathcal{Z}$ form a real Lie algebra trivializing $TM$.

REMARK 12.3. If a complex Lie algebra $\mathcal{Z}$ of complex vector fields on a manifold $M$ trivializes its complexified tangent bundle and $\dim_{\mathbf{C}}\mathcal{Y} = \dim M$ for the Lie algebra $\mathcal{Y}$ of all $C^\infty$ complex vector fields $w$ on $M$ with $[w,\mathcal{Z}] = \{0\}$, then

(i) $\mathcal{Y}$ trivializes the complexified tangent bundle of $M$.

(ii) $\mathcal{Z}$ admits a real form whenever $\mathcal{Y}$ does.

To see this, first note that Lemma 12.1 yields (i). Next, let $\mathcal{V}$ be a real form of $\mathcal{Y}$, and let a complex vector field $w$ commute with $\mathcal{Y}$, so that $[w,\mathcal{Y}]=\{0\}$. Since $\mathcal{V}\subset\mathcal{Y}$, we have $[w,\mathcal{V}]=\{0\}$. Therefore $[\mathrm{Re}\,w,\mathcal{V}]=\{0\}$, as $\mathcal{V}$ consists of real vector fields and $[\,,\,]$ is complex-bilinear; this and relation $\mathcal{Y}=\mathcal{V}+i\mathcal{V}$ now give $[\mathrm{Re}\,w,\mathcal{Y}]=\{0\}$. The Lie algebra $\mathcal{Z}'$ of all complex vector fields commuting with $\mathcal{Y}$ thus is closed under the real-part operator Re. However, $\mathcal{Z}\subset\mathcal{Z}'$ and, by Lemma 12.1, $\dim_{\mathbf{C}}\mathcal{Z}'\le\dim M=\dim_{\mathbf{C}}\mathcal{Z}$, so that $\mathcal{Z}'=\mathcal{Z}$, which proves (ii).

LEMMA 12.4. *Let $\mathcal{Z}$ be a complex Lie algebra of complex vector fields on a pseudo-Riemannian manifold $(M,g)$, trivializing the complexified tangent bundle of $M$ and such that $g(u,v)$ is constant for any $u,v\in\mathcal{Z}$. Then*

(a) *$(M,g)$ is locally homogeneous.*

(b) *Under the additional assumption that $[u,v]=0$ for every $u\in\mathcal{Z}$ and every real Killing field $v$ defined on any open subset of $M$, we have $\mathrm{Re}\,w\in\mathcal{Z}$ whenever $w\in\mathcal{Z}$, i.e., $\mathcal{Z}$ admits a real form.*

In fact, by Lemma 12.1 and Remark 12.2, every vector in $T_xM$, $x\in M$, is the value at $x$ of some real Killing field on a neighborhood of $x$, which proves (a) (cf. [**10**], p 546). Now let us fix $x\in M$ and choose $U,\mathcal{Y}$ for $x,\mathcal{Z}$ as in Lemma 12.1. Remark 12.2 and our hypothesis show that $\mathcal{Y}$ then is *precisely* the Lie algebra of all complex Killing fields on $U$. Thus, $\mathcal{Y}$ is closed under the real-part operator Re, i.e., admits a real form, and Remark 12.3(ii) yields (b). □

## §13. Real forms of some specific complex Lie algebras

We use the standard notation Ad for the adjoint representation of any given Lie algebra $\mathcal{X}$, so that $\mathrm{Ad}\,v:\mathcal{X}\to\mathcal{X}$ is, for any $v\in\mathcal{X}$, given by $(\mathrm{Ad}\,v)w=[v,w]$.

LEMMA 13.1. *Let a basis $w,v_1,v_2,v_3$ of a four-dimensional complex Lie algebra $\mathcal{Z}$ satisfy conditions* (18) *for some complex-bilinear symmetric form $g$ on $\mathcal{Z}$ and a complex number $\gamma\ne 0$, where $[\,,\,]$ is the Lie-algebra multiplication of $\mathcal{Z}$ and $\rho_1,\rho_2,\rho_3$ are the three cubic roots of $\gamma^2$. Also, let $\mathcal{X}\subset\mathcal{Z}$ be a four-dimensional real Lie subalgebra with $\mathcal{Z}=\mathcal{X}+i\mathcal{X}$ and $g(\mathcal{X},\mathcal{X})\subset\mathbf{R}$. In other words, $\mathcal{X}$ spans $\mathcal{Z}$ as a complex space and the form $g$ restricted to $\mathcal{X}$ is real-valued.*

*Then $w\in\mathcal{X}$ and there exist a three-dimensional real vector subspace $V$ of $\mathcal{X}$, a linear operator $F:V\to V$, and a real-valued bilinear form $\langle\,,\,\rangle$ on $V$, satisfying conditions* (4) *with $u=|\gamma|^{-1/2}w$ and $\delta=\mathrm{sgn}\,\gamma$, and such that $\langle\,,\,\rangle,F$ are, for a suitable isomorphic identification $V=\mathbf{C}\times\mathbf{R}$, given by* (a),(b) *in* §4 *with some sign $\pm$ and some $p\in\mathbf{R}\smallsetminus\{0\}$.*

PROOF. We set $V=\mathcal{X}\cap\mathrm{Ker}\,\Psi$, where $\Psi:\mathcal{Z}\to\mathbf{C}$ is the $\mathbf{C}$-linear functional with $\Psi(w)=1$ and $\Psi(v_j)=0$, $j=1,2,3$. For any $u\in\mathcal{Z}\smallsetminus\mathrm{Ker}\,\Psi$,

(a) $\mathrm{Ad}\,u$ has the characteristic roots 0 and $\Psi(w)\rho_j$, $j=1,2,3$.

(b) $\dim_{\mathbf{R}}V=3$ and $\mathrm{Span}_{\mathbf{C}}V=\mathrm{Ker}\,\Psi$.

In fact, by (18), $\mathrm{Ad}\,u:\mathcal{Z}\to\mathcal{Z}$ is diagonalizable with the eigenvalues as in (a) for the eigenvectors $u$ and $v_j$, which proves (a). Also, as $\dim_{\mathbf{C}}[\mathrm{Ker}\,\Psi]=3$, our $\mathcal{X}$ cannot be contained in $\mathrm{Ker}\,\Psi$, and so the image $\Psi(\mathcal{X})$ is a nontrivial real vector subspace of $\mathbf{C}$. For any fixed $u\in\mathcal{X}\smallsetminus\mathrm{Ker}\,\Psi$, (a) gives $\Psi(w)\rho_j\in\mathbf{R}$ for some $j\in\{1,2,3\}$. (In fact, as $\mathcal{X}$ contains a basis of $\mathcal{Z}$, the characteristic roots of $\mathrm{Ad}\,w:\mathcal{Z}\to\mathcal{Z}$ coincide with those of $\mathrm{Ad}\,w:\mathcal{X}\to\mathcal{X}$, so that the number of nonreal ones among them is 0 or 2.) Thus, $\Psi(\mathcal{X})$ is contained in the union of the real lines $\mathbf{R}\bar{\rho}_j\subset\mathbf{C}$, $j=1,2,3$, i.e., must coincide with one of them, and we may fix $j \in \{1,2,3\}$ with $\Psi(\mathcal{X}) = \mathbf{R}\overline{\rho_j}$. Now $\dim_{\mathbf{R}} V = 3$, since $V = \mathcal{X} \cap \operatorname{Ker}\Psi$ is the kernel of $\Psi : \mathcal{X} \to \mathbf{C}$. Also, $\mathcal{X}$ spans $\mathcal{Z}$, so that vectors in $\mathcal{X}$, linearly independent over $\mathbf{R}$, are also linearly independent over $\mathbf{C}$ in $\mathcal{Z}$. This implies (b): $\operatorname{Span}_{\mathbf{C}} V = \operatorname{Ker}\Psi$ as $\operatorname{Span}_{\mathbf{C}} V \subset \operatorname{Ker}\Psi$ and $\dim_{\mathbf{C}}[\operatorname{Span}_{\mathbf{C}} V] = \dim_{\mathbf{C}}[\operatorname{Ker}\Psi] = 3$.

Since $\dim_{\mathbf{R}} V = 3$, we may choose $u \in \mathcal{X} \smallsetminus \{0\}$ which is $g$-orthogonal to $V$. By (b), $u$ then is also $g$-orthogonal to $\operatorname{Ker}\Psi$. Thus, in view of (18), $u \in \mathbf{C}w$, i.e., $u = \Psi(u)w$ with $\Psi(u) \neq 0$. Also, $\Psi(u)\rho_j$ is real, for $j$ chosen above (as $\Psi(u) \in \mathbf{R}\overline{\rho_j}$), and hence so is its cube $[\Psi(u)]^3\gamma^2$. On the other hand, (18) gives $[\Psi(u)]^2\gamma = g(u,u) \in \mathbf{R}$. Consequently, the numbers $\Psi(u)\gamma$, $\Psi(u)$, $\rho_j$ and $\gamma$ are all real, while $w \in \mathcal{X}$, as $\mathcal{X}$ contains $u = \Psi(u)w$ and $\Psi(u) \in \mathbf{R} \smallsetminus \{0\}$.

As $\gamma \in \mathbf{R} \smallsetminus \{0\}$, replacing such $u$ by $|\gamma|^{-1/2}w$ and letting $\langle\,,\rangle$ stand for the restriction of $g$ to $V$, we now obtain $\langle u,u\rangle = \delta$ with $\delta = \operatorname{sgn}\gamma \in \{1,-1\}$.

The real 3-space $V = \mathcal{X} \cap \operatorname{Ker}\Psi$ is $(\operatorname{Ad}u)$-invariant, since so are $\mathcal{X}$ (as $u \in \mathcal{X}$) and $\operatorname{Ker}\Psi$ (by (18) with $u = |\gamma|^{-1/2}w$). The restriction $F : V \to V$ of $\operatorname{Ad}u$ is self-adjoint, since that is the case for $F, V, \langle\,,\rangle$ replaced by $\operatorname{Ad}u, \mathcal{Z}, g$ (as $\operatorname{Ad}u : \mathcal{Z} \to \mathcal{Z}$ is diagonalized by the $g$-orthogonal basis $w, v_1, v_2, v_3$, cf. (18)). Combining (a) with our assumptions about the cubes $\rho_j^3$ and the fact that $\dim_{\mathbf{R}} V = 3$ is odd, we see that $F$ has the characteristic roots $p, pq, p\overline{q}$, where $q = e^{2\pi i/3}$ and $p \in \mathbf{R} \smallsetminus \{0\}$, and we may choose $\xi, \eta, \zeta \in V$ such that $\zeta$ and $\xi + i\eta$ are eigenvectors of $\operatorname{Ad}u : \mathcal{Z} \to \mathcal{Z}$ for the eigenvalues $p$ and $pq$. (Since $pq \notin \mathbf{R}$, this implies that $\xi, \eta$ are linearly independent over $\mathbf{R}$.) By (18), $\zeta$ and $\xi + i\eta$ are complex multiples of $v_j, v_k$ for some $j, k$. Thus, $g(\zeta,\zeta) \neq 0$, i.e., $\zeta$ may be normalized so that $\langle\zeta,\zeta\rangle = \pm 1$ for some sign $\pm$, while $g(\zeta, \xi + i\eta) = 0$, and so $\langle\zeta,\xi\rangle = \langle\zeta,\eta\rangle = 0$, as $g$ is real-valued on $V \subset \mathcal{X}$. Next, $\langle F\xi,\eta\rangle = \langle\xi, F\eta\rangle$ since $F$ is self-adjoint, so that $\langle\xi,\xi\rangle + \langle\eta,\eta\rangle = 0$ in view of the eigenvector relation $F\xi + iF\eta = pq(\xi + i\eta)$ with $q = (\sqrt{3}i - 1)/2$. Finally, let $c$ be a complex number with $2\overline{c}^2 = -g(\xi + i\eta, \xi + i\eta)$. Thus, $c \neq 0$, since $g(v_k, v_k) \neq 0$, and it is easy to verify that the isomorphism $V \to \mathbf{C} \times \mathbf{R}$ sending the basis $\xi, \eta, \zeta$ onto $(c,0), (-ic,0), (0,1)$ has the required properties. This completes the proof. $\square$

PROOFS OF THEOREMS 5.1, 6.1 AND 7.2. In all three cases, $R - W$ is a constant multiple of the identity (Remark 1.1), and so the hypotheses of Theorem 11.1 are satisfied (cf. Remark 9.1). If $g$ is Riemannian, Theorem 11.1(i) yields Theorem 7.2. If $g$ is Lorentzian and $\nabla W = 0$, i.e., $\nabla R = 0$, Theorem 41.5 of [**10**] (pp. 662–663) implies (a) or (b) in Theorem 5.1, as the diagonalizability condition excludes option (c) in [**10**] on p. 663. The only remaining cases now are those named in (ii) of Theorem 11.1, the conclusion of which shows that Lemma 13.1 can be applied to the Lie algebra $\mathcal{Z} = \operatorname{Span}_{\mathbf{C}}\{w, v_1, v_2, v_3\}$ and its real form $\mathcal{X}$ which exists in view of Lemma 12.4(b). As a result, $(M,g)$ is obtained as in Example 4.2(i) or (ii); the situation where $\delta = -1$ and $\pm$ is $-$ cannot occur, as it would lead to the sign pattern $---+$, which is not one of (1). $\square$

## Appendix. Simply transitive Lie algebras of vector fields

In this section we prove Corollary A.3, a well-known result included here to provide a convenient, self-contained reference for a conclusion in Example 4.2.

A *simply transitive Lie algebra of vector fields* on a manifold $M$ is any vector space $\mathcal{X}$ of $C^\infty$ (real) vector fields on $M$, closed under the Lie bracket and such that the evaluation operator $\mathcal{X} \ni w \mapsto w(x) \in T_x M$ is bijective for every $x \in M$. An example is the Lie algebra of left-invariant vector fields on a Lie group.

Given a simply transitive Lie algebra $\mathcal{X}$ of vector fields on a manifold $M$ and a fixed point $y \in M$, the *exponential mapping* $E : U_y \to M$ *for* $\mathcal{X}$, *centered at* $y$, is given by $E(v) = x(1)$, where $U_y$ is the set of all $v \in \mathcal{X}$ for which an integral curve $t \mapsto x(t)$ of $v$ with $x(0) = y$ can be defined on the whole interval $[0,1]$. It is clear that $U_y$ is a neighborhood of 0 in $\mathcal{X}$ and, for every $v \in U_y$ and $t \in [0,1]$, we have $tv \in U_y$ and $x(t) = E(tv)$, with $x(t)$ as above.

Let $Q : \mathbf{C} \to \mathbf{C}$ be the entire function with $Q(z) = (1 - e^{-z})/z$ if $z \neq 0$ and $Q(0) = 1$. Its Maclaurin series defines $Q(A)$ for any linear operator $A : V \to V$ in a vector space $V$ with $\dim V < \infty$. Thus, with Ad as in §13, $Q(\operatorname{Ad} v) = \sum_{k=0}^{\infty}(-\operatorname{Ad} v)^k/[(k+1)!]$ for a Lie algebra $\mathcal{X}$ with $\dim \mathcal{X} < \infty$ and $v \in \mathcal{X}$.

PROPOSITION A.1. *Let* $\mathcal{X}$ *be a simply transitive Lie algebra of vector fields on a manifold* $M$, *and let* $dE_v : \mathcal{X} \to T_{E(v)}M$ *be the differential at* $v \in U_y$ *of the exponential mapping of* $\mathcal{X}$ *centered at a point* $y \in M$, *with the usual identification* $T_v\mathcal{X} = \mathcal{X}$. *Then* $dE_v$ *equals the composite mapping in which* $Q(\operatorname{Ad} v) : \mathcal{X} \to \mathcal{X}$, *defined above, is followed by the evaluation isomorphism* $\mathcal{X} \to T_{E(v)}M$.

PROOF. For any $C^\infty$ mapping $(s,t) \mapsto x(s,t) \in M$ of a rectangle $K \subset \mathbf{R}^2$, let $u_s, u_t : K \to \mathcal{X}$ assign to $(s,t)$ the unique elements of $\mathcal{X}$ which coincide, at $x(s,t)$, with $\partial x/\partial s$ and, respectively, $\partial x/\partial t$ (that is, with the velocity at $s$, or $t$, of the curve $s \mapsto x(s,t)$ or $t \mapsto x(s,t)$). Using subscripts for partial derivatives of $u_s, u_t$ we thus have $u_{st}, u_{ts}, u_{stt} : K \to \mathcal{X}$ with $u_{st} = \partial u_s/\partial t$, etc.; we also let $[u_s, u_t] : K \to \mathcal{X}$ stand for the valuewise bracket of the Lie-algebra valued functions $u_s, u_t$. In local coordinates $x^j$ at any given $x_0 = x(s_0, t_0)$, the vector fields $u_s(s,t), u_t(s,t)$ have some component functions $u_s^j(s,t,x), u_t^j(s,t,x)$, also depending on a point $x$ near $x_0$. Thus, $u_s^j(s,t,x(s,t)) = \partial[x^j(s,t)]/\partial s$ and $u_t^j(s,t,x(s,t)) = \partial[x^j(s,t)]/\partial t$. Applying $\partial/\partial t$ to the first relation, $\partial/\partial s$ to the second, and using equality of mixed partial derivatives for the $x^j(s,t)$, we get $\partial u_s^j/\partial t - \partial u_t^j/\partial s = u_s^k \partial_k u_t^j - u_t^k \partial_k u_s^j$, with $\partial_k = \partial/\partial x^k$, which is the coordinate form of the identity $u_{st} - u_{ts} = [u_s, u_t]$. If $u_{tt} = 0$ for all $(s,t) \in K$, taking $\partial/\partial t$ of that identity, we obtain the *Jacobi equation* $u_{stt} = [u_{st}, u_t]$ (as $u_{tst} = u_{tts} = 0$).

It is clear that $u_{tt} = 0$ identically if and only if $t \mapsto x(s,t)$ is, for each fixed $s$, an integral curve of some vector field $v(s) \in \mathcal{X}$. Then, obviously, $u_t(s,t) = v(s)$.

Now let $u_{tt} = 0$ for all $(s,t)$, and let $K$ intersect the $s$-axis $\mathbf{R} \times \{0\}$. The Jacobi equation (see above) reads $\partial u_{st}/\partial t = -[\operatorname{Ad} v(s)]\, u_{st}$, with $v(s) = u_t(s,t)$, and so $u_{st}(s,t) = e^{-t \operatorname{Ad} v(s)}\, w(s)$, where $w(s) = u_{st}(s,0)$. Since $d[tQ(t \operatorname{Ad} v)]/dt = e^{-t \operatorname{Ad} v}$ (cf. our formula for $Q(\operatorname{Ad} v)$), we get $u_s(s,t) = u_s(s,0) + tQ(t \operatorname{Ad} v(s))w(s)$, as both sides satisfy the same initial value problem in the variable $t$.

Finally, let $K = I \times [0,1]$ and $x(s,t) = E(tv(s))$ for some interval $I$ and some $C^\infty$ curve $I \ni s \mapsto v(s) \in U_y$. Thus, $u_{tt} = 0$ identically and $u_t(s,t) = v(s)$, so that $u_{ts}(s,t) = \dot{v}(s)$, with $\dot{v} = dv/ds$. Also, $x(s,0) = y$, and hence $u_s(s,0) = 0$. Evaluating at $(s,0)$ the identity $u_{st} - u_{ts} = [u_s, u_t]$, established above, and setting $w(s) = u_{st}(s,0)$ as in the preceding paragraph, we thus get $w(s) = u_{ts}(s,0) = \dot{v}(s)$. Writing $v, \dot{v}$ instead od $v(s), dv/ds$ we now see that $u_s(s,1)$ equals the preimage of $dE_v\dot{v}$ under the evaluation isomorphism $\mathcal{X} \to T_{E(v)}M$ (cf. the definition of $u_s$) while $u_s(s,1) = Q(\operatorname{Ad} v)\dot{v}$, as one sees setting $t = 1$ in $u_s(s,t) = u_s(s,0) + tQ(t \operatorname{Ad} v(s))w(s)$. This completes the proof. □

COROLLARY A.2. *Given a simply transitive Lie algebra $\mathcal{X}$ of vector fields on a manifold $M$ and a point $y \in M$, there exists a neighborhood $U$ of $0$ in $\mathcal{X}$ such that $U \subset U_y$ and the exponential mapping $E : U_y \to M$ sends $U$ diffeomorphically onto an open subset of $M$. For any $U$ with this property, $Q(\operatorname{Ad} v) : \mathcal{X} \to \mathcal{X}$ is an isomorphism for every $v \in U$, and the pullback under $E$ of any vector field $w \in \mathcal{X}$ is the vector field on $U$ given by $U \ni v \mapsto [Q(\operatorname{Ad} v)]^{-1} w$.*

In fact, $Q(\operatorname{Ad} v)$ is an isomorphism by Proposition A.1, since $dE_v$ is. □

By Corollary A.2, the local diffeomorphism type of a simply transitive Lie algebra of vector fields is determined by its Lie-algebra isomorphism type. Since every finite-dimensional Lie algebra is the Lie algebra of some Lie group, this yields

COROLLARY A.3. *Given a simply transitive Lie algebra $\mathcal{X}$ of vector fields on a manifold $M$, there exists a Lie group $G$ with the following property: Every point of $M$ has a neighborhood $U$ which may be diffeomorphically identified with an open set $U' \subset G$ so as to make $\mathcal{X}$ restricted to $U$ appear as the Lie algebra of the restrictions to $U'$ of all left-invariant vector fields on $G$.* □

## References


1. A. L. Besse, *Einstein Manifolds*, Ergebnisse, ser. 3, vol. 10, Springer-Verlag, 1987.
2. E. Boeckx, O. Kowalski, L. Vanhecke, *Riemannian Manifolds of Conullity Two*, World Scientific, 1996.
3. C. H. Brans, *Complex 2-form representation of the Einstein equations: The Petrov type III solutions*, J. Math. Phys. **12** (1971), 1616–1619.
4. P. Bueken, *On curvature homogeneous three-dimensional Lorentzian manifolds*, J. Geom. Phys. **22** (1997), 349–362.
5. P. Bueken, *Three-dimensional Lorentzian manifolds with constant principal Ricci curvatures* $\rho_1 = \rho_2 \neq \rho_3$, J. Math. Phys. **38** (1997), 1000–1013.
6. P. Bueken, L. Vanhecke, *Examples of curvature homogeneous Lorentz metrics*, Class. Quantum Grav. **14** (1997), L93–L96.
7. M. Cahen, J. Leroy, M. Parker, F. Tricerri, L. Vanhecke, *Lorentz manifolds modelled on a Lorentz symmetric space*, J. Geom. Phys. **7** (1990), 571–581.
8. M. Cahen, M. Parker, *Pseudo-riemannian symmetric spaces*, Mem. AMS **229** (1980), 1–108.
9. M. Cahen, N. Wallach, *Lorentzian symmetric spaces*, Bull. AMS **76** (1970), 585–591.
10. F. J. E. Dillen, L. C. A. Verstraelen (eds.), *Handbook of Differential Geometry I*, Elsevier, 2000.
11. D. Ferus, H. Karcher, H. F. Münzner, *Cliffordalgebren und neue isoparametrische Hyperflächen*, Math. Z. **177** (1981), 479–502.
12. G. R. Jensen, *Homogeneous Einstein spaces of dimension* 4, J. Diff. Geom. **3** (1969), 309–349.
13. A. Koutras, C. McIntosh, *A metric with no symmetries or invariants*, Class. Quantum Grav. **13** (1996), L47–L49.
14. O. Kowalski, F. Prüfer, *Curvature tensors in dimension four which do not belong to any curvature homogeneous space*, Arch. Math. (Brno) **30** (1994), 45–57.
15. A. Z. Petrov, *Einstein Spaces*, English translation of *Prostranstva Eynshteyna* (Fizmatlit, Moscow, 1961), Pergamon Press, Oxford-New York, 1969.
16. I. M. Singer, J. A. Thorpe, *The curvature of 4-dimensional Einstein spaces*, Global Analysis, Papers in Honor of K. Kodaira, Princeton (1969), 355–365.
17. H. Takagi, *On curvature homogeneity of Riemannian manifolds*, Tôhoku Math. J. **26** (1974), 581–585.


DEPARTMENT OF MATHEMATICS, OHIO STATE UNIVERSITY, COLUMBUS, OHIO 43210
*E-mail address*: andrzej@math.ohio-state.edu

Contemporary Mathematics
Volume **337**, 2003

# Jacobians and Volume Comparison for Lorentzian Warped Products

PAUL E. EHRLICH, YOON-TAE JUNG,
JEONG-SIK KIM AND SEON-BU KIM

ABSTRACT. After a review of Jacobi field techniques for submanifold volume comparison, the particular case of the level sets of the space-time distance function compared with a warped product is discussed.

## 1. Introduction

The volume of Riemannian manifolds, or of geodesic balls in them, is one of the geometric invariants for which Bishop-Gromov's comparison theorem is well known [7, 8, 9, 22, 23]. E. Heintze and H. Karcher[24], moreover, generalized the volume comparisons for submanifolds. In those references, Jacobi fields and index form arguments are the main technical tools, even though H. Karcher and J.-H. Eschenburg[16, 17, 18, 19, 20, 27] used the Riccati equation technique associated with Jacobi tensors to obtain the Bishop-Gromov comparison theorem of a complete Riemannian manifold. On the other hand, in [26] Y. Itokawa obtained a volume comparison of a complete Riemannian manifold with a Riemannian geometric model called the Darboux-Cheeger-Yau model to generalize the Myers' Theorem. Inspired by this work, J.-S. Kim studied volume calculation for a Lorentzian warped product in his Ph. D. Thesis[28] by a representation of the linearly independent Jacobi fields in terms of the orthonormal parallel fields along a unit timelike geodesic perpendicular to a spacelike submanifold. Indeed, warped product metrics have played a prominent role in modeling the end structure of Riemannian manifolds and the asymptotic behavior of space-times (cf. Anderson [1]). On complete Riemannian manifolds, since any pair of points may be joined by a minimal geodesic segment by the Hopf-Rinow Theorem, the volume of any subregion may be calculated as an iterated integral over the unit sphere of an appropriate tangent space after lifting to $T_pM$ via $(\exp_p)^{-1}$. In Lorentzian case, however, the

2000 Mathematics Subject Classification: 53C50

The second author was supported by KOSEF R02-2000-00012-0
The fourth author was supported by BK21-2002, Chonnam National University

completeness does not guarantee the existence of a maximal nonspacelike geodesic segment between any causally related pair of points while global hyperbolicity does imply this[3, 4, 5]. In [14], P. Ehrlich, Y.-T. Jung and S.-B. Kim considered a globally hyperbolic space-time instead of a (geodesically)complete space-time in order to use the iterated integrals for a compact subset of $I^+(p)$. Moreover, they employed Riccati equation techniques to compare the volume of the compact set with the volume of a corresponding compact region in a given Lorentzian space-form. In [15], adapting the Riemannian machinery as presented in [10], P. Ehrlich and M. Sanchez obtained comparison results for starshaped subsets of space-time corresponding to the Riemannian comparison theorems of Günther, Bishop and Bishop-Gromov. Also, local results were obtained for the non-Lorentzian, pseudo-Riemannian case.

This paper is organized as follows. In Section 2, we prepare notations and some preliminary results for Lorentzian manifolds. Especially, we recall the notion of a type of a compact subregion called the $K$-distance wedge given in [14]. In Section 3, the Jacobian comparison for $r$-dimensional volume using $K$-Jacobi fields along the timelike geodesic initially perpendicular to a spacelike submanifold $K$ is given in Theorem 3.1, cf. [24, 32] for treatments of this topic in the Riemannin case. Especially when $r = 1$ or $\dim K = 0$, Theorem 3.1 may be interpreted as reducing to the Rauch comparison theorem and a space-time analogue of the Bishop volume comparison respectively. In Section 4, we study the volume computation for a compact subregion in a Lorentzian warped product manifold, and a comparison of volumes to a Lorentzian warped product under assumptions on the Ricci and mean curvatures.

## 2. Preliminaries and $D$-distance Wedges

Let $(M, g)$ be an arbitrary space-time of dimension $m \geq 2$ and let $K$ be a spacelike submanifold of dimension $k \geq 0$.

Given a timelike curve $\gamma$ from $K$ to $q$, suppose we have a variation $\alpha$ of $\gamma(t)$ with $s \to \alpha(0, s)$ in $K$ for all $s$ and with $\alpha(b, s) = q$ for all $s$. Then if $V(t)$ denotes the variation vector field of $\alpha$ along $\gamma$, $V(t) = \frac{d}{ds}\alpha(t, s)_{|s=0}$, then we have $V(b) = 0$ and $V(0) \in T_{\gamma(0)}K$.

If $\gamma : [0, b] \to (M, g)$ is a unit speed timelike geodesic from $K$ to $q$, then $L'(0) = g(V(0), \gamma'(0))$. Thus, $\gamma$ is extremal iff $\gamma$ is orthogonal at $\gamma(0)$ to $K$. Moreover, if $\gamma : [0, b] \to (M, g)$ is a unit timelike geodesic which is orthogonal at $\gamma(0)$ to the spacelike submanifold $K$ and assuming that $V$ is a piecewise smooth vector field along $\gamma$ orthogonal to $\gamma'$, then

$$L''(0) = g(S_{\gamma'(0)}V(0), V(0)) + I(V, V)$$

where $I(V, V) = -\int_0^b [g(V', V') - g(R(V, \gamma')\gamma', V)]dt$ and $S_{\gamma'(0)}$ is the second fundamental tensor given by $S_{\gamma'(0)}x = -(\nabla_x \gamma'(0))^T$ for $x \in T_pK$ where $T$ means "tangential part".

Let $V^\perp(\gamma)$ denote the vector space of piecewise smooth vector fields $Y$ orthogonal to $\gamma'$ and set $V^\perp(\gamma, K) = \{Y \in V^\perp(\gamma) | Y(0) \in T_{\gamma(0)}K\}$. Then the *Lorentzian submanifold index form* $I_{(b,K)} : V^\perp(\gamma, K) \times V^\perp(\gamma, K) \to \mathbf{R}$ for $X, Y \in V^\perp(\gamma, K)$, is defined by

$$I_{(b,K)}(X, Y) = g(S_{\gamma'(0)}X(0), Y(0)) + I(X, Y)$$

where $I$ is the index form on $V^\perp(\gamma)$.

Now a smooth vector field $J \in V^\perp(\gamma, K)$ is called a *$K$-Jacobi field along* $\gamma$ if $J$ satisfies

(1) $J'' + R(J, \gamma')\gamma' = 0$.

(2) $J'(0) + S_{\gamma'(0)}J(0) \in (T_{\gamma(0)}K)^\perp$.

Hence, $\gamma(t_0)$, $t_0 \in (0, b]$ is said to be a *$K$-focal point* if there is a nontrivial $K$-Jacobi field with $J(t_0) = 0$. Note that if $J$ is a $K$-Jacobi field, then $I_{(b,K)}(J, J) = -g(J'(b), J(b))$.

Now we recall a class of compact subregions from [14]. For $p \in M$, let $Fut(T_pM)$ denote the set of all future directed timelike vectors $w \in T_pM$ such that $\exp_p w$ is defined. Then for any unit vector $v \in Fut(T_pM)$ we have a timelike **radial** geodesic $\gamma_v(t) = \exp_p tv$ with $\gamma_v(0) = p, \gamma_v'(0) = v$. Put

$$cut_v(p) = \sup\{t \geq 0 \mid L(\gamma_v|_{[0,t]}) = d(p, \gamma_v(t))\} > 0$$

where $d : M \times M \to \mathbf{R}$ denotes the Lorentzian distance function of $(M, g)$, which is continuous and finite valued since $(M, g)$ is globally hyperbolic. Denote by $\delta : \exp_p(Fut(T_pM)) \to [0, +\infty)$ the distance function given by $\delta(q) := d(p, q)$. Fix $t_0 > 0$ in $\delta(\exp_p(Fut(T_pM)))$ such that for some unit timelike vector $v$ in $Fut(T_pM)$ we have $t_0 < cut_v(p)$. Since $v \longrightarrow cut_v(p)$ is lower semi-continuous for unit timelike vectors in a globally hyperbolic space-time (cf. [4, Prop. 8.7] or [5, Prop. 9.7]), it makes sense to consider the following geodesic variation of $\gamma_v$. Let $w \in T_{\gamma_v(t_0)}(\delta^{-1}(t_0))$. Extend to a parallel field $W$ along $\gamma_v$; thus $W(0) \in T_pM$ and we may put $\alpha(t, s) := \exp_p(t(v + sW(0)))$ with $0 \leq t \leq t_0$, $-\epsilon < s < \epsilon$ for some $\epsilon > 0$. Then

$$\alpha_{*|(t,0)}\frac{\partial}{\partial s} = (d\exp_p)\tau_{tv}(tW(0)),$$

where $\tau_v : T_pM \to T_v(T_pM)$ is the canonical isomorphism[5, p. 337]. Denoting $(d\exp_p)\tau_{tv}(tW(0))$ by $(d\exp_p)_{tv}tW(0)$, for $0 \leq t \leq t_0$, then $J(t) = (d\exp_p)_{tv}tW(0)$ is the Jacobi field along the timelike geodesic $\gamma_v(t)$ such that $J(0) = 0$, $J'(0) = W(0)$.

Now, let $\mathcal{C}^+(p)$ denote the following subset of $I^+(p)$:

$$\mathcal{C}^+(p) := \{\gamma_w(t) | g(w, w) = -1, w \in Fut(T_pM), \text{ and } 0 < t < cut_w(p)\}.$$

One way to calculate the volume of subsets of $\mathcal{C}^+(p)$ is an iterated integral, with one integration ranging over the unit future timelike vectors. (Unlike the unit sphere bundle for the Riemannian case, the set of unit timelike vectors is noncompact but is closed, and has infinite volume in the Minkowski metric induced on $T_pM$ by $g_{|p}$).

P. Ehrlich, S.-B. Kim and Y.-T. Jung[14] used a compact subset $D$ of $\delta^{-1}(t_0)$ to obtain finite integrals over distance wedges. Similarly, let

$$S := \{v \in Fut(T_pM) \mid g(v,v) = -1, \exp_p(v) \text{ is defined}\}$$

and let $C$ be a compact subset of $S$. Since $v \longrightarrow cut_v(p)$ is lower semicontinuous, if we put $inj_C(p) = \inf\{cut_v(p) \mid v \in C\}$, there exists $w$ in $C$ with $cut_w(p) = inj_C(p)$. Hence, $inj_C(p) > 0$. Also, since $\exp_p$ is continuous, $D = \exp_p(t_0C)$ is compact in $\delta^{-1}(t_0)$ for any $t_0$.

DEFINITION 2.1. The $D$-**distance wedge** $B_p^D(t_0)$ is defined as

$$B_p^D(t_0) = \{\exp_p(tv) \mid v \in C,\ 0 \le t \le t_0\}.$$

If $\{v, e_2, e_3, \cdots, e_m\}$ is an orthonormal basis of $T_pM$, then we have $(m-1)$ linearly independent Jacobi fields $J_2, J_3, \cdots, J_m$ along the timelike geodesic $\gamma_v = \exp_p(tv)$ where $J_i(t) = (d\exp_p)_{tv}tE_i(0)$ for the parallel fields $E_i$ along $\gamma_v$ with $E_i(0) = e_i$. Thus, the volume distortion by the exponential map $\exp_p$ along $\gamma_v$ is given by

$$\begin{aligned} |\det(d\exp_p)_{tv}| &= \frac{||(d\exp_P)_{tv}tE_2(0) \wedge \cdots \wedge (d\exp_P)_{tv}tE_m(0)||}{||tE_2(0) \wedge \cdots \wedge tE_m(0)||} \\ &= \frac{||J_2(t) \wedge \cdots \wedge J_m(t)||}{t^{m-1}}. \end{aligned} \tag{2.1}$$

A customary notation is $\theta(t,v) = ||J_2(t) \wedge \cdots \wedge J_m(t)||$. Then the following result on change of variables on $T_pM$ may be obtained([14]).

LEMMA 2.2. *$du = t^{m-1}dtdv$, where $dv$, $du$ are the volume elements of $S$ and $Fut(T_pM)$ respectively.*

Recalling that $J_i(t) = \alpha_*|_{(t,0)}(\frac{d}{ds}) = (d\exp_p)_{tv}te_i$ for the normal variation $\alpha(t,s) = \exp_p(t(v+se_i))$ and applying (2.1) and Lemma 2.2, we obtain

$$\begin{aligned} V_p^D(r) = Vol(B_p^D(r)) &= \int\int_{(\exp_p)^{-1}(B_p^D(r))} |\det(d\exp_p)_{tv}|du \\ &= \int_C\int_0^r ||J_2(t) \wedge \cdots \wedge J_m(t)||dtdv = \int_C\int_0^r \theta(t,v)dtdv. \end{aligned} \tag{2.2}$$

## 3. Jacobians and Comparison Theorems

Let $M$ be a globally hyperbolic space-time of dimension $m$ and $K$ a spacelike submanifold of dimension $k$, $0 \le k < m$. Let $\pi : TK^\perp \to K$ be the normal bundle with the normal connection $\nabla^\perp$. Then, for $p \in K$, we have a (normal) exponential map $\exp_p : (T_pK)^\perp \to M$ defined by $\exp_p(v) = \gamma(1)$ where $\gamma_v$ is the timelike geodesic in $M$ of initial velocity $v$.

Let $\xi \in TK^\perp$ be a normal field along $K$ with $\xi(0) = \xi_p$. Then, as in [24, pp. 456-457], $T_{\xi_p}(TK^\perp)$ may be split into the horizontal tangent space isometrically identified with $T_pK$ by $\pi_*$ and the vertical tangent space isometrically identified with $(T_pK)^\perp$ by $\nabla^\perp$-parallel translation.

Thus, if $U(t) = \frac{d}{ds}t\xi(s)|_{s=0}$ for a vector field $\xi(s)$ along $K$ with $\xi(0) = \xi_p$ at $p$, then $U(t)$ is a linear vector field along a curve $t \to t\xi_p$ in $TK^\perp$ given by $U(t) = A(t) + tB(t)$ where $A(t)$ is a horizontal field such that $A(0) \in T_pK$ and $B(t)$ is a vertical field such that $B(0) \in (T_pK)^\perp$ along the curve $t \to t\xi_p$.

Let $\gamma(t)$ be a timelike geodesic orthogonal at $p = \gamma(0)$ to the spacelike submanifold $K$ as $\gamma(t) = \exp_p(t\xi_p)$. Then we have a normal variation $\alpha(t,s) = \exp_p(t\xi(s))$ and we have a $K$-Jacobi field $J(t) = (d\exp_p)_{t\xi_p}U(t)$ along $\gamma$ as follows;

$$J(t) = \frac{d}{ds}\exp_p(t\xi(s))|_{s=0} = (d\exp_p)_{t\xi_p}\frac{d}{ds}(t\xi(s))|_{s=0} = (d\exp_p)_{t\xi_p}U(t)$$

with the initial conditions:

(1) $J(0) = U(0) = A(0) \in T_pK$.

(2) $J'(0) + S_{\xi_p}J(0) = B(0) \in (T_pK)^\perp$, because

$$\begin{aligned} J'(0) &= \nabla_{\frac{\partial}{\partial t}}(\frac{\partial}{\partial s}\alpha(t,s)|_{s=0})|_{t=0} = \nabla_{\frac{\partial}{\partial s}}(\frac{\partial}{\partial t}\alpha(t,s)|_{t=0})|_{s=0} \\ &= \nabla_{\frac{\partial}{\partial s}}\xi(s)|_{s=0} = \tan(\nabla_{\frac{\partial}{\partial s}}\xi(s)|_{s=0}) + \nabla^\perp_{\frac{\partial}{\partial s}}\xi(s)|_{s=0} \\ &= \tan(\nabla_{A(0)}\xi(s)|_{s=0}) + B(s)|_{s=0} = -S_{\xi_p}J(0) + B(0). \end{aligned}$$

Note that $\pi_* A(t) = J(0)$ and $||B(t)|| = ||J'(0)^\perp||$.

Now, for $r$ linearly independent linear vector fields $U_1, U_2, \cdots, U_r$, $0 < r < \dim M$ we may obtain $r$ linearly independent $K$-Jacobi fields $J_i(t)$ along $\gamma$ given by

$$J_i(t) = (d\exp_p)_{t\xi_p}U_i(t)$$

for $i = 1, 2, 3, \cdots, r$. To integrate the distortion of $r$-dimensional volumes, we need to compute the Jacobian (determinant) of the map $\exp_p$ as follows;

$$\begin{aligned} |\det(d\exp_p)_{t\xi_p}| &= \frac{||(d\exp_p)_{t\xi_p}U_1(t) \wedge \cdots \wedge (d\exp_p)_{t\xi_p}U_r(t)||}{||U_1(t) \wedge \cdots \wedge U_r(t)||} \\ &= \frac{||J_1(t) \wedge \cdots \wedge J_r(t)||}{||U_1(t) \wedge \cdots \wedge U_r(t)||} \end{aligned} \tag{3.1}$$

which is independent of any choice of the linearly independent fields $U_1, \cdots, U_r$, cf. [32].

Let $t_0(K) > 0$ denote the first focal value of $K$ along $\gamma$. For $0 < t < t_0(K)$, set $V_t = Span_{\mathbf{R}}\{J_1(t), \cdots, J_r(t)\}$ and let

$$\eta(t) := \log\{||J_1(t) \wedge \cdots \wedge J_r(t)||/||U_1(t) \wedge \cdots \wedge U_r(t)||\}$$

which is independent of basis of $V_t$. Then, $\lim_{t\to 0}\eta(t) = 0$. In fact, changing the basis of $V_t$ if necessary, we may assume that $A_{s+1}(0) = \cdots = A_r(0) = 0$ and $\{A_1(0), \cdots, A_s(0), B_{s+1}(0), \cdots, B_r(0)\}$ are linearly independent, where $s = \dim Span_{\mathbf{R}}\{A_1(0), \cdots, A_r(0)\} = \dim\{J(0) \mid J \in V_t\}$. Using $\lim_{t\to 0} J_{s+1}(t)/t =$

$\nabla J_{s+1}(0) = B_{s+1}(0), \cdots, \lim_{t\to 0} J_r(t)/t = \nabla J_r(0) = B_r(0)$, we get

$$\begin{aligned}
&\lim_{t\to 0} ||J_1(t) \wedge \cdots \wedge J_r(t)||/t^{r-s} \\
&= \lim_{t\to 0} ||J_1(t) \wedge \cdots \wedge J_s(t) \wedge \frac{J_{s+1}(t)}{t} \wedge \cdots \wedge \frac{J_r(t)}{t}|| \\
&= ||A_1(0) \wedge \cdots \wedge A_s(0) \wedge B_{s+1}(0) \wedge \cdots \wedge B_r(0)|| \\
&= \lim_{t\to 0} ||U_1(t) \wedge \cdots \wedge U_r(t)||/t^{r-s}
\end{aligned}$$

using the properties of determinants, which implies the desired limit for $\eta(t)$. If $\dim K = 0$, then $S_{\gamma'(0)} = 0$ and $A_i(t) = 0$. Thus, $U_i(t) = tB_i(t)$, $i = 2, \cdots, m$. If $\dim K = m-1$, then $J_i'(0) + S_{\gamma'(0)}J_i(0) = 0$ and $B_i(t) = 0$ for $i = 1, \cdots, m-1$. So, $U_i(t) = A_i(t)$ for $i = 1, \cdots, m-1$. Hence,

$$(3.2) \qquad ||U_1(t) \wedge \cdots \wedge U_r(t)|| = \begin{cases} ||J_1(0) \wedge \cdots \wedge J_r(0)||, & \text{if } \dim K = m-1 \\ t^r||J_1'(0) \wedge \cdots \wedge J_r'(0)||, & \text{if } \dim K = 0. \end{cases}$$

We now compare the Jacobians along timelike geodesics satisfying some appropriate initial conditions in two globally hyperbolic space-times $\bar{M}$ and $M$. Let $K$ be an $k$-dimensional spacelike submanifold of $M$ and $\gamma : [0, b] \to M$ a unit speed timelike geodesic perpendicular at $p = \gamma(0)$ to $K$. Then, for $0 < t < t_0(K)$, we set

$$\begin{aligned}
\Lambda(t) &:= \max\{g(R(v, \gamma'(t))\gamma'(t), v) \mid v \perp \gamma'(t),\ g(v, v) = 1\}, \\
\lambda(t) &:= \min\{g(R(v, \gamma'(t))\gamma'(t), v) \mid v \perp \gamma'(t),\ g(v, v) = 1\}, \\
S(t) &:= \max\{g(S_{\gamma'(t)}w, w) \mid g(w, w) = 1 \text{ for all } w \in T_{\gamma(t)}K\}, \\
s(t) &:= \min\{g(S_{\gamma'(t)}w, w) \mid g(w, w) = 1 \text{ for all } w \in T_{\gamma(t)}K\}.
\end{aligned}$$

Moreover, in $\bar{M}$ we may consider the corresponding notions $\bar{K}$, $\bar{\gamma}$, $\bar{\Lambda}(t)$, $\bar{\lambda}(t)$, $\bar{S}(t)$, $\bar{s}(t)$, $t_0(\bar{K})$, $\bar{V}_t$, $V^{\perp}(\bar{\gamma}, \bar{K})$, $\bar{I}_{(t_1,\bar{K})}$, etc.

If we take linearly independent spacelike $\bar{K}$-Jacobi fields $\bar{J}_1, \cdots, \bar{J}_r$ on $\bar{M}$ along the timelike geodesic $\bar{\gamma}$ perpendicular at $\bar{p} = \bar{\gamma}(0)$ to $\bar{K}$ and define $\bar{\eta}(t)$ as above, then similarly $\lim_{t\to 0} \bar{\eta}(t) = 0$.

Now, assume that

$$(3.3) \qquad \dim\bar{M} \le \dim M, \ \dim\bar{K} = \dim K = k.$$

And suppose

$$(3.4) \qquad \bar{\Lambda}(t) \le \lambda(t) \text{ for } 0 \le t \le \min\{t_0(\bar{K}), t_0(K)\}, \text{ and } \bar{S}(0) \le s(0).$$

Let $\imath : T_{\bar{p}}\bar{M} \to T_pM$ be a linear isometry with $\imath(\bar{\gamma}'(0)) = \gamma'(0)$, and $\imath(T_p\bar{K}) = T_pK$. Further construct the usual family of linear isometries $\phi_t : T_{\bar{\gamma}(t)}\bar{M} \to T_{\gamma(t)}M$ given by $\phi_t = p_t^{-1} \circ \imath \circ \bar{p}_t$ where $\bar{p}_t : T_{\bar{\gamma}(t)}\bar{M} \to T_{\bar{\gamma}(0)}\bar{M}$ and $p_t : T_{\gamma(t)}M \to T_{\gamma(0)}M$ denote the parallel translations along $\bar{\gamma}$ and $\gamma$ respectively. Now take $r$ linearly independent spacelike $K$-Jacobi fields $J_1, J_2, \cdots, J_r$ in $M$ along the timelike geodesic $\gamma$ and also $r$ linearly independent spacelike $\bar{K}$-Jacobi fields $\bar{J}_1, \bar{J}_2, \cdots, \bar{J}_r$ in $\bar{M}$ along the timelike geodesic $\bar{\gamma}$. Similarly to the setup in [24] and [32], make the further assumption on the two sets of Jacobi fields that

$$(3.5) \qquad \phi_t < \bar{J}_1, \bar{J}_2, \cdots, \bar{J}_r > = < J_1, J_2, \cdots, J_r >$$

for $t$ with $0 \le t < \min\{t_0(\bar{K}), t_0(K)\}$, which is a further condition on the above linear isometry $\imath : T_{\bar{p}}\bar{M} \to T_pM$. Fix any $t_1$ with $0 < t_1 < \min\{t_0(\bar{K}), t_0(K)\}$. By (3.5) we have $\phi_{t_1}(\bar{V}_{t_1}) = V_{t_1}$. Making a change of basis if necessary (cf. [32, p. 144]), we may suppose that $\{\bar{J}_i(t_1)\}$ forms an orthonormal basis of $\bar{V}_{t_1}$. Then $W_j(t) = \phi_t(\bar{J}_j(t))$ along $\gamma$ belong to $V^\perp(\gamma, K)|_{[0,t_1]}$ and $\{W_j(t_1)\}$ forms an orthonormal basis for $V_{t_1}$. Thus we may assume that $J_j(t_1) = W_j(t_1)$ for $i = 1, 2, \cdots, r$ so that $\{J_j(t_1)\}$ forms an orthonormal basis for $V_{t_1}$. Note that $g(W_j(t), W_j(t)) = \bar{g}(\bar{J}_j(t), \bar{J}_j(t))$, $g(W_j'(t), W_j'(t)) = \bar{g}(\bar{J}_j'(t), \bar{J}_j'(t))$ for $0 \le t \le t_1$ as in [4, 5, 29].

By the timelike index theorem for the spacelike submanifold $K$ and the maximality theorem of $K$-Jacobi fields in [29], we have the following inequalities for the submanifold index form:

$$\bar{I}_{(t_1,\bar{K})}(\bar{J}_j(t), \bar{J}_j(t)) \le I_{(t_1,K)}(W_j(t), W_j(t)) \le I_{(t_1,K)}(J_j(t), J_j(t))$$

for $j = 1, 2, \cdots, r$. Hence, we may now show the Lorentzian version of the comparison result

$$\eta(t) \le \bar{\eta}(t) \tag{3.6}$$

for $0 \le t < \min\{t_0(\bar{K}), t_0(K)\}$ along similar lines as the Riemannian proof in [24], cf. [32], under the further condition employed in these references that

$$\{\log ||\bar{U}_1(t) \wedge \cdots \wedge \bar{U}_r(t)|| - \log ||U_1(t) \wedge \cdots \wedge U_r(t)||\}' = 0. \tag{3.7}$$

This condition (3.7) typically holds in the usual comparison theorems because the $U_j$'s and $\bar{U}_j$'s may be chosen appropriately, cf. (3.2). Indeed, in view of (3.1) and (3.7) one should first consider separately

$$h(t) = \frac{d}{dt}\{\log ||\bar{J}_1(t) \wedge \cdots \wedge \bar{J}_r(t)|| - \log ||J_1(t) \wedge \cdots \wedge J_r(t)||\} \tag{3.8}$$

and show $h(t_1) \ge 0$ en route to deriving (3.6). Recalling that $\{\bar{J}_i(t_1)\}$ is orthonormal, we have

$$\begin{aligned}
\frac{d}{dt}\log ||\bar{J}_1(t) \wedge \cdots \wedge \bar{J}_r(t)||_{t=t_1} &= \frac{\frac{d}{dt}||\bar{J}_1 \wedge \cdots \wedge \bar{J}_r||(t_1)}{||\bar{J}_1 \wedge \cdots \wedge \bar{J}_r||(t_1)} \\
&= \sum_{i=1}^r \bar{g}(\bar{J}_1(t_1) \wedge \cdots \wedge \bar{J}_i'(t_1) \wedge \cdots \wedge \bar{J}_r(t_1), \bar{J}_1(t_1) \wedge \cdots \wedge \bar{J}_r(t_1)) \\
&= \sum_{i=1}^r \bar{g}(\bar{J}_1(t_1) \wedge \cdots \wedge \sum_{j=1}^r \bar{g}(\bar{J}_i'(t_1), \bar{J}_j(t_1))\bar{J}_j(t_1) \wedge \cdots \wedge \bar{J}_r(t_1), \bar{J}_1(t_1) \wedge \cdots) \\
&= \sum_{i=1}^r \bar{g}(\bar{J}_i'(t_1), \bar{J}_i(t_1)) = -\sum_{i=1}^r \bar{I}_{(t_1,\bar{K})}(\bar{J}_i(t), \bar{J}_i(t)) \ge -\sum_{i=1}^r I_{(t_1,K)}(J_i(t), J_i(t)) \\
&= \frac{d}{dt}\log ||J_1(t) \wedge \cdots \wedge J_r(t)||_{t=t_1}.
\end{aligned}$$

as desired.

Thus the Jacobi comparison theorem may now be obtained.

THEOREM 3.1. *Under conditions (3.3), (3.4) and (3.5),*

*(1)* $F(t) = \dfrac{||\bar{J}_1(t) \wedge \cdots \wedge \bar{J}_r(t)||}{||J_1(t) \wedge \cdots \wedge J_r(t)||}$

*is monotone increasing on* $(0, \min\{t_0(\bar{K}), t_0(K)\})$. *In fact,* $F'(t) \geq 0$ *and in particular,* $F(t) \geq F(0)$ *for* $0 \leq t < \min\{t_0(\bar{K}), t_0(K)\}$.

*(2) If in addition (3.7) holds for* $0 \leq t < \min\{t_0(\bar{K}), t_0(K)\}$, *then*

$$\frac{||J_1(t) \wedge \cdots \wedge J_r(t)||}{||U_1(t) \wedge \cdots \wedge U_r(t)||} \leq \frac{||\bar{J}_1(t) \wedge \cdots \wedge \bar{J}_r(t)||}{||\bar{U}_1(t) \wedge \cdots \wedge \bar{U}_r(t)||}. \tag{3.9}$$

*for* $0 \leq t < \min\{t_0(\bar{K}), t_0(K)\}$.

PROOF. For $t \in (0, \min\{t_0(K), t_0(\bar{K})\})$, inequality (1) follows from $\{\log F(t)\}' = h(t) \geq 0$. Then (2) follows assuming (3.7) because $\lim_{t \to 0} \eta(t) = \lim_{t \to 0} \bar{\eta}(t) = 0$ and $\{\bar{\eta}(t) - \eta(t)\}' = h(t) \geq 0$ provided (3.7) is assumed.

REMARK 3.2. If $r = 1$, the timelike Rauch Comparison Theorem for $K$-Jacobi fields as given in [29] is recovered. In the case that $\dim K = 0$, then the $U_j$'s and $J_j$'s may be chosen so that

$$\frac{||J_2(t) \wedge \cdots \wedge J_m(t)||}{||U_2(t) \wedge \cdots \wedge U_m(t)||} = \frac{||J_2(t) \wedge \cdots \wedge J_m(t)||}{t^{m-1}} = \frac{\theta(t,v)}{t^{m-1}} \tag{3.10}$$

recalling (2.1). Hence, in this case inequality (3.9) may be interpreted as a comparison

$$\theta(t,v) \leq \bar{\theta}(t,\bar{v}) \tag{3.11}$$

and thus Theorem 3.1 may be viewed as implying the space-time versions of the Bishop Volume Comparison and the Bishop-Gromov Volume Comparison given in [14], (cf. [13, 16, 17, 20]) for $D$-distance wedges as given in Definition 2.1.

For use in Section 4, we should also note that in the hypersurface case $\dim K = m-1$, the expression (3.9) for the Jacobian of the exponential map $\exp_p$ at $t\gamma'(0)$ becomes

$$\frac{||J_2(t) \wedge \cdots \wedge J_m(t)||}{||J_2(0) \wedge \cdots \wedge J_m(0)||} \tag{3.12}$$

in view of (3.2).

Finally, since the index comparison theorem and the maximality of $K$-Jacobi vector classes for the quotient index form associated to a null geodesic initially perpendicular to a spacelike submanifold has been obtained in [12, Theorem 2.3 and 2.1 respectively], results analogous to the infinitesimal Theorem 3.1 should be valid for the case of a null geodesic initially perpendicular to a spacelike submanifold.

## 4. A Volume Comparison for Lorentzian Warped Products

A way to approach volume calculations is from the viewpoint of the family of level sets $\{\delta^{-1}(t) \mid 0 < t \leq cut_w(p)\}$ of the local space-time distance function $\delta(q) = d(p,q)$. In this notion, it will be helpful to recall the concept of a "variation tensor field associated to a normal variation of a hypersurfaces $H$" as formulated in Eschenburg and O'Sullivan [20, pp. 6-7] for Riemannian manifolds.

Let $H \subset \mathcal{C}^+(p)$ be a local smooth spacelike hypersurface which is perpendicular to $\gamma_v'(r)$ at $q := \gamma_v(r)$. Let $\xi$ be a future directed unit timelike vector field along $H$ with $\xi_q = \gamma_v'(r)$, $r \in (0, cut_v(p))$. Define a variation $\psi : I \times H \to M$ by $\psi(t,q) = \exp_q((t-r)\xi_q)$, where we suppose $I$ and $H$ to be suitably chosen such that $\psi$ is a diffeomorphism onto its image. Such a mapping $\psi : I \times H \to M$ will be called a normal geodesic variation of $\gamma$ along the hypersurface $H$. Further, for each $t \in I$, define a differentiable map $\psi : H \to M$ by $\psi_t(q) = \psi(t,q)$.

Now, we specialize to the case that $H = H_t(p) = \delta^{-1}(t) \cap \mathcal{C}^+(p)$, or $H$ is a portion of a future pseudo-hyperbolic space which is contained in a suitably small neighborhood of $\gamma(t)$ for $t \in (0, cut_w(p))$, $w \in Fut(T_pM)$.

And, let $S = \{v \in Fut(T_pM) \mid g(v,v) = -1, \exp_p(v) \text{ is defined}\}$ and let $C \subset S$ be a compact subset. Then we denote $H_t^*(p) = H_t(p) \cap D$, where $D = \exp_p(tC)$ is compact in $\delta^{-1}(t)$ for any $t \in (0, cut_w(p))$.

Let $\xi$ be a future directed unit timelike vector field along $H_{r_0}^*(p)$ with $\xi_q = \gamma_w'(r_0)$, $r_0 \in (0, cut_w(p))$. As in (3.2), we have $m-1$ linearly independent $H_{r_0}^*(p)$-Jacobi fields $J_2, J_3, \cdots, J_m$ along the unit timelike geodesic $\gamma(t) = \exp_q((t-r_0)\xi_q)$ and we may define the area of $H_t^*(p)$ by

$$A(t) = \int_{H_{r_0}^*(p)} \frac{||J_2(t) \wedge J_3(t) \wedge \cdots \wedge J_m(t)||}{||J_2(r_0) \wedge J_3(r_0) \wedge \cdots \wedge J_m(r_0)||} d(Area_g H_{r_0}^*(p))(q), \tag{4.1}$$

for $q \in H_{r_0}^*(p)$.

In obtaining the volume estimate on distance wedges in Theorem 4.4, a comparison with a warped product space-time will be used inspired by the Riemannian results obtained in Itokawa[26]. Here the Riemannian factor will be taken to be the lift $S(p) = \exp_p^{-1}(H_{t_0}(p))$ to $T_pM$ in the induced metric. However, en route to Theorem 4.4, calculations are made which are valid for a general warped product equally well. Hence we will first treat this topic in the more general setting.

Thus let $(\bar{K}, h)$ be a Riemannian manifold of dimension $m-1$ and suppose $r_0$ and $r_1$ with $0 < r_0 < r_1 < inj_C(p)$ together with a warping function $f : [r_0, r_1) \to (0, \infty)$ may be found so that $f$ is a $C^2$ solution to the Jacobi differential equation

$$f''(t) + \mu(t) f(t) = 0 \tag{4.2}$$

with initial conditions $f(r_0) = f_0$, $f'(r_0) = f_1$. Here $\mu : [r_0, r_1) \to \mathbf{R}$ is a continuous function which may be regarded as prescribing the curvature, in view of Proposition 4.1. Form the generalized Robertson-Walker space-time

$$\bar{M} = [r_0, r_1) \times_f \bar{K}$$

with warped product metric

$$\bar{g} = -dt^2 + (f(t))^2 h. \tag{4.3}$$

In particular, $\bar{M}$ contains the class of unit timelike geodesics $\bar{\gamma}(t) = (t, \bar{q})$, $\bar{q} \in \bar{K}$, sometimes called "stationary", which are always globally maximal regardless of choice of warping function $f(t)$ or Riemannian factor $(\bar{K}, h)$.

We will use the notation $\bar{K}_t = \{t\} \times \bar{K}$ below. A second important property of this class of warped products is that $\bar{K}_{r_0}$ (or more generally, any $\bar{K}_t$) is free of focal points along the normal timelike geodesics $\bar{\gamma}(t) = (t, \bar{q})$.

Let $\bar{\nabla}$ denote the Levi-Civita connection on $(\bar{M}, \bar{g})$. The following basic curvature calculation is well known, cf. [31].

PROPOSITION 4.1. *Let $\bar{M}$ be as above and $\bar{q} \in \bar{K}$. Suppose that $\bar{u}$ is a unit timelike vector at $(t, \bar{q})$ normal to $\bar{K}_t$ and $\bar{v}, \bar{w}$ are unit spacelike vectors tangent to $\bar{K}_t$ at $(t, \bar{q})$. Then we have*

*1) the sectional curvature:* $K(\bar{u}, \bar{v}) = -\mu(t) = \dfrac{f''(t)}{f(t)}$,

*2) the Ricci curvature:* $Ric(\bar{u}, \bar{u}) = (m-1)\mu(t)$,

*3) the mean curvature:* $H(t, \bar{q}) = -(m-1)\dfrac{f'(t)}{f(t)}$.

With the above calculations in hand, we may also obtain the $\bar{K}_{r_0}$-Jacobi fields in $\bar{M}$. Note that for $\bar{\gamma}(t) = (t, \bar{q}), \bar{q} \in \bar{K}$, we have

$$S_{\bar{\gamma}'(r_0)}\bar{v} = -\frac{f'(r_0)}{f(r_0)}\bar{v}$$

from (3) of Proposition 4.1. Thus we find that a Jacobi field $\bar{J}(t)$ along $\bar{\gamma}(t)$ satisfies the $\bar{K}_{r_0}$-initial conditions provided $\bar{J}(r_0) \in T_{(r_0, \bar{q})}\bar{K}_{r_0}$ and

$$\bar{J}'(r_0) = \frac{f'(r_0)}{f(r_0)}\bar{J}(r_0). \tag{4.4}$$

The Riemannian analogue of (4.4) motivates Itokawa's [26] concept of the $\mathbb{M}(r_0, r_1, f, f_0, f_1/f_0)$-Darboux-Cheeger-Yau model. With this initial condition translated into our context, the $\bar{K}_{r_0}$-Jacobi fields may now be easily obtained.

LEMMA 4.2. *Given a spacelike vector $\bar{v}$ tangent to $\bar{K}_{r_0}$ at $(r_0, \bar{q})$, let $\bar{E}(t)$ be the parallel field along $\bar{\gamma}(t) = (t, \bar{q})$ with $\bar{E}(r_0) = \bar{v}$. Then the $\bar{K}_{r_0}$-Jacobi field along $\bar{\gamma}(t)$ with initial condition $\bar{J}(r_0) = \bar{v}$ (and hence also $\bar{J}'(r_0) = \dfrac{f_1}{f_0}\bar{v}$) is given by*

$$\bar{J}(t) = \frac{f(t)}{f(r_0)}\bar{E}(t). \tag{4.5}$$

PROOF. From the curvature calculations in the proof of Prop. 4.1, we have

$$R(\bar{J}(t), \bar{\gamma}'(t))\bar{\gamma}'(t) = \mu(t)\bar{J}(t) = \frac{\mu(t)f(t)}{f(r_0)}\bar{E}(t).$$

On the other hand, differentiating (4.5), we have $\bar{J}''(t) = \dfrac{f''(t)}{f(r_0)}\bar{E}(t)$, Hence,

$$\bar{J}''(t) + R(\bar{J}(t), \bar{\gamma}'(t))\bar{\gamma}'(t) = \frac{f''(t) + \mu(t)f(t)}{f(r_0)}\bar{E}(t) = 0.$$

Remark that for a frame $\bar{U}_2, \bar{U}_3, \cdots, \bar{U}_m$ on $\bar{K}$ near $\bar{q}$ and the two volume forms $d(Area_{\bar{g}}\bar{K}_{r_0})$ and $d(Area_h\bar{K})$ on $\bar{K}_{r_0}$ and $\bar{K}$ respectively, we have

$$\begin{aligned} & d(Area_{\bar{g}}\bar{K}_{r_0})((0,\bar{U}_2) \wedge \cdots \wedge (0,\bar{U}_m)) \\ &= |\det(\bar{g}((0,\bar{U}_i),(0,\bar{U}_j)))|^{1/2} \\ &= f(r_0)^{m-1}|\det(h(\bar{U}_i,\bar{U}_j))|^{1/2} \\ &= f(r_0)^{m-1} d(Area_h\bar{K})(\bar{U}_2 \wedge \cdots \wedge \bar{U}_m). \end{aligned}$$

Thus,

$$d(Area_h\bar{K}) = \left(\frac{1}{f(r_0)}\right)^{m-1} d(Area_{\bar{g}}\bar{K}_{r_0}). \tag{4.6}$$

With (4.6) and Lemma 4.2, the warped product version of the integrand in (4.1) takes a simple form: for linearly independent $\bar{K}_{r_0}$-Jacobi fields $\bar{J}_2(t), \cdots, \bar{J}_m(t)$ in $\bar{M}$ along the stationary timelike geodesic $\bar{\gamma}(t) = (t, \bar{q})$,

$$\frac{||\bar{J}_2(t) \wedge \cdots \wedge \bar{J}_m(t)||}{||\bar{J}_2(r_0) \wedge \cdots \wedge \bar{J}_m(r_0)||} d(Area_{\bar{g}}\bar{K}_{r_0})(r_0,\bar{q}) = (f(t))^{m-1} d(Area_h\bar{K})(\bar{q}). \tag{4.7}$$

With these properties of the warped product comparison model established, we may now obtain Theorem 4.4 with the results of section 3. Here we will now specialize to a warped product comparison model obtained by lifting the future pseudo-hyperbolic space to $T_pM$. For sake of convenience, assume that

$$S(p) = \exp_p^{-1}(H_1(p))$$

is defined and also that $\exp_p S(p)$ is a diffeomorphism onto $H_1(p)$. Also, let $S^*(p) = C \cap S(p)$, where $C$ is the compact subset of $S$ as discussed at the beginning of the section. Let $h$ be the Riemannian metric induced on $S(p)$ as a subset of $T_pM$. We consider a warped product comparison model

$$\bar{M} = [r_0, r_1) \times_f S(p)$$

with Lorentzian metric $\bar{g} = -dt^2 \oplus (f(t))^2 h$ where $0 < r_0 < r_1 < inj_C(p)$ and $f$ satisfies the Jacobi differential equation

$$f''(t) + \mu(t) f(t) = 0, \; f(r_0) = f_0, \; f'(r_0) = f_1$$

with $f(t) > 0$ on $[r_0, r_1)$, $\mu : [r_0, r_1) \to \mathbf{R}$ continuous. Finally, let $S_t(p) = \{t\} \times S^*(p)$.

Applying Theorem 3.1 (2) and (3.12), we have the following proposition.

PROPOSITION 4.3. *Let $H^*_{r_0}(p)$ be the compact subset as above of a future pseudo-hyperbolic space $H_{r_0}(p)$ in a globally hyperbolic space-time $(M, g)$ with $S(p) = \exp_p^{-1}(H_1(p))$ as above and suppose the mean curvature of $H^*_{r_0}(p)$ at $q$ satisfies*

$$H(q) \geq -(m-1)\frac{f'(r_0)}{f(r_0)}.$$

*Let $J_2(t), J_3(t), \cdots, J_m(t)$ be linearly independent $H^*_{r_0}(p)$-Jacobi fields along the geodesic $\gamma(t) = \exp_q((t - r_0)\xi_q)$ where $\xi_q$ is a unit timelike vector orthogonal to*

*$H^*_{r_0}(p)$ at $q$. Let $\bar{J}_2(t), \bar{J}_3(t), \cdots, \bar{J}_m(t)$ be linearly independent $S^*(p)$-Jacobi fields along the timelike geodesic $\bar{\gamma}(t) = (t, \bar{q})$ in $M$. Assume that*

$$Ric(\gamma'(t), \gamma'(t)) \geq (m-1)\mu(t)$$

*for all $t \in [r_0, r_1)$. Then,*

$$\frac{||J_2(t) \wedge J_3(t) \wedge \cdots \wedge J_m(t)||}{||J_2(r_0) \wedge J_3(r_0) \wedge \cdots \wedge J_m(r_0)||} \leq \frac{||\bar{J}_2(t) \wedge \bar{J}_3(t) \wedge \cdots \wedge \bar{J}_m(t)||}{||\bar{J}_2(r_0) \wedge \bar{J}_3(r_0) \wedge \cdots \wedge \bar{J}_m(r_0)||}. \tag{4.8}$$

*for $r_0 \leq t < r_1 < inj_C(p)$.*

Now we may obtain the volume estimate for a distance wedge region (recall Definition 2.1) which correspond to Itokawa's[26] estimates for the distance spheres and metric balls in a complete Riemannian manifold. First, fix $q \in H^*_{r_0}(p)$, $\bar{q} \in S^*(p)$ and $t_0$ with $r_0 < t_0 < r_1$. Applying (4.7) in the context of Proposition 4.3,

$$\begin{aligned}
&\int_{r_0}^{t_0} \int_{H^*_{r_0}(p)} \frac{||J_2(t) \wedge J_3(t) \wedge \cdots \wedge J_m(t)||}{||J_2(r_0) \wedge J_3(r_0) \wedge \cdots \wedge J_m(r_0)||} d(Area_g H^*_{r_0}(p))(q)dt \\
&\leq \int_{r_0}^{t_0} \int_{S^*_{r_0}(p)} \frac{||\bar{J}_2(t) \wedge \bar{J}_3(t) \wedge \cdots \wedge \bar{J}_m(t)||}{||\bar{J}_2(r_0) \wedge \bar{J}_3(r_0) \wedge \cdots \wedge \bar{J}_m(r_0)||} d(Area_{\bar{g}} S^*_{r_0}(p))(r_0, \bar{q})dt \\
&= Area_h(S^*(p)) \int_{r_0}^{t_0} (f(t))^{m-1} dt.
\end{aligned}$$

THEOREM 4.4. *Let $(M, g)$ be a globally hyperbolic space-time of dimension $m$ with $Ric(\gamma'(t), \gamma'(t)) \geq (m-1)\mu(t)$ for $t$ in $[r_0, r_1)$ along all unit timelike geodesics $\gamma(t) = \exp_q((t - r_0)\xi_q)$ where $\xi_q$ is a future directed timelike vector normal to the compact subset $H^*_{r_0}(p) = H_{r_0}(p) \cap \exp_p(r_0 C)$ of a pseudo-hyperbolic space $H_{r_0}(p)$ in $M$, where $f(t)$ is a positive $C^2$-solution to the Jacobi differential equation $f''(t) + \mu(t)f(t) = 0$, for $t$ in $[r_0, r_1)$ with $0 < r_0 < r_1 < inj_C(p)$. Suppose further that the mean curvature of $H^*_{r_0}(p)$ in $M$ satisfies*

$$H(q) \geq -(m-1)\frac{f'(r_0)}{f(r_0)}$$

*for all $q$ in $H^*_{r_0}(p)$. Suppose that $\exp_p$ is a diffeomorphism of $S(p) = \exp_p^{-1}(H_1(p))$ onto $H_1(p)$ and $S^*(p) = S(p) \cap C$. Then*

$$Vol(B) \leq Area_h(S^*(p)) \int_{r_0}^{r_1} (f(t))^{m-1} dt$$

*where $B = B^D_p(r_1) \backslash B^D_p(r_0)$.*

For a concluding example, we consider a class of warped products that have played a role in Andersson and Howard[2, Example 2.7]. Let $f(t) = \cos(t)$ on $[r_0, r_1) = [\pi/6, \pi/2)$, and suppose $\dim H = 3$. Now $Ric(T, T) = 3$ for $T = \left(\frac{d}{dt}, 0\right)$ regardless of the curvature of the Riemannian factor. But, for $w = \left(c\frac{d}{dt}, v\right) \in$

$T_{(t,p)}M$ with $g(w,w) = -1$, if $x = \dfrac{v}{||v||_h}$, then

$$Ric(w,w) = 3 + (c^2 - 1)\sec^2(t)[Ric_H(x,x) + 2].$$

Hence, as $|c| \geq 1$, if $K_H \geq -1$, then $Ric_H(x,x) \geq -2$ and $Ric(w,w) \geq 3$.

In the special case that $K_{S(p)} = -1$ considered above, then $Ric_H(x,x) = -2$, so that $Ric(w,w) = 3$ on all unit timelike vectors. Moreover, since the mean curvature of $S^*_{\pi/6}(p)$ is $-(4-1)\dfrac{f'(\pi/6)}{f(\pi/6)} = \sqrt{3}$, we obtain the following corollary to Theorem 4.4.

COROLLARY 4.5. *Let $(M,g)$ be a globally hyperbolic space-time of dimension 4 with $Ric(\gamma'(t),\gamma'(t)) \geq 3$ for any unit timelike geodesic with $\gamma'(\pi/6)$ perpendicular to $H^*_{\pi/6}(p)$, for $t \in \left[\frac{\pi}{6}, \frac{\pi}{2}\right)$. Suppose also $\frac{\pi}{2} < inj_C(p)$ and the mean curvature of $H^*_{\pi/6}(p)$ satisfies $H(q) \geq \sqrt{3}$ for all $q$. Let $S^*_{\pi/6}(p)$ be a compact subset of a pseudo-hyperbolic space $S_{\pi/6}(p)$ in a Lorentzian warped product $\bar{M} = \left[\frac{\pi}{6}, \frac{\pi}{2}\right) \times_{\cos(t)} S(p)$ with $S(p) = \exp_p^{-1}(H_1(p))$. Then*

$$Vol(B) \leq \frac{5}{24} Area_h(S^*(p))$$

*where $B = B_p^D\left(\frac{\pi}{2}\right) \backslash B_p^D\left(\frac{\pi}{6}\right)$.*

## REFERENCES


1. M. T. Anderson, *On Long-Time Evolution in General Relativity and Geometrization of 3-Manifolds*, Commun. Math Phys. **222** (2001), 533-567.
2. L. Andersson and R. Howard, *Comparison and rigidity theorems in semi-Riemannian geometry*, Communications in Analysis and Geometry **6** (1998), 819-877.
3. J. K. Beem and P. E. Ehrlich, *Cut points, conjugate points and Lorentzian comparison theorems*, Math. Proc. Camb. Phil. Soc. **86** (1979), 365-384.
4. J. K. Beem and P. E. Ehrlich, *Global Lorentzian Geometry*, vol. 67, Marcel Dekker Pure and Applied Mathematics, 1981.
5. J. Beem, P. Ehrlich, and K. Easley, *Global Lorentzian Geometry, Second Revised Edition*, vol. 202, Marcel Dekker Pure and Applied Mathematics, 1996.
6. J. Beem, P. Ehrlich, S. Markvorsen and G. Galloway, *Decomposition theorems for Lorentzian manifolds with nonpositive curvature*, J. Diff. Geom. **22** (1985), 29-42.
7. R. Bishop, *A relation between volume, mean curvature, and diameter*, Notices of Amer. Math. Soc. 10 (1963), 364-375.
8. R. Bishop and R. Crittenden, *Geometry of Manifolds*, vol. 15, Academic Press Pure and Applied Mathematics, 1964.
9. Y. D. Burago and V. A. Zalgaller, *Geometric Inequalities*, Spring - Verlag, 1980.
10. I. Chavel, *Riemannian Geometry: A Modern Introduction*, Cambridge Univ. Press, 1993.
11. J. Cheeger and D. G. Ebin, *Comparison theorems in Riemannian Geometry*, North Holl. Publ. Co., 1975.
12. P. E. Ehrlich and S.-B. Kim, *A focal comparison theorem for null geodesics*, Communications of the Korean Mathematical Society **6** (1991), 73-87.
13. P. E. Ehrlich and S.-B. Kim, *From the Riccati equation to the Raychaudhuri inequality*, Contemporary Math. of AMS **170** (1994), 65-78.
14. P. Ehrlich, S.-B. Kim, Y.-T. Jung, *Volume comparison theorems for Lorentzian manifolds*, Geo. Dedicata **73** (1998), 39-56.
15. P. Ehrlich and M. Sanchez, *Some semi-Riemannian volume comparison theorems*, Tohoku Math. J. **52** (2000), 331-348.

16. J.-H. Eschenburg, *Comparison theorems and hypersurfaces*, Manuscripta Math. **59** (1987), 295-323.
17. J.-H. Eschenburg, *Maximum principle for hypersurfaces*, Manuscripta Math. **64** (1989), 55-75.
18. J.-H. Eschenburg and E. Heintze, *Comparison theory for Riccati equations*, Manuscripta Math. **68** (1990), 209-214.
19. J.-H. Eschenburg and J. O'Sullivan, *Growth of Jacobi fields and divergence of geodesics*, Math. Z. **150** (1976), 221-237.
20. J.-H. Eschenburg and J. O'Sullivan, *Jacobi tensors and Ricci curvature*, Math. Ann. **252** (1980), 1-26.
21. G. J. Galloway, *A generalization of Myers theorem and an application to relativistic cosmology*, J. of Differential Geometry **14** (1979), 105-116.
22. S. Gallot, D. Hulin, and J. Lafontaine, *Riemannian Geometry, Second Edition*, Springer Verlag Universitext, 1990.
23. K. Grove, *Metric Differential Geometry, Nordic Summer School, Lyngby 1985*, Springer Lecture Notes 1263, 1987.
24. E. Heintze and H. Karcher, *A general comparison theorem with application to volume estimates for submanifolds*, Ann. Sci. Ecole Norm. Sup. **11** (1978), 459-470.
25. S. W. Hawking and G. F. R. Ellis, *The large scale structure of space-time*, Cambridge Monographs on Mathematical Physics, Cambridge University Press, 1973.
26. Y. Itokawa, *Distance spheres and Myers type theorems for manifolds with lower bounds on the Ricci curvature*, Illinois J. of Math 34(4) (1990), 693-705.
27. H. Karcher, *Riemannian Comparison Constructions*, M.A.A. Studies in Mathematics Vol. 27, Global Differential Geometry, S. S. Chern, ed. (1989), 170-222.
28. J.-S. Kim, *Volumes and Ricci Curvatures on Lorentzian Warped Products*, Ph.D. Thesis, Chonnam National University (1998).
29. S.-B. Kim, *K-Jacobi fields and Lorentzian comparison theorems*, Ph.D. Thesis, University of Missouri-Columbia (1986).
30. S.-B. Kim and D.-S. Kim, *A focal Myers-Galloway theorem on space-times*, J.of the Korean Math. Soc. **31** (1994), 97-110.
31. B. O'Neill, *Semi-Riemannian Geometry*, Academic Press Inc, 1983.
32. T. Sakai, *Riemannian Geometry*, vol. 149, A.M.S. Math. Monographs, 1996.

Department of Mathematics, The University of Florida,
Gainesville, FL 32611-8105.

*E-mail address*: *ehrlich@math.ufl.edu*

Department of Mathematics, Chosun University, Kwangju,
501-759 S. Korea.

*E-mail address*: *ytajung@mail.chosun.ac.kr*

Department of Mathematics, Chonnam National University,
Kwangju, 500-757 S. Korea.

*E-mail address*: *jskim01@hanmir.com*

Department of Mathematics, Chonnam National University,
Kwangju, Korea 500-757

*E-mail address*: *sbk@chonnam.ac.kr*

Contemporary Mathematics
Volume **337**, 2003

# Nilpotent Szabó, Osserman and Ivanov-Petrova pseudo-Riemannian manifolds

B. Fiedler and P. Gilkey

ABSTRACT. We exhibit pseudo Riemannian manifolds which are Szabó nilpotent of arbitrary order, or which are Osserman nilpotent of arbitrary order, or which are Ivanov-Petrova nilpotent of order 3.

## 1. Introduction

Let $R$ be the Riemann curvature tensor of a pseudo-Riemannian manifold $(M, g)$ of signature $(p, q)$. The *Szabó operator* $\mathcal{S}$ is the self-adjoint linear map which is characterized by the identity:

$$g(\mathcal{S}(x)y, z) = \nabla R(y, x, x, z; x).$$

One says that $(M, g)$ is *Szabó* if the eigenvalues of $\mathcal{S}(x)$ are constant on the pseudo-spheres of unit timelike and spacelike vectors:

$$S^{\pm}(M, g) := \{x \in TM : g(x, x) = \pm 1\}.$$

Szabó [**22**] used techniques from algebraic topology to show in the Riemannian setting ($p = 0$) that any such metric is locally symmetric. He used this observation to give a simple proof that any 2 point homogeneous space is either flat or is a rank 1 symmetric space. Subsequently Gilkey and Stavrov [**15**] extended his results to show that any Szabó Lorentzian ($p = 1$) manifold has constant sectional curvature. By replacing $g$ by $-g$, one can interchange the roles of $p$ and of $q$, thus these results apply to the cases $q = 0$ and $q = 1$ as well.

The eigenvalue zero is distinguished. One says that $(M, g)$ is *Szabó nilpotent of order* $n$ if $\mathcal{S}(x)^n = 0$ for every $x \in TM$ and if there exists a point $P_0 \in M$ and a tangent vector $x_0 \in T_{P_0}M$ so that $\mathcal{S}(x_0)^{n-1} \neq 0$. One says that $(M, g)$ is *Szabó nilpotent* if $(M, g)$ is Szabó nilpotent of order $n$ for some $n$. Note that $(M, g)$ is Szabó nilpotent if and only if 0 is the only eigenvalue of $\mathcal{S}$; consequently any Szabó nilpotent manifold is Szabó. There is some evidence [**12, 21**] to suggest, conversely, that any Szabó manifold is Szabó nilpotent.

*Key words and phrases.* Szabó operator, Jacobi operator, Osserman conjecture, Higher order Jacobi operator, Ivanov-Petrova manifolds
2000 *Mathematics Subject Classification.* 53B20
Research of P. Gilkey partially supported by the NSF (USA) and MPI (Germany).

If $(M,g)$ is Szabó nilpotent of order 1, then $\mathcal{S}(x)=0$ for all $x\in TM$. This implies [**15**] that $\nabla R=0$ so $(M,g)$ is a local symmetric space; this is to be regarded, therefore, as a trivial case. Gilkey, Ivanova, and Zhang [**13**] have constructed pseudo-Riemannian manifolds of any signature $(p,q)$ with $p\geq 2$ and $q\geq 2$ which are Szabó nilpotent of order 2; these were the only previously known examples of Szabó manifolds which were not local symmetric spaces. In this brief note, we shall construct pseudo-Riemannian metrics $g_n$ on $\mathbb{R}^{n+2}$ which are Szabó nilpotent of order $n\geq 2$; the metric will be *balanced* (i.e. $p=q$) if $n$ is even and *almost balanced* (i.e. $p=q\pm 1$) if $n$ is odd. By taking an isometric product with a suitable flat manifold, the signature can be increased without changing the order of nilpotency.

THEOREM 1.1. *Let $n\geq 2$. There exists a pseudo-Riemannian metric $g_n$ on $\mathbb{R}^{n+2}$ which is Szabó nilpotent of order $n$. If $n=2p$, $g_n$ has signature $(p+1,p+1)$; if $n=2p+1$, $g_n$ has signature $(p+1,p+2)$.*

The *Jacobi operator* is defined analogously; it is characterized by the identity:

$$g(J(x)y,z)=R(y,x,x,z).$$

One says that $(M,g)$ is *Osserman* if the eigenvalues of $J$ are constant on $S^{\pm}(M)$. In the Riemannian setting, Osserman wondered [**19**] if this implied $(M,g)$ was a 2 point homogeneous space. This question has been answered in the affirmative in the Riemannian setting [**5, 18**] for dimensions $\neq 8,16$, and in all dimensions in the Lorentzian setting [**1, 8**].

We shall say that $(M,g)$ is *Osserman nilpotent of order n* if $J(x)^n=0$ for every $x\in TM$ and if there exists a point $P_0\in M$ and a tangent vector $x_0\in T_{P_0}M$ so that $J(x_0)^{n-1}\neq 0$, i.e. 0 is the only eigenvalue of $J$. Such manifolds are necessarily Osserman. Osserman nilpotent manifolds of orders 2 and 3 have been constructed previously [**2, 10, 9, 11**]. These manifolds need not be homogeneous, thus the question Osserman raised has a negative answer in the higher signature setting. A byproduct of our investigation of Szabó manifolds yields new examples of Osserman manifolds; again, the signature can be increased by taking isometric products with flat factors.

THEOREM 1.2. *Let $n\geq 2$. There exists a pseudo-Riemannian metric $\tilde{g}_n$ on $\mathbb{R}^{n+2}$ which is Osserman nilpotent of order $n$. If $n=2p$, then $\tilde{g}_n$ has signature $(p+1,p+1)$; if $n=2p+1$, then $\tilde{g}_n$ has signature $(p+1,p+2)$.*

If $\{f_1,f_2\}$ is an oriented orthonormal basis for a non-degenerate oriented 2 plane $\pi$, we define the skew-symmetric curvature operator by setting $\mathcal{R}(\pi):=R(f_1,f_2)$. We say $(M,g)$ is Ivanov-Petrova nilpotent of order $n$ if $\mathcal{R}(\pi)^n=0$ for any non-degenerate oriented 2 plane $\pi$ and if there exists $\pi$ so $\mathcal{R}(\pi)^{n-1}\neq 0$. We refer to [**11**] for further details concerning Ivanov-Petrova manifolds. Another byproduct of our investigation yields new examples of these manifolds:

THEOREM 1.3. *There exist Ivanova-Petrova pseudo-Riemannian manifolds which are nilpotent of order 2 and of order 3.*

Here is a brief outline to the paper. In Section 2, we give a general procedure for constructing pseudo-Riemannian manifolds with certain kinds of curvature and covariant derivative curvature tensors. We apply this procedure in Section 3 to complete the proof of Theorem 1.1. Lemma 3.1 deals with the cases $n=2$ and $n=3$, Lemma 3.2 deals with the case $n=2\ell+1\geq 5$, and Lemma 3.3 deals with the case $n = 2\ell + 2 \geq 4$. In Section 4, we prove Theorem 1.2 and in Section 5, we prove Theorem 1.3.

One can also work with the Jordan normal form; one says $(M, g)$ is Jordan Szabo (resp. Jordan Osserman or Jordan IP) if the Jordan normal form of $\mathcal{S}$ (resp. $J$ or $\mathcal{R}$) is constant on the appropriate domains of definition. The examples constructed in this paper do **not** fall into this framework; in particular, there are no known Jordan Szabo pseudo-Riemannian manifolds which are not locally symmetric.

It is a pleasant task to acknowledge helpful discussions with Prof. S. Nikčević concerning these matters.

## 2. A family of pseudo-Riemannian manifolds

We introduce the following notational conventions. Let $(x, u_1, ..., u_\nu, y)$ be coordinates on $\mathbb{R}^{\nu+2}$. We shall use several different notations for the coordinate frame:

$$\mathcal{B} = \{e_0, e_1, ..., e_{\nu+1}\} = \{X, U_1, ..., U_\nu, Y\} := \{\partial_x, \partial_{u_1}, ..., \partial_{u_\nu}, \partial_y\}.$$

Let indices $i, j, ...$ range from 0 through $\nu+1$ and index the full coordinate frame. Let indices $a, b$ range from 1 through $\nu$ and index the tangent vectors $\{U_1, ...., U_\nu\}$. In the interests of brevity, we shall give non-zero entries in a metric $g$, curvature tensor $R$, and covariant derivative curvature tensor $\nabla R$ up to the obvious $\mathbb{Z}_2$ symmetries.

Lemma 2.1. *Let $f = f(u)$ be a smooth function on $\mathbb{R}^\nu$ and let $\Xi$ be a constant invertible symmetric $\nu \times \nu$ matrix. Define a metric $g_f$ on $\mathbb{R}^{\nu+2}$ by setting:*

$$g_f(X, X) = f(u), \quad g_f(X, Y) = 1, \quad \text{and} \quad g_f(U_a, U_b) = \Xi_{ab}.$$

*All other scalar products equals zero. The non-zero entries in $R_{g_f}$ and $\nabla R_{g_f}$ are:*

(1) $R_{g_f}(X, U_a, U_b, X) = -\frac{1}{2} U_a U_b(f)$.
(2) $\nabla R_{g_f}(X, U_a, U_b, X; U_c) = -\frac{1}{2} U_a U_b U_c(f)$.

Proof. Since $d\Xi = 0$, the non-zero Christoffel symbols of the first kind are:

$$\Gamma_{a00} = \Gamma_{0a0} = -\Gamma_{00a} = \tfrac{1}{2} U_a(f). \tag{2.a}$$

Let $\Xi^{ab}$ be the inverse matrix. We adopt the Einstein convention and sum over repeated indices to compute:

$$\begin{aligned}
&\Gamma_{ijb} = g(\nabla_{e_i} e_j, e_b) = g(\Gamma_{ij}{}^k e_k, e_b) = \Gamma_{ij}{}^a \Xi_{ab} && \text{so } \Gamma_{ij}{}^a = \Xi^{ab}\Gamma_{ijb},\\
&\Gamma_{ij\nu+1} = g(\nabla_{e_i} e_j, e_{\nu+1}) = g(\Gamma_{ij}{}^k e_k, e_{\nu+1}) = \Gamma_{ij}{}^0 && \text{so } \Gamma_{ij}{}^0 = 0,\\
&\Gamma_{ij0} = g(\nabla_{e_i} e_j, e_0) = g(\Gamma_{ij}{}^k e_k, e_0) = f\Gamma_{ij}{}^0 + \Gamma_{ij}{}^{\nu+1} && \text{so } \Gamma_{ij}{}^{\nu+1} = \Gamma_{ij0}.
\end{aligned}$$

Thus the non-zero Christoffel symbols of the second kind are:

$$\Gamma_{a0}{}^{\nu+1} = \Gamma_{0a}{}^{\nu+1} = \tfrac{1}{2} U_a(f) \quad \text{and} \quad \Gamma_{00}{}^a = -\tfrac{1}{2} \textstyle\sum_b \Xi^{ab} U_b(f). \tag{2.b}$$

The components of the curvature tensor relative to the coordinate frame are:

$$R_{ijkl} = e_i\Gamma_{jkl} - e_j\Gamma_{ikl} + \textstyle\sum_n \{\Gamma_{inl}\Gamma_{jk}{}^n - \Gamma_{jnl}\Gamma_{ik}{}^n\}. \tag{2.c}$$

By equation (2.b), $\Gamma_{ik}{}^0 = \Gamma_{jk}{}^0 = 0$. By equation (2.a), $\Gamma_{i,\nu+1,k} = \Gamma_{j,\nu+1,k} = 0$. Thus the index $n$ in equation (2.c) is neither 0 nor $\nu+1$. Thus by equation (2.a) and equation (2.b), $i = j = k = l = 0$. This shows that the terms which are quadratic in $\Gamma$ play no role in equation (2.c). Assertion (1) then follows from equation (2.a).

The covariant derivative of the curvature tensor is given by:

(2.d) $R_{ijkl;n} = e_n R_{ijkl} - \sum_p \{\Gamma_{ni}{}^p R_{pjkl} + \Gamma_{nj}{}^p R_{ipkl} + \Gamma_{nk}{}^p R_{ijpl} + \Gamma_{nl}{}^p R_{ijkp}\}.$

By equation (2.b) $\Gamma_{**}{}^0 = 0$. Thus we may assume $p \neq 0$ in equation (2.d). Furthermore, by assertion (1), $R_{\nu+1***} = R_{*\nu+1**} = R_{**\nu+1*} = R_{***\nu+1} = 0$ so we may also assume $p \neq \nu + 1$ in equation (2.d). Thus $\Gamma_{ni}{}^p R_{pjkl} = 0$ unless $i = j = 0$ and similarly $\Gamma_{nj}{}^p R_{ipkl} = 0$ unless $i = j = 0$. Thus these two terms cancel. Similarly $\Gamma_{nk}{}^p R_{ijpl}$ cancels $\Gamma_{nl}{}^p R_{ijkp}$. Thus $R_{ijkl;n} = e_n R_{ijkl}$ and assertion (2) follows. □

REMARK 2.2. Such metrics are called *plane wave metrics* in the Lorentzian setting in the physics literature. Let $\rho$ be the associated Ricci tensor; we have $\rho(\xi, \xi) = \text{Trace}(J(\xi))$. We have $\rho(e_i, e_j) = \sum_{kl} g^{kl} R(e_i, e_k, e_l, e_j)$. Since $R$ vanishes on $e_{\nu+1}$, we may sum over $k, l \leq \nu$. Since $g^{0k} = g^{k0} = 0$ for $k \leq \nu$, we have $\rho(e_i, e_j) = \sum_{ab} g^{ab} R(e_i, e_a, e_b, e_j)$. Thus $\rho(e_i, e_j) = 0$ for $(i, j) \neq (0, 0)$ and the only non-zero entry of the Ricci tensor is $\rho(e_0, e_0) = -\frac{1}{2} \sum_{ab} \Xi^{ab} \partial_a \partial_b f$. The associated Jacobi operator will be nilpotent if and only if this sum vanishes. Raising indices yields a Ricci operator $\hat{\rho}$ with the property that $\hat{\rho}(e_0) = -\frac{1}{2} \sum_{ab} \Xi^{ab} \partial_a \partial_b f$ and $\hat{\rho}(e_i) = 0$ for $i > 0$. Thus $\hat{\rho}^2 = 0$ so the Ricci operator is nilpotent of order 2 and non-trivial if and only if $g$ is not Osserman. We refer to R. Bryant [**4**] and to I. Kath [**17**] where more general examples of metrics with nilpotent Ricci operators are constructed in the context of an investigation of parallel spinors.

If $f$ is quadratic, then $R$ is constant on the coordinate frame; if $f$ is cubic, then $\nabla R$ is constant on the coordinate frame. However, these tensors are not curvature homogeneous in the sense of Kowalski, Tricerri, and Vanhecke [**16**] since the metric relative to the coordinate frames is not constant.

The tensors of Lemma 2.1 are related to hypersurface theory. Let $M$ be a non-degenerate hypersurface in $\mathbb{R}^{(a,b)}$; we assume $M$ is spacelike but similar remarks hold in the timelike setting. Let $L$ be the associated second fundamental form and let $S = \nabla L$ be the covariant derivative of $L$; $L$ is a totally symmetric 2 form and $S$ is a totally symmetric 3 form. We may then, see for example [**11**], express:

$$\begin{aligned} R_L(x_1, x_2, x_3, x_4) &= L(x_1, x_4)L(x_2, x_3) - L(x_1, x_3)L(x_2, x_4), \\ \nabla R_{L,S}(x_1, x_2, x_3, x_4; x_5) &= S(x_1, x_4, x_5)L(x_2, x_3) + L(x_1, x_4)S(x_2, x_3, x_5) \\ &- S(x_1, x_3, x_5)L(x_2, x_4) - L(x_1, x_3)S(x_2, x_4, x_5). \end{aligned} \tag{2.e}$$

If $L$ is an arbitrary symmetric 2 tensor and if $S$ is an arbitrary totally symmetric 3 tensor, then we may use equation (2.e) to define tensors we continue to denote by $R_L$ and $\nabla R_{L,S}$. We refer to [**6**] for the proof of assertion (1) and to [**7**] for the proof of assertion (2) in the following result:

THEOREM 2.3.

(1) *The tensors $R_L$ which are defined by a symmetric 2 form $L$ generate the space of all algebraic curvature tensors.*
(2) *The tensors $\nabla R_{L,S}$ which are defined by a symmetric 2 form $L$ and by a totally symmetric 3 form $S$ generate the space of all algebraic covariant derivative curvature tensors.*

The tensors of Lemma 2.1 (2) are of this form. Let

$$f(u) := -\tfrac{1}{3} \sum_{a,b,c} c_{a,b,c} u_a u_b u_c$$

be a cubic polynomial in the $u$ variables which is independent of $x$ and $y$. Then:

$$\nabla R = \nabla R_{L,S} \quad \text{for} \quad L(\partial_i, \partial_j) := \delta_{0,i}\delta_{0,j} \quad \text{and} \quad S(\partial_i, \partial_j, \partial_k) := -\tfrac{1}{2}\partial_i\partial_j\partial_k f.$$

## 3. Nilpotent Szabó manifolds

In this section we will use Lemma 2.1 to prove Theorem 1.1 by choosing $\Xi$ and $f$ appropriately. We shall consider metrics of the form:

$$g(X,X) = f(t,u,v),\ g(X,Y) = 1,\ g(T,T) = 1,\ g(U_a, V_b) = \delta_{ab};$$

the spacelike vector $T$ will not be present in some cases. The vectors $\{U_a, V_a\}$ are a hyperbolic pair.

We begin by discussing the cases $n = 2$ and $n = 3$.

LEMMA 3.1.

(1) *Let $\mathcal{B}_2 := \{X, U, V, Y\} = \{\partial_x, \partial_u, \partial_v, \partial_y\}$ be the coordinate frame on $\mathbb{R}^4$ relative to the coordinate system $(x, u, v, y)$. Define a metric $g_2$ by:*

$$g_2(X,X) = -\tfrac{1}{3}u^3,\ g_2(X,Y) = 1,\ g_2(U,V) = 1.$$

*Then $g_2$ has signature $(2,2)$ on $\mathbb{R}^4$ and $g_2$ is Szabó nilpotent of order 2.*

(2) *Let $\mathcal{B}_3 := \{X, T, U, V, Y\} = \{\partial_x, \partial_t, \partial_u, \partial_v, \partial_y\}$ be the coordinate frame on $\mathbb{R}^5$ relative to the coordinate system $(x, t, u, v, y)$. Define a metric $g_3$ by:*

$$g_3(X,X) = -tu^2,\ g_3(T,T) = 1,\ g_3(U,V) = 1,\ g_3(X,Y) = 1.$$

*Then $g_3$ has signature $(2,3)$ on $\mathbb{R}^5$ and $g_3$ is Szabó nilpotent of order 3.*

PROOF. Let $\mathcal{B}^* = \{e^0, \dots, e^{\nu+1}\}$ be the corresponding *dual basis* of $\mathcal{B}$; it is characterized by the relations $g(e_i, e^j) = \delta_i^j$. For example, we have

$$\mathcal{B}_2^* = \{Y, V, U, X - fY\} \quad \text{and} \quad \mathcal{B}_3^* = \{Y, T, V, U, X - fY\} \tag{3.a}$$

By Lemma 2.1, the only non-zero component of $\nabla R_{g_2}$ is

$$\nabla R_{g_2}(X, U, U, X; U) = 1.$$

Let $\xi = \xi_0 X + \xi_1 U + \xi_2 V + \xi_3 Y$ be a tangent vector. We use equation (3.a) to raise indices and conclude:

$$\mathcal{S}_{g_2}(\xi)X = \xi_1^3 Y - \xi_0\xi_1^2 V,\quad \mathcal{S}_{g_2}(\xi)U = -\xi_0\xi_1^2 Y + \xi_0^2\xi_1 V,\quad \mathcal{S}_{g_2}(\xi)Y = \mathcal{S}(\xi)V = 0.$$

Thus $\mathcal{S}_{g_2}(\xi)^2 = 0$ for all $\xi$ while $\mathcal{S}_{g_2}(\xi) \neq 0$ for generic $\xi$. Assertion (1) now follows.

Similarly, the only non-zero components of $\nabla R_{g_3}$ are

$$\nabla R_{g_3}(X, U, U, X; T) = \nabla R_{g_3}(X, U, T, X; U) = 1.$$

We use equation (3.a) to raise indices and compute:

$$\begin{aligned} &\mathcal{S}_{g_3}(\xi)X = \star T + \star Y + \star V, \quad \mathcal{S}_{g_3}(\xi)Y = 0,\\ &\mathcal{S}_{g_3}(\xi)T = \star Y + \star V,\\ &\mathcal{S}_{g_3}(\xi)U = \star T + \star Y + \star V, \quad \mathcal{S}_{g_3}(\xi)V = 0 \end{aligned}$$

where $\star = \star(\xi)$ denotes suitably chosen cubic polynomials in the coefficients of $\xi$ that is generically non-zero; as the precise value of this coefficient is not important, we shall suppress it in the interests of notational simplicity. It is now clear that $\mathcal{S}_{g_3}(\xi)^3 = 0$ for all $\xi$ while $\mathcal{S}_{g_3}(\xi)^2$ is generically non-zero. □

Next we consider the case $n = 2\ell + 1 \geq 5$. Let $(x, t, u_2, ..., u_{\ell+1}, v_2, ..., v_{\ell+1}, y)$ be coordinates on $\mathbb{R}^{2\ell+3}$ which define the associated coordinate frame:

$$\mathcal{B} := \{X, T, U_2, ..., U_{\ell+1}, V_2, ..., V_{\ell+1}, Y\} = \{\partial_x, \partial_t, \partial_{u_2}, ..., \partial_{u_{\ell+1}}, \partial_{v_2}, ..., \partial_{v_{\ell+1}}, \partial_y\}.$$

LEMMA 3.2. *Let $\ell \geq 2$. Define a metric $g_{2\ell+1}$ on $\mathbb{R}^{2\ell+3}$ by setting:*

$$g_{2\ell+1}(X,X) = -tu_2^2 - \textstyle\sum_{2\leq a\leq\ell}(u_a + v_a)u_{a+1}^2,$$

$$g_{2\ell+1}(X,Y) = 1, \quad g_{2\ell+1}(T,T) = 1, \quad g_{2\ell+1}(U_a, V_b) = \delta_{ab}.$$

*Then $g_{2\ell+1}$ is a metric of signature $(\ell+1, \ell+2)$ and Szabó nilpotent of order $2\ell+1$.*

PROOF. Let $2 \leq a \leq \ell$. By Lemma 2.1, the non-zero components of $\nabla R_{g_{2\ell+1}}$ are:

$$\begin{array}{rclcl}
\nabla R_{g_{2\ell+1}}(X, U_2, U_2, X; T) &=& \nabla R_{g_{2\ell+1}}(X, T, U_2, X; U_2) &=& 1,\\
\nabla R_{g_{2\ell+1}}(X, U_{a+1}, U_{a+1}, X; U_a) &=& \nabla R_{g_{2\ell+1}}(X, U_{a+1}, U_a, X; U_{a+1}) &=& 1,\\
\nabla R_{g_{2\ell+1}}(X, U_{a+1}, U_{a+1}, X; V_a) &=& \nabla R_{g_{2\ell+1}}(X, U_{a+1}, V_a, X; U_{a+1}) &=& 1.
\end{array}$$

The dual basis is $\mathcal{B}^* = \{Y, T, V_2, ..., V_{\ell+1}, U_2, ..., U_{\ell+1}, X - fY\}$. Let $\xi$ be an arbitrary tangent vector. We raise indices and compute:

$$\begin{array}{lcl}
\mathcal{S}_{g_{2\ell+1}}(\xi)X &\in& \mathrm{Span}\{Y, T, U_2, ..., U_\ell, V_2, ..., V_{\ell+1}\},\\
\mathcal{S}_{g_{2\ell+1}}(\xi)Y &=& 0,\\
\mathcal{S}_{g_{2\ell+1}}(\xi)T &=& \star Y + \star V_2,\\
\mathcal{S}_{g_{2\ell+1}}(\xi)U_2 &=& \star T + \star Y + \star V_2 + \star V_3,\\
\mathcal{S}_{g_{2\ell+1}}(\xi)U_a &=& \star U_{a-1} + \star Y + \star V_{a-1} + \star V_a + \star V_{a+1} \quad \text{for} \quad 3 \leq a \leq \ell,\\
\mathcal{S}_{g_{2\ell+1}}(\xi)U_{\ell+1} &=& \star U_\ell + \star Y + \star V_\ell + \star V_{\ell+1}\\
\mathcal{S}_{g_{2\ell+1}}(\xi)V_a &=& \star Y + \star V_{a+1} \quad \text{for} \quad 2 \leq a \leq \ell,\\
\mathcal{S}_{g_{2\ell+1}}(\xi)V_{\ell+1} &=& 0
\end{array}$$

where $\star$ is a coefficient that is non-zero for generic $\xi$. If $\mathcal{E}$ is a subspace, let $\alpha = \beta + \mathcal{E}$ mean that $\alpha - \beta \in \mathcal{E}$. We compute:

$$\begin{array}{lcll}
\mathcal{S}_{g_{2\ell+1}}(\xi)^\mu U_{\ell+1} &=& \star U_{\ell+1-\mu} + \mathrm{Span}\{V_2, ..., V_{\ell+1}, Y\}, & 1 \leq \mu \leq \ell - 1\\
\mathcal{S}_{g_{2\ell+1}}(\xi)^\ell U_{\ell+1} &=& \star T + \mathrm{Span}\{V_2, ..., V_{\ell+1}, Y\}, &\\
\mathcal{S}_{g_{2\ell+1}}(\xi)^\mu U_{\ell+1} &=& \star V_{\mu+1-\ell} + \mathrm{Span}\{V_{\mu+2-\ell}, ..., V_{\ell+1}, Y\}, & \ell + 1 \leq \mu \leq 2\ell - 1\\
\mathcal{S}_{g_{2\ell+1}}(\xi)^{2\ell} U_{\ell+1} &=& \star V_{\ell+1} + \mathrm{Span}\{Y\}. &
\end{array}$$

Thus $\mathcal{S}_{g_{2\ell+1}}(\xi)^{2\ell} \neq 0$ for generic $\xi$. One shows similarly $\mathcal{S}_{g_{2\ell+1}}(\xi)^{2\ell+1} = 0$ for every $\xi$ by:

$$\begin{array}{lcll}
\mathcal{S}_{g_{2\ell+1}}(\xi)^\mu \mathcal{B} &\subseteq& \mathrm{Span}\{T, U_2, ..., U_{\ell+1-\mu}, V_2, ..., V_{\ell+1}, Y\}, & 1 \leq \mu \leq \ell - 1\\
\mathcal{S}_{g_{2\ell+1}}(\xi)^\ell \mathcal{B} &\subseteq& \mathrm{Span}\{T, V_2, ..., V_{\ell+1}, Y\}, &\\
\mathcal{S}_{g_{2\ell+1}}(\xi)^\mu \mathcal{B} &\subseteq& \mathrm{Span}\{V_{\mu+1-\ell}, ..., V_{\ell+1}, Y\}, & \ell + 1 \leq \mu \leq 2\ell
\end{array}$$

and $\mathcal{S}_{g_{2\ell+1}}(\xi)^{2\ell+1}\mathcal{B} = \{0\}$. □

We complete the proof of Theorem 1.1 by considering the case $n = 2\ell + 2$ for $\ell \geq 1$. Let $(x, u_1, ..., u_{\ell+1}, v_1, ..., v_{\ell+1}, y)$ be coordinates on $\mathbb{R}^{2\ell+4}$ which define the associated coordinate frame:

$$\mathcal{B} := \{X, U_1, ..., U_{\ell+1}, V_1, ..., V_{\ell+1}, Y\} = (\partial_x, \partial_{u_1}, ..., \partial_{u_{\ell+1}}, \partial_{v_1}, ..., \partial_{v_{\ell+1}}, \partial_y\}.$$

LEMMA 3.3. *Let $\ell \geq 1$. Define a metric $g_{2\ell+2}$ on $\mathbb{R}^{2\ell+4}$ by setting:*

$$g_{2\ell+2}(X,X) = -\sum_{1\leq a\leq \ell}(u_a+v_a)u_{a+1}^2 - \tfrac{1}{3}u_1^3,$$
$$g_{2\ell+2}(X,Y)=1, \quad g_{2\ell+2}(U_a,V_b)=\delta_{ab}.$$

*Then $g_{2\ell+2}$ is a metric of signature $(\ell+2,\ell+2)$ and Szabó nilpotent of order $2\ell+2$.*

PROOF. Let $2 \leq a \leq \ell+1$. The non-zero components of $\nabla R_{g_{2\ell+2}}$ are:

$$\begin{array}{lclcl}
\nabla R_{g_{2\ell+2}}(X,U_1,U_1,X;U_1) & = & 1, & & \\
\nabla R_{g_{2\ell+2}}(X,U_a,U_a,X;U_{a-1}) & = & \nabla R_{g_{2\ell+2}}(X,U_a,U_{a-1},X;U_a) & = & 1, \\
\nabla R_{g_{2\ell+2}}(X,U_a,U_a,X;V_{a-1}) & = & \nabla R_{g_{2\ell+2}}(X,U_a,V_{a-1},X;U_a) & = & 1.
\end{array}$$

We compute:

$$\begin{array}{lcl}
\mathcal{S}_{g_{2\ell+2}}(\xi)X & = & \star U_1+\ldots+\star U_\ell+\star Y+\star V_1+\ldots+\star V_{\ell+1}, \\
\mathcal{S}_{g_{2\ell+2}}(\xi)U_1 & = & \star Y+\star V_1+\star V_2, \\
\mathcal{S}_{g_{2\ell+2}}(\xi)U_a & = & \star U_{a-1}+\star Y+\star V_{a-1}+\star V_a+\star V_{a+1} \quad \text{for} \quad 2\leq a\leq \ell, \\
\mathcal{S}_{g_{2\ell+2}}(\xi)U_{\ell+1} & = & \star U_\ell+\star Y+\star V_\ell+\star V_{\ell+1}, \\
\mathcal{S}_{g_{2\ell+2}}(\xi)Y & = & 0, \\
\mathcal{S}_{g_{2\ell+2}}(\xi)V_a & = & \star Y+\star V_{a+1} \quad \text{for} \quad 1\leq a\leq \ell, \\
\mathcal{S}_{g_{2\ell+2}}(\xi)V_{\ell+1} & = & 0.
\end{array}$$

We may then show $\mathcal{S}_{g_{2\ell+2}}(\xi)^{2\ell+1}$ is generically non-zero by computing:

$$\begin{array}{lcll}
\mathcal{S}_{g_{2\ell+2}}(\xi)^\mu U_{\ell+1} & = & \star U_{\ell+1-\mu}+\mathrm{Span}\{Y,V_1,\ldots,V_{\ell+1}\}, & 1\leq\mu\leq\ell \\
\mathcal{S}_{g_{2\ell+2}}(\xi)^\mu U_{\ell+1} & = & \star V_{\mu-\ell}+\mathrm{Span}\{Y,V_{\mu+1-\ell},\ldots,V_{\ell+1}\}, & \ell+1\leq\mu\leq 2\ell \\
\mathcal{S}_{g_{2\ell+2}}(\xi)^{2\ell+1}U_{\ell+1} & = & \star V_{\ell+1}+\mathrm{Span}\{Y\}. &
\end{array}$$

A similar argument shows $\mathcal{S}_{g_{2\ell+2}}(\xi)^{2\ell+2}=0$ for all $\xi$. We can write

$$\begin{array}{lcll}
\mathcal{S}_{g_{2\ell+2}}(\xi)^\mu\mathcal{B} & \subseteq & \mathrm{Span}\{U_1,\ldots,U_{\ell+1-\mu},V_1,\ldots,V_{\ell+1},Y\}, & 1\leq\mu\leq\ell, \\
\mathcal{S}_{g_{2\ell+2}}(\xi)^\mu\mathcal{B} & \subseteq & \mathrm{Span}\{V_{\mu-\ell},\ldots,V_{\ell+1},Y\}, & \ell+1\leq\mu\leq 2\ell+1
\end{array}$$

and consequently we have that $\mathcal{S}_{g_{2\ell+2}}(\xi)^{2\ell+2}\mathcal{B}=\{0\}$. □

REMARK 3.4. One can also consider the purely pointwise question. We shall say that $(M,g)$ is Szabó nilpotent of order $n$ at $P\in M$ if $\mathcal{S}(x)^n=0$ for all $x\in T_PM$ and if $\mathcal{S}(x_0)^{n-1}\neq 0$ for some $x_0\in T_PM$. Throughout Section 3, we considered cubic functions to ensure that $\nabla R$ was constant on the coordinate frames; thus the point in question played no role. However, had we replaced $u^3$ by $u^4$, $tu^2$ by $tu^3$, $u_au_{a+1}^2$ by $u_au_{a+1}^3$, and $v_au_{a+1}^2$ by $v_au_{a+1}^3$, then we would have constructed metrics $g_n$ which were Szabó nilpotent of order $n$ on $T_P\mathbb{R}^{n+2}$ for generic points $P\in\mathbb{R}^{n+2}$, but where $\nabla R$ vanishes at the origin $0\in\mathbb{R}^{n+2}$. Since the order of nilpotency would vary with the point of the manifold, these metrics clearly are not homogeneous.

## 4. Nilpotent Osserman manifolds

In Section 3, we used cubic expressions to define our metrics to ensure the tensors $R_{ijkl;n}$ were constant on the coordinate frame. To discuss the Jacobi operator, we use the corresponding quadratic polynomials. We adopt the notation of Section

3 to define metrics:

$$\begin{aligned}
&\tilde g_2(X,X) = -u^2,\ \tilde g_2(X,Y) = 1,\ \tilde g_2(U,V) = 1,\\
&\tilde g_3(X,X) = -2tu - u^2,\ \tilde g_3(T,T) = 1,\ \tilde g_3(U,V) = 1,\ \tilde g_3(X,Y) = 1,\\
&\tilde g_{2\ell+1}(X,X) = -2tu_2 - u_2^2 - \textstyle\sum_{2\le a\le \ell}\{2(u_a+v_a)u_{a+1} + u_{a+1}^2\},\\
&\qquad \tilde g_{2\ell+1}(X,Y) = 1,\ \tilde g_{2\ell+1}(T,T) = 1,\ \tilde g_{2\ell+1}(U_a,V_b) = \delta_{uv}, \quad (\ell \ge 2)\\
&\tilde g_{2\ell+2}(X,X) = -\textstyle\sum_{1\le a\le \ell}\{2(u_a+v_a)u_{a+1} + u_{a+1}^2\} - u_1^2,\\
&\qquad \tilde g_{2\ell+2}(X,Y) = 1, \quad \tilde g_{2\ell+2}(U_a,V_b) = \delta_{ab} \qquad (\ell \ge 1).
\end{aligned}$$

LEMMA 4.1.

(1) $\tilde g_2$ *has signature* $(2,2)$ *and is Osserman nilpotent of order* $2$.
(2) $\tilde g_3$ *has signature* $(2,3)$ *and is Osserman nilpotent of order* $3$.
(3) $\tilde g_{2\ell+1}$ *has signature* $(\ell+1,\ell+2)$ *and is Osserman nilpotent of order* $2\ell+1$.
(4) $\tilde g_{2\ell+2}$ *has signature* $(\ell+2,\ell+2)$ *and is Osserman nilpotent of order* $2\ell+2$.

PROOF. By Lemma 2.1, the non-zero components of $R_{\tilde g_2}$ are

$$R_{\tilde g_2}(X,U,U,X) = 1. \tag{4.a}$$

Assertion (1) now follows since:

$$J_{\tilde g_2}(\xi)X = \star Y + \star V, \quad J_{\tilde g_2}(\xi)U = \star Y + \star V, \quad J_{\tilde g_2}(\xi)Y = J_{\tilde g_2}(\xi)V = 0$$

where $\star$ denotes suitably chosen quadratic polynomials in the components of $\xi$ which are non-zero for generic $\xi$.

Similarly, the only non-zero component of $\nabla R_{\tilde g_3}$ are

$$R_{\tilde g_3}(X,U,U,X) = 1 \quad \text{and} \quad R_{\tilde g_3}(X,U,T,X) = 1. \tag{4.b}$$

Assertion (2) now follows since:

$$\begin{aligned}
&J_{\tilde g_3}(\xi)X = \star T + \star Y + \star V, \quad J_{\tilde g_3}(\xi)Y = 0,\\
&J_{\tilde g_3}(\xi)T = \star Y + \star V,\\
&J_{\tilde g_3}(\xi)U = \star T + \star Y + \star V, \quad J_{\tilde g_3}(\xi)V = 0.
\end{aligned}$$

We take $\ell \ge 2$ to prove assertion (3). Let $2 \le a \le \ell$. The non-zero components of $R_{\tilde g_{2\ell+1}}$ are:

$$\begin{aligned}
1 &= R_{\tilde g_{2\ell+1}}(X,U_2,U_2,X) = R_{\tilde g_{2\ell+1}}(X,T,U_2,X)\\
&= R_{\tilde g_{2\ell+1}}(X,U_{a+1},U_{a+1},X) = R_{\tilde g_{2\ell+1}}(X,U_{a+1},U_a,X)\\
&= R_{\tilde g_{2\ell+1}}(X,U_{a+1},V_a,X).
\end{aligned} \tag{4.c}$$

Assertion (3) follows from the same argument as that used to prove Lemma 3.2 as:

$$\begin{aligned}
&J_{\tilde g_{2\ell+1}}(\xi)X \in \operatorname{Span}\{Y,T,U_2,...,U_\ell,V_2,...,V_{\ell+1}\},\\
&J_{\tilde g_{2\ell+1}}(\xi)Y = 0,\\
&J_{\tilde g_{2\ell+1}}(\xi)T = \star Y + \star V_2,\\
&J_{\tilde g_{2\ell+1}}(\xi)U_2 = \star T + \star Y + \star V_2 + \star V_3,\\
&J_{\tilde g_{2\ell+1}}(\xi)U_a = \star U_{a-1} + \star Y + \star V_{a-1} + \star V_a + \star V_{a+1} \quad \text{for} \quad 3 \le a \le \ell,\\
&J_{\tilde g_{2\ell+1}}(\xi)U_{\ell+1} = \star U_\ell + \star Y + \star V_\ell + \star V_{\ell+1},\\
&J_{\tilde g_{2\ell+1}}(\xi)V_a = \star Y + \star V_{a+1} \quad \text{for} \quad 2 \le a \le \ell,\\
&J_{\tilde g_{2\ell+1}}(\xi)V_{\ell+1} = 0
\end{aligned}$$

To prove assertion (4), we take $\ell \geq 1$. Let $2 \leq a \leq \ell + 1$. The non-zero components of $R_{\tilde{g}_{2\ell+2}}$ are:

$$\begin{aligned}1 &= R_{\tilde{g}_{2\ell+2}}(X, U_1, U_1, X) = R_{\tilde{g}_{2\ell+2}}(X, U_a, U_a, X) \\ &= R_{\tilde{g}_{2\ell+2}}(X, U_a, U_{a-1}, X) = R_{\tilde{g}_{2\ell+2}}(X, U_a, V_{a-1}, X).\end{aligned} \tag{4.d}$$

We may then compute:

$$\begin{aligned}
J_{\tilde{g}_{2\ell+2}}(\xi)X &= \star U_1 + \ldots + \star U_\ell + \star Y + \star V_1 + \ldots + \star V_{\ell+1}, \\
J_{\tilde{g}_{2\ell+2}}(\xi)U_1 &= \star Y + \star V_1 + \star V_2, \\
J_{\tilde{g}_{2\ell+2}}(\xi)U_a &= \star U_{a-1} + \star Y + \star V_{a-1} + \star V_a + \star V_{a+1} \quad \text{for} \quad 2 \leq a \leq \ell, \\
J_{\tilde{g}_{2\ell+2}}(\xi)U_{\ell+1} &= \star U_\ell + \star Y + \star V_\ell + \star V_{\ell+1}, \\
J_{\tilde{g}_{2\ell+2}}(\xi)Y &= 0, \\
J_{\tilde{g}_{2\ell+2}}(\xi)V_a &= \star Y + \star V_{a+1} \quad \text{for} \quad 1 \leq a \leq \ell, \\
J_{\tilde{g}_{2\ell+2}}(\xi)V_{\ell+1} &= 0.
\end{aligned}$$

Assertion (4) now follows from the argument used to establish Lemma 3.3. □

REMARK 4.2. Again, one can consider pointwise questions. We shall say that $(M, g)$ is Ossersman nilpotent of order $n$ at $P \in M$ if $J(x)^n = 0$ for all $x \in T_P M$ and if $J(x_0)^{n-1} \neq 0$ for some $x_0 \in T_P M$. By replacing $u^2$ by $u^3$, $tu$ by $tu^2$, $u_a u_{a+1}$ by $u_a u_{a+1}^2$, and $v_a u_{a+1}$ by $v_a u_{a+1}^2$, we could construct metrics $\tilde{g}_n$ on $\mathbb{R}^{n+2}$ which are Osserman of order $n$ on $T_P\mathbb{R}^{n+2}$ for generic points $P \in \mathbb{R}^{n+2}$, but where $R$ vanishes at the origin $0 \in \mathbb{R}^{n+2}$. This gives rise to metrics where the order of nilpotency varies with the point of the manifold; such examples, clearly, are neither symmetric nor homogeneous.

REMARK 4.3. Stanilov and Videv [**20**] defined a higher order analogue of the Jacobi operator in the Riemannian setting which was subsequently extended to arbitrary signature. Let $\text{Gr}_{r,s}(M, g)$ be the Grassmannian bundle of all non-degenerate subspaces of $TM$ of signature $(r, s)$. We assume $0 \leq r \leq p$, $0 \leq s \leq q$, and $0 < r+s < p+q$ to ensure $\text{Gr}_{r,s}(M, g)$ is non-empty and does not consist of a single point; such a pair $(r, s)$ will be said to be *admissible*. Let $\mathcal{B} = \{e_1^+, ..., e_r^+, e_1^-, ..., e_s^-\}$ be an orthonormal basis for $\pi \in \text{Gr}_{r,s}(M, g)$. Then

$$J(\pi) := J(e_1^+) + \ldots + J(e_r^+) - J(e_1^-) - \ldots - J(e_s^-)$$

is independent $\mathcal{B}$ and depends only on $\pi$. Following Stanilov, one says that $(M, g)$ is Osserman of type $(r, s)$ if the eigenvalues of $J(\pi)$ are constant on $\text{Gr}_{r,s}(M, g)$. Let $J_n$ be defined by the metric $\tilde{g}_n$ defined in Lemma 4.1. The discussion given above then implies $J_n(\pi)^n = 0$ for all $\pi$ and thus $(\mathbb{R}^{n+2}, \tilde{g}_n)$ is Osserman of type $(r, s)$ for all admissible $(r, s)$. We refer to the discussion in [**3, 14**] for other examples of higher order Osserman manifolds.

## 5. Ivanov-Petrova manifolds

LEMMA 5.1. *The pseudo-Riemannian manifold* $(\mathbb{R}^{n+2}, \tilde{g}_n)$ *defined in* Lemma 4.1 *is nilpotent Ivanov-Petrova of order* 2 *if* $n = 2$ *and nilpotent Ivanova-Petrova of order* 3 *if* $n \geq 3$.

PROOF. Suppose first $n = 2$. We use equation (4.a) to see:

$$\mathcal{R}_{\tilde{g}_2}(\pi)X = \star V,\ \mathcal{R}_{\tilde{g}_2}(\pi)U = \star Y,\ \mathcal{R}_{\tilde{g}_2}(\pi)V = \mathcal{R}_{\tilde{g}_2}(\pi)Y = 0,$$

where $\star$ are suitably chosen quadratic polynomials in the components of the generating vectors of $\pi = \text{Span}\{f_1, f_2\}$ which are non-zero for generic $f_i$. Thus $\mathcal{R}_{\tilde{g}_2}(\pi) \neq 0$ for generic $\pi$ while $\mathcal{R}_{\tilde{g}_2}(\pi)^2 = 0$ for all $\pi$.

We use equations (4.b), (4.c), and (4.d) to compute $\mathcal{R}_{\tilde{g}_n}(\pi)Y = 0$ and:

$$\begin{array}{ll}
\mathcal{R}_{\tilde{g}_3}(\pi)X \in \text{Span}\{V, T\}, & \mathcal{R}_{\tilde{g}_3}(\pi)T \in \text{Span}\{Y\}, \\
\mathcal{R}_{\tilde{g}_3}(\pi)V = 0, & \mathcal{R}_{\tilde{g}_3}(\pi)U \in \text{Span}\{Y\}, \\
\mathcal{R}_{g_{2\ell+1}}(\pi)X \in \text{Span}\{T, U_a, V_a\}, & \mathcal{R}_{g_{2\ell+1}}(\pi)T \in \text{Span}\{Y\}, \\
\mathcal{R}_{g_{2\ell+1}}(\pi)U_a \in \text{Span}\{Y\}, & \mathcal{R}_{g_{2\ell+1}}(\pi)V_a \in \text{Span}\{Y\}, \\
\mathcal{R}_{\tilde{g}_{2\ell+2}}(\pi)X \in \text{Span}\{U_a, V_a\}, & \mathcal{R}_{\tilde{g}_{2\ell+2}}(\pi)U_a \in \text{Span}\{Y\}, \\
\mathcal{R}_{\tilde{g}_{2\ell+2}}(\pi)V_a \in \text{Span}\{Y\}. &
\end{array}$$

This shows $\mathcal{R}^3_{\tilde{g}_n}(\pi) = 0 \ \forall \ \pi$ and $\mathcal{R}^2_{\tilde{g}_n}(\pi) \neq 0$ for generic $\pi$. □

## References


[1] N. Blažic, N. Bokan, and P. Gilkey, A note on Osserman Lorentzian manifolds, *Bull. London Math. Soc.*, **29**, (1997), 227–230.

[2] N. Blažić, N. Bokan, P. Gilkey and Z. Rakić, Pseudo-Riemannian Osserman manifolds, *Balkan J. Geom. Appl.*, **2**, (1997), 1–12.

[3] A. Bonome, P. Castro, E. Garcia-Rio, Generalized Osserman four-dimensional manifolds, *Classical and Quantum Gravity*, **18** (2001), 4813–4822.

[4] R. Bryant, Pseudo-Riemannian metrics with parallel spinor fields and vanishing Ricci tensor, *Global analysis and harmonic analysis (Marseille-Luminy, 1999)*, Sémin. Congr. 4, Soc. Math. France, Paris 2000, 53–94.

[5] Q.-S. Chi, A curvature characterization of certain locally rank-one symmetric spaces, *J. Differential Geom.*, **28**, (1988), 187–202.

[6] B. Fiedler, Determination of the structure of algebraic curvature tensors by means of Young symmetrizers, (preprint). math.CO/0212278

[7] B. Fiedler, On the symmetry classes of the first covariant derivatives of tensor fields, (preprint).

[8] E. García-Rio, D. Kupeli, and M. Vázquez-Abal, On a problem of Osserman in Lorentzian geometry, *Differential Geom. Appl.*, **7**, (1997), 85–100.

[9] E. García-Rió, M. E. Vázquez-Abal and R. Vázquez-Lorenzo, Nonsymmetric Osserman pseudo-Riemannian manifolds, *Proc. Amer. Math. Soc.*, **126**, (1998), 2771–2778.

[10] E. Garciá-Rió, D. Kupeli, and R. Vázquez-Lorenzo, **Osserman Manifolds in Semi-Riemannian Geometry**, Lecture notes in Mathematics 1777, Springer Verlag, Berlin, (2002), ISBN 3-540-43144-6.

[11] P. Gilkey, **Geometric properties of natural operators defined by the Riemann curvature tensor**, World Scientific Publishing Co., (2001), ISBN 981-02-4752-4.

[12] P. Gilkey, R. Ivanova, and I. Stavrov, Spacelike Jordan Szabó algebraic covariant curvature tensors in the higher signature setting, preprint http://arXiv.org/abs/math.DG/0211089.

[13] P. Gilkey, R. Ivanova, and T. Zhang, Szabó Osserman IP Pseudo-Riemannian manifolds, preprint http://arXiv.org/abs/math.DG/0205085

[14] —, Higher order Jordan Osserman pseudo-Riemannian manifolds, preprint http://arXiv.org/abs/math.DG/0205269

[15] P. Gilkey and I. Stavrov, Curvature tensors whose Jacobi or Szabó operator is nilpotent on null vectors, *Bull. London Math. Soc.*, to appear; preprint http://arXiv.org/abs/math.DG/0205074

[16] O. Kowalski, F. Tricerri and L. Vanhecke, Curvature homogeneous Riemannian manifolds, *J. Math. Pures Appl.*, **71** (1992), 471–501.

[17] I. Kath, *Killing spinors on pseudo-Riemannian manifolds, Habilitation*, Humboldt Universität zu Berlin (2000).

[18] Y. Nikolayevsky, Osserman Conjecture in dimension $n \neq 8, 16$, preprint: http://arXiv.org/abs/math.DG/0204258.

[19] R. Osserman, Curvature in the eighties, *Amer. Math. Monthly*, **97**, (1990) 731–756.

[20] G. Stanilov and V. Videv, Four dimensional pointwise Osserman manifolds, *Abh. Math. Sem. Univ. Hamburg*, **68**, (1998), 1–6.
[21] I. Stavrov, Ph. D. Thesis, University of Oregon (2003).
[22] Z. I. Szabó, A short topological proof for the symmetry of 2 point homogeneous spaces, *Invent. Math.*, **106**, (1991), 61–64.

BF: Mathematics Institute, University of Leipzig, Augustusplatz 10/11, 04109 Leipzig, Germany. Email: *bernd.fiedler.roschstr.leipzig@t-online.de*

PG: Mathematics Department, University of Oregon, Eugene Or 97403 USA. Email: *gilkey@darkwing.uoregon.edu*

Contemporary Mathematics
Volume **337**, 2003

# Jordan Szabó algebraic covariant derivative curvature tensors

Peter B. Gilkey, Raina Ivanova, and Iva Stavrov

ABSTRACT. We show that if $\mathfrak{R}$ is a Jordan Szabó algebraic covariant derivative curvature tensor on a vector space of signature $(p, q)$, where $q \equiv 1 \bmod 2$ and $p < q$ or $q \equiv 2 \bmod 4$ and $p < q - 1$, then $\mathfrak{R} = 0$. This algebraic result yields an elementary proof of the geometrical fact that any pointwise totally isotropic pseudo-Riemannian manifold with such a signature $(p, q)$ is locally symmetric.

## 1. Introduction

Let $V$ be a vector space which is equipped with a non-degenerate inner product $(\cdot, \cdot)$ of signature $(p, q)$. Let $m := p+q = \dim\{V\}$. Let $S^{\pm}(V)$ be the pseudo-spheres of unit spacelike $(+)$ and timelike $(-)$ vectors in $V$:

$$S^{\pm}(V) := \{v \in V : (v, v) = \pm 1\}.$$

We now review briefly some facts from linear algebra which we shall need in what follows. Let $A$ be a linear map from $V$ to $V$. We say $A$ is *self-adjoint* if $(Av, w) = (v, Aw)$ for all $v, w \in V$; let $\mathcal{A}(V)$ be the vector space of all self-adjoint linear maps from $V$ to $V$. If $A$ is self-adjoint and if $\lambda \in \mathbb{C}$, then define real operators $A_\lambda$ on $V$ and associated generalized subspaces $E_\lambda$ by setting:

$$A_\lambda := \begin{cases} A - \lambda \cdot \mathrm{Id} & \text{if } \lambda \in \mathbb{R}, \\ (A - \lambda \cdot \mathrm{Id})(A - \bar{\lambda} \cdot \mathrm{Id}) & \text{if } \lambda \in \mathbb{C} - \mathbb{R}, \end{cases}$$

$$E_\lambda := \ker\{A_\lambda^m\}. \tag{1.1}$$

Since $A_\lambda = A_{\bar{\lambda}}$, $E_\lambda = E_{\bar{\lambda}}$. Both $A$ and $A_\lambda$ preserve each generalized eigenspace $E_\lambda$. The operator $A$ is said to be *Jordan simple* if $A_\lambda = 0$ on $E_\lambda$ for all $\lambda$. We say $\lambda$ is an *eigenvalue of* $A$ if $\dim\{E_\lambda\} > 0$; the *spectrum* Spec $\{A\} \subset \mathbb{C}$ is the set of complex eigenvalues of $A$.

LEMMA 1.1. *Let $A$ be a self-adjoint linear map of a vector space of signature $(p, q)$.*

---

*Key words and phrases.* algebraic covariant derivative curvature tensor, Szabó tensor, totally isotropic pseudo-Riemannian manifold

2000 *Mathematics Subject Classification.* 53B20.

Research of PG partially supported by the NSF (USA) and the MPI (Leipzig).
Research of RI and IS partially supported by the NSF (USA).

(1) *We may decompose* $V = \oplus_{\mathrm{Im}(\lambda)\geq 0} E_\lambda$.
(2) *We have* $E_\lambda \perp E_\mu$ *if* $\lambda \neq \mu$ *and* $\lambda \neq \bar{\mu}$.
(3) *The spaces* $E_\lambda$ *inherit non-degenerate metrics of signature* $(p_\lambda, q_\lambda)$.
(4) *If* $\lambda \notin \mathbb{R}$, *then* $p_\lambda = q_\lambda$.

We say that $R \in \otimes^4 V^*$ is an *algebraic curvature tensor* on $V$ if $R$ satisfies the symmetries:

$$\begin{aligned} &R(x, y, z, w) = R(z, w, x, y) = -R(y, x, z, w), \\ &R(x, y, z, w) + R(y, z, x, w) + R(z, x, y, w) = 0. \end{aligned} \tag{1.2}$$

We say that $\mathfrak{R} \in \otimes^5 V^*$ is an *algebraic covariant derivative curvature tensor* on $V$ if $\mathfrak{R}$ satisfies the symmetries:

$$\begin{aligned} &\mathfrak{R}(x, y, z, w; v) = \mathfrak{R}(z, w, x, y; v) = -\mathfrak{R}(y, x, z, w; v), \\ &\mathfrak{R}(x, y, z, w; v) + \mathfrak{R}(y, z, x, w; v) + \mathfrak{R}(z, x, y, w; v) = 0, \\ &\mathfrak{R}(x, y, z, w; v) + \mathfrak{R}(x, y, w, v; z) + \mathfrak{R}(x, y, v, z; w) = 0. \end{aligned} \tag{1.3}$$

Let $(M, g)$ be a pseudo-Riemannian manifold of signature $(p, q)$. The Riemann curvature tensor $R_g$ is an algebraic curvature tensor on the tangent space $T_P M$ for every point $P \in M$. Similarly, the covariant derivative of the curvature tensor $\nabla R_g$ is an algebraic covariant derivative curvature tensor on $T_P M$. Conversely, such tensors are geometrically realizable. Thus tensors with the symmetries of equations (1.2) and (1.3) are important in differential geometry.

Let $R$ be an algebraic curvature tensor on $V$. The associated *Jacobi operator* $J$ is the linear map of $V$ characterized by the identity:

$$(J(v)y, z) = R(y, v, v, z);$$

$J(v)$ is self-adjoint by equation (1.2). We say that $R$ is a *spacelike Osserman* (resp. *timelike Osserman*) tensor if $R$ is an algebraic curvature tensor and if $\mathrm{Spec}\{J\}$ is constant on $S^+(V)$ (resp. on $S^-(V)$); these two notions are equivalent [**4**] and such a tensor is said to be an *Osserman* tensor.

Analogously, let $\mathfrak{R}$ be an algebraic covariant derivative curvature tensor on $V$. The associated *Szabó operator* $\mathfrak{S}$ is the linear map of $V$ characterized by the identity:

$$(\mathfrak{S}(v)y, z) = \mathfrak{R}(y, v, v, z; v);$$

the symmetries of equation (1.3) show $\mathfrak{S}$ is self-adjoint. We say that $\mathfrak{R}$ is a *spacelike Szabó* (resp. *timelike Szabó*) tensor if $\mathfrak{R}$ is an algebraic covariant derivative curvature tensor and if $\mathrm{Spec}\{\mathfrak{S}\}$ is constant on $S^+(V)$ (resp. $S^-(V)$); we denote this common spectrum by $\mathrm{Spec}^+\{\mathfrak{S}\}$ (resp. $\mathrm{Spec}^-\{\mathfrak{S}\}$). The notions spacelike Szabó and timelike Szabó are equivalent [**4**]; such a tensor is said to be a *Szabó tensor*.

Osserman [**11**] and Szabó [**14**] wrote the original papers concerning the spectral properties of the operators $J$ and $\mathfrak{S}$ in the Riemannian setting. That is why their names have become associated with the subject. The Jacobi operator has been studied extensively in this context; we refer to [**3, 4**] for a more complete bibliography. By contrast, the Szabó operator has received considerably less attention and the present paper is devoted to the study of the spectral properties of $\mathfrak{S}$ in the pseudo-Riemannian setting. Although there are certain formal parallels between the Jacobi operator and the Szabó operator, the fact that $J(-v) = J(v)$ while $\mathfrak{S}(v) = -\mathfrak{S}(-v)$ plays a crucial role.

In the higher signature setting, the eigenvalue structure does not control the Jordan normal form (i.e. the conjugacy class) of a self-adoint map. We shall say that $\mathfrak{R}$ is a *spacelike* (resp. *timelike*) *Jordan Szabó tensor* if $\mathfrak{R}$ is an algebraic covariant derivative curvature tensor and if the Jordan normal form of $\mathfrak{S}$ is constant on $S^+(V)$ (resp. $S^-(V)$).

The following result is due in the Riemannian setting ($p = 0$) to Szabó [**14**] and in the Lorentzian setting ($p = 1$) to Gilkey and Stavrov [**7**]:

THEOREM 1.2. *Let $\mathfrak{R}$ be a Szabó tensor on a vector space of signature $(p, q)$. If $p = 0$ or if $p = 1$, then $\mathfrak{R} = 0$.*

By replacing $g$ by $-g$, one can interchange the roles of $p$ and $q$. Consequently Theorem 1.2, implies there are no non-trivial Szabó tensors if $q = 0$ or if $q = 1$ as well. Therefore, we shall assume $p \geq 2$ and $q \geq 2$ henceforth. Although we are primarily interested in spacelike or timelike Jordan Szabó tensors, many of our results hold under the weaker assumption that the tensor is Szabó. We shall establish the following result in Section 2.

THEOREM 1.3. *Let $\mathfrak{R}$ be a Szabó tensor on a vector space $V$ of signature $(p, q)$, where $p \geq 2$ and $q \geq 2$. Then*

(1) $\operatorname{Spec}^{\pm}\{\mathfrak{S}\} = -\operatorname{Spec}^{\pm}\{\mathfrak{S}\} = \sqrt{-1}\operatorname{Spec}^{\mp}\{\mathfrak{S}\}$;
(2) $\operatorname{Spec}^{\pm}\{\mathfrak{S}\} \subset \mathbb{R} \cup \sqrt{-1}\,\mathbb{R}$;
(3) *If $p < q$, then* $\operatorname{Spec}^{+}\{\mathfrak{S}\} \subset \sqrt{-1}\,\mathbb{R}$ *and* $\operatorname{Spec}^{-}\{\mathfrak{S}\} \subset \mathbb{R}$;
(4) *If $q < p$, then* $\operatorname{Spec}^{+}\{\mathfrak{S}\} \subset \mathbb{R}$ *and* $\operatorname{Spec}^{-}\{\mathfrak{S}\} \subset \sqrt{-1}\,\mathbb{R}$.

A pseudo-Riemannian manifold $(M, g)$ of signature $(p, q)$ is said to be a *Szabó* manifold if $\nabla R_g$ is a Szabó tensor on $T_PM$ for all $P \in M$. If $p \geq 2$ and if $q \geq 2$, then there exist Szabó pseudo-Riemannian manifolds and algebraic covariant derivative curvature tensors of signature $(p, q)$ which are neither locally symmetric, nor locally homogeneous, nor pointwise totally isotropic [**2, 6**].

Let $S^{q-1}$ be the unit sphere in $\mathbb{R}^q$ with the usual positive definite inner product. If $q$ is odd, then there are no non-vanishing vector fields on $S^{q-1}$. This well known result has been generalized by Adams [**1**]. The *Adams number* $\nu(q)$ is defined as follows. Let $q = 2^\ell q_0$, where $q_0$ is odd. Then $\nu(q) = \nu(2^\ell)$, where

$$\nu(1) = 0, \quad \nu(2) = 1, \quad \nu(4) = 3, \ \nu(8) = 7, \quad \text{and} \quad \nu(2^{\ell+4}) = \nu(2^\ell) + 8. \tag{1.4}$$

Let $\{v_1, ..., v_\ell\}$ be linearly independent vector fields on $S^{q-1}$. Adams showed that $\ell \leq \nu(q)$; thus $\nu(q)$ provides an upper bound to the number of linearly independent vector fields that can exist on $S^{q-1}$. Furthermore, the estimate is sharp. If $\nu(q) > 0$, then there exist $\nu(q)$ linearly independent vector fields on $S^{q-1}$. Since $\nu(q) = 0$ if $q$ is odd, the result of Adams contains as a special case the original observation that there does not exist a nowhere vanishing vector field on an even dimensional sphere.

If $p \geq 2$ and if $q \geq 2$ , then there non-trivial nilpotent Szabó pseudo-Riemannian manifolds [**2, 6**]. However, there are as yet no known examples of spacelike Jordan Szabó tensors $\mathfrak{R}$, where $\mathfrak{R} \neq 0$, and we conjecture none exist. We have some partial results in this direction. In section 3, we shall establish the following result:

THEOREM 1.4. *Let $\mathfrak{R}$ be a spacelike Jordan Szabó tensor on a vector space $V$ of signature $(p, q)$, where $p < q$. Let $v \in S^+(V)$. Then:*

(1) *$\mathfrak{S}(v)$ is Jordan simple;*

(2) *If* $p < q - \nu(q)$, *then* $\operatorname{rank}\{\mathfrak{S}(v)\} \leq 2\nu(q)$;
(3) *If* $q$ *is odd, then* $\mathfrak{R} = 0$.

If $\mathfrak{R}$ is spacelike and timelike Jordan Szabó, then we let $r_\pm$ be the rank of $\mathfrak{S}$ on the pseudo-spheres $S^\pm(V)$. In Section 4, we shall prove:

THEOREM 1.5. *Let* $\mathfrak{R} \neq 0$ *be an algebraic covariant derivative curvature tensor on a vector space* $V$ *of signature* $(p, q)$ *which is both spacelike and timelike Jordan Szabó. Then:*
(1) $r_+ = r_-$.
(2) *If* $p \neq q$, *then* $\mathfrak{S}(v)$ *is Jordan simple for* $v \in S^\pm(V)$.

We say $\mathfrak{R}$ is *null Jordan Szabó* if the Jordan normal form of $\mathfrak{S}$ is constant on the null cone

$$\mathcal{N} := \{\{v \in V : (v, v) = 0\} - \{0\}\};$$

this implies $\mathfrak{S}$ is nilpotent and has constant rank, which we shall denote by $r_0$, on $\mathcal{N}$. We say that $\mathfrak{R}$ is *Jordan Szabó* if $\mathfrak{R}$ is spacelike, timelike, and null Jordan Szabó. In Section 5, we shall prove:

THEOREM 1.6. *Let* $\mathfrak{R}$ *be an algebraic covariant derivative curvature tensor on a vector space* $V$ *of signature* $(p, q)$ *which is Jordan Szabó. Then*
(1) *If* $\mathfrak{R} \neq 0$, *then* $r_0 < r_+$.
(2) *If* $q \equiv 2$ *mod* 4 *and if* $p < q - 1$, *then* $\mathfrak{R} = 0$.

Following Wolf [**15**], we say that a pseudo-Riemannian manifold $(M, g)$ is *locally isotropic* if given a point $P \in M$ and nonzero tangent vectors $X$ and $Y$ in $T_P M$ with $(X, X) = (Y, Y)$, there is a local isometry of $(M, g)$, fixing $P$, which sends $X$ to $Y$. Wolf showed that such a manifold is necessarily locally symmetric, see [**15**] (Theorem 12.3.1). The Szabó operator of a locally isotropic pseudo-Riemannian manifold is necessarily Jordan Szabó. Thus Wolf's result in certain special cases can be derived from Theorems 1.4 and 1.5:

COROLLARY 1.7. *Let* $(M, g)$ *be a locally isotropic pseudo-Riemannian manifold of signature* $(p, q)$. *If* $q \equiv 1$ *mod* 2 *and if* $p < q$ *or if* $q \equiv 2$ *mod* 4 *and if* $p < q - 1$, *then* $(M, g)$ *locally symmetric.*

## 2. Spacelike Szabó Tensors

In this section we prove Theorem 1.3. Let $V$ be a vector space which is equipped with a non-degenerate inner product of signature $(p, q)$, where $p \geq 2$ and $q \geq 2$. We can choose a non-canonical decomposition

$$V = V^+ \oplus V^-, \tag{2.1}$$

where $V^+$ is a maximal spacelike subspace and where $V^- := (V^+)^\perp$ is the complementary maximal timelike subspace. Let $\pi^\pm$ be orthogonal projection on $V^\pm$. The following is a useful technical observation.

LEMMA 2.1. *Let* $v \in S^+(V^+)$. *If* $w \perp v$, *then* $\pi^+ w \in V^+ \cap v^\perp = T_v(S^+(V^+))$.

PROOF. We may decompose $w = cv + v^+ + v^-$, where $v^\pm \in V^\pm$ and $v \perp v^+$. Because $c = (v, w) = 0$, we have $\pi^+ w = v^+$. □

The next result is well known, see, for example [**5**], and generalizes the decomposition of equation (2.1) to the bundle setting:

LEMMA 2.2. *Let $E$ be a vector bundle over a smooth manifold $M$ which is equipped with a non-degenerate fiber metric. Then we can decompose $E$ as a direct sum $E^+ \oplus E^-$ of complementary orthogonal subbundles, where $E^+$ is a maximal spacelike subbundle and $E^- = (E^+)^\perp$ is a maximal timelike subbundle.*

Let $\mathfrak{R}$ be a Szabó tensor on $V$. Since $\mathfrak{R}(-v) = -\mathfrak{R}(v)$,

$$\mathrm{Spec}\{\mathfrak{S}(v)\} = -\mathrm{Spec}\{\mathfrak{S}(-v)\} \quad \text{so} \quad \mathrm{Spec}^{\pm}\{\mathfrak{S}\} = -\mathrm{Spec}^{\pm}\{\mathfrak{S}\}.$$

We now establish Theorem 1.3 (1). Let $\mathfrak{R}$ be a Szabó tensor on a vector space of signature $(p,q)$ for $p \geq 2$ and $q \geq 2$. Let $V_{\mathbb{C}} := V \otimes \mathbb{C}$ be the complexification of $V$. We extend the inner product and the tensor $\mathfrak{R}$ to be complex multilinear on $V_{\mathbb{C}}$. Let

$$\begin{aligned} \mathcal{O}_{\mathbb{C}} : &= \{v \in V_{\mathbb{C}} : (v,v) \neq 0\}, \\ \mathfrak{P}(v) : &= (v,v)^{-3}\mathfrak{S}(v)^2 \text{ for } v \in \mathcal{O}_{\mathbb{C}}, \text{ and} \\ p(t,v) : &= \det\{\mathfrak{P}(v) - t \cdot \mathrm{Id}\} = c_0(v) + \ldots + c_m(v)t^m \text{ for } v \in \mathcal{O}_{\mathbb{C}}. \end{aligned}$$

The set $\mathcal{O}_{\mathbb{C}}$ is a connected open subset of $V_{\mathbb{C}}$ and the coefficients $c_i(v)$ are holomorphic functions on $\mathcal{O}_{\mathbb{C}}$. The intersection of $\mathcal{O}_{\mathbb{C}}$ with the real subspace $V$ is the disjoint union of the non-empty sets $\mathcal{O}_{\pm}$ of spacelike and timelike vectors in $V$. Since $\mathfrak{R}$ is a Szabó tensor, $\mathrm{Spec}\{\mathfrak{P}\}$, the characteristic polynomial $p$, and hence the functions $c_i$ are constant on $\mathcal{O}_{+}$. By the identity theorem, the functions $c_i$ are constant on $\mathcal{O}_{\mathbb{C}}$. Let $v_{\pm} \in S^{\pm}(V)$. Because $p(t,v^-) = p(t,v^+)$, the operators $\mathfrak{P}(v^-) = -\mathfrak{S}(v^-)^2$ and $\mathfrak{P}(v^+) = \mathfrak{S}(v^+)^2$ have the same spectrum. This completes the proof of assertion (1) of Theorem 1.3 by establishing the identity:

$$\mathrm{Spec}\{\mathfrak{S}(v^-)\} = \sqrt{-1}\,\mathrm{Spec}\{\mathfrak{S}(v^+)\}.$$

Before continuing to prove the other assertions of Theorem 1.3, we must recall some facts from algebraic topology. Assertion (1) in the following Lemma follows from results of Szabó [**14**], and assertion (2) follows from the Borsuk-Ulam theorem (see for example Spanier [**12**], Corollary 8 p. 266). They concern $\mathbb{Z}_2$ equivariant vector fields and vector valued functions on spheres.

LEMMA 2.3. *Let $S^{q-1}$ be the unit sphere in a vector space of signature $(0,q)$.*

(1) *There does not exist a continuous nowhere vanishing vector field $F$ on $S^{q-1}$ so $F(v) = F(-v)$ for every $v \in V$.*
(2) *If $\dim\{V^-\} = p < q$, then there does not exist a continuous nowhere vanishing map $F$ from $S^{q-1}$ to $V^-$ so that $F(v) = -F(-v)$ for every $v \in S^{q-1}$.*

Recall that $\mathcal{A}(V)$ is the set of all self-adjoint linear maps of $V$. We adopt the notation of Lemma 1.1. Let $V = V^+ \oplus V^-$ be the decomposition described in equation (2.1). The remaining assertions of Theorem 1.3 will be proved, as has become traditional in this subject, using methods from algebraic topology. The essential point is to abstract precisely those properties relevant to the investigation. What will be crucial for us are the facts that:

$$\begin{aligned} &\mathfrak{S}(-v) = -\mathfrak{S}(v), \quad \mathfrak{S}(v)v = 0, \quad \text{and} \\ &\mathfrak{S}(v) \in \mathcal{A}(V) \quad \text{for all} \quad v \in S^{q-1} := S^+(V^+). \end{aligned} \tag{2.2}$$

LEMMA 2.4. *Let $V$ be a vector space of signature $(p, q)$. Let $V^+$ be a maximal spacelike subspace of $V$ and let $S^{q-1} := S^+(V^+)$. Let $A : v \to A_v$ be a continuous map from $S^{q-1}$ to $\mathcal{A}(V)$ with $A_{-v} = -A_v$ and with $A_v v = 0$ for all $v \in S^{q-1}$. Assume that $A$ has constant spectrum. Then:*

(1) *If $\lambda \in \operatorname{Spec}\{A\}$ with $\lambda \notin \sqrt{-1}\,\mathbb{R}$, then $q_\lambda = 0$;*
(2) *$\operatorname{Spec}\{A\} \subset \mathbb{R} \cup \sqrt{-1}\,\mathbb{R}$;*
(3) *If $p < q$, then $\operatorname{Spec}\{A\} \subset \sqrt{-1}\,\mathbb{R}$.*

PROOF. Let $\pi^\pm$ be orthogonal projection on $V^\pm$. Consider the following vector bundles on $S^{q-1}$:

$$\begin{aligned} TS^{q-1} &:= \{(v_1, v_2) \in S^{q-1} \times V^+ : (v_1, v_2) = 0\}, \text{ and} \\ TS^{q-1} \oplus V^- &:= \{(v_1, v_2) \in S^{q-1} \times V : (v_1, v_2) = 0\}. \end{aligned}$$

Let $A : S^{q-1} \to \mathcal{A}(V)$ satisfy the hypothesis of the Lemma. Since $A$ has constant spectrum, the generalized eigenspaces $E_{\lambda,v}$ of $A_v$, as described in Lemma 1.1, have constant rank and fit together to define smooth vector bundles $E_\lambda$ over $S^{q-1}$.

Suppose $\lambda \in \operatorname{Spec}\{A\}$ and $\lambda \notin \sqrt{-1}\,\mathbb{R}$. We apply Lemma 1.1; $0 \neq \lambda$, $-\lambda \neq \lambda$ and $-\lambda \neq \bar{\lambda}$. Thus $E_{0,v}$, $E_{\lambda,v}$, and $E_{-\lambda,v}$ are mutually orthogonal vector spaces which inherit non-degenerate metrics of signatures $(p_0, q_0)$, $(p_\lambda, q_\lambda)$, and $(p_{-\lambda}, q_{-\lambda})$, respectively. As $A_v v = 0$, $v \in E_{0,v}$ so $v \perp E_{\lambda,v}$. By Lemma 2.2, there exist maximal orthogonal timelike and spacelike sub-bundles $E_\lambda^\pm$ of $E_\lambda$ so:

$$E_\lambda = E_\lambda^+ \oplus E_\lambda^-.$$

Since $A_{-v} = -A_v$, we have $E_{\lambda,-v} = E_{-\lambda,v}$. Thus we may obtain a similar splitting of the bundle $E_{-\lambda}$ by setting:

$$E^\pm_{-\lambda,v} := E^\pm_{\lambda,-v}.$$

Let $\varepsilon$ be a choice of signs $\pm$. We suppose $\dim\{E_\lambda^\varepsilon\} > 0$; this means $p_\lambda > 0$ if $\varepsilon$ is $-$ and $q_\lambda > 0$ if $\varepsilon$ is $+$. Let $N$ be the north pole of $S^{q-1}$. Since $S^{q-1} - \{N\}$ is contractable, there exists a global section $s_\lambda$ to $E_\lambda^\varepsilon$ which only vanishes at $N$. Let $s_{-\lambda}(v) := s_\lambda(-v)$. Then $s_{-\lambda}$ is a section to $E^\varepsilon_{-\lambda}$ which only vanishes at the south pole $S := -N$ and which satisfies $s_\lambda(v) \perp s_{-\lambda}(v)$. Let

$$F := s_\lambda + \varepsilon s_{-\lambda}.$$

If $\varepsilon$ is $+$, then $s_\lambda$ and $s_{-\lambda}$ are spacelike; if $\varepsilon$ is $-$, then $s_\lambda$ and $s_{-\lambda}$ are timelike. Since $s_\lambda \perp s_{-\lambda}$, $F$ is a nowhere vanishing section to $E_\lambda^\varepsilon \oplus E^\varepsilon_{-\lambda}$.

To prove assertion (1) of Lemma 2.4, we suppose that $q_\lambda > 0$ and argue for a contradiction. We take $\varepsilon$ to be $+$. Then $F(v)$ is nowhere vanishing, spacelike, and perpendicular to $v$ so Lemma 2.1 shows that $\pi^+ F(v)$ is a nowhere vanishing vector field on $S^{q-1}$ with the equivariance property $\pi^+ F(v) = \pi^+ F(-v)$. This contradicts Lemma 2.3 (1) and proves Lemma 2.4 (1).

If $\lambda \in \operatorname{Spec}\{A\}$ and if $\lambda \notin \mathbb{R}$, then $p_\lambda = q_\lambda \neq 0$. Assertion (1) then implies $\lambda \in \sqrt{-1}\,\mathbb{R}$. Thus $\operatorname{Spec}\{A\} \subset \mathbb{R} \cup \sqrt{-1}\,\mathbb{R}$ which establishes assertion (2).

Let $p < q$. To prove assertion (3), we assume that there exists an eigenvalue $\lambda$ belonging to $\mathbb{R} - \{0\}$ and argue for a contradiction. We take $\varepsilon$ equal to $-$. As $F(v)$ is timelike and non-vanishing, $\pi^- F(v)$ is a nowhere vanishing continuous map from $S^{q-1}$ to $V^-$ with the equivariance property $\pi^- F(v) = -\pi^- F(-v)$. This contradicts Lemma 2.3 (2) and proves assertion (3) of Lemma 2.4. □

Proof of Theorem 1.3. Let $\mathfrak{R}$ be a spacelike Szabó tensor on $V$. We have already established assertion (1) of the Theorem. Assertions (2), (3), and (4) follow from Lemma 2.4, where, if necessary, we change the sign of the quadratic form to interchange the roles of $p$ and $q$ and the notions spacelike and timelike. □

## 3. Spacelike Jordan Szabó Tensors

We begin the preparation of Theorem 1.4 by recalling some topological results. Let $\mathbb{RP}^{q-1}$ be real projective space; this is the space of lines thru the origin in $\mathbb{R}^q$. If $v \in S^{q-1}$, then we denote the corresponding element of $\mathbb{RP}^{q-1}$ by $\langle v\rangle = v \cdot \mathbb{R}$. We may identify $\mathbb{RP}^{q-1} = S^{q-1}/\mathbb{Z}_2$. The classifying line bundle and orthogonal complement bundle over projective space are defined by

$$\begin{aligned} L &:= \{(\langle v\rangle, w) \in \mathbb{RP}^{q-1} \times \mathbb{R}^q : w \in \langle v\rangle\} \\ L^\perp &:= \{(\langle v\rangle, w) \in \mathbb{RP}^{q-1} \times \mathbb{R}^q : v \perp w\}. \end{aligned}$$

We identify $T(\mathbb{RP}^{q-1})$ with $L \otimes L^\perp$. Let $\nu(q)$ be the Adams number which was defined in equation (1.4). We refer to [**5**] for the proof of the first assertion and to Adams [**1**] for the proof of the second assertion in the following Lemma:

Lemma 3.1.

(1) *Let $U_i$ be non-trivial vector bundles over $\mathbb{RP}^{q-1}$. If $U_1$ is a sub-bundle of $L^\perp$ and if $U_2$ is a sub-bundle of a trivial bundle of dimension $p < q$, then $U_1$ is not isomorphic to $U_2$.*

(2) *Let $S^{q-1}$ be the unit sphere in $\mathbb{R}^q$. Let $\nu(q)$ be the Adams number. Then:*
   (a) *If $E$ is a rank $r$ sub-bundle of $TS^{q-1}$, then either $q - 1 - \nu(q) \le r$ or $r \le \nu(q)$;*
   (b) *If $\{\chi_1, ..., \chi_\ell\}$ are linearly independent vector fields on $S^{q-1}$, then we have $\ell \le \nu(q)$.*

Again, the properties of equation (2.2) are crucial. We begin the proof of Theorem 1.4 by using Lemma 3.1 to establish a related result in the abstract setting. As in Lemma 2.4, we decompose $V = V^- \oplus V^+$ and identify $S^{q-1}$ with the unit sphere in $V^+$.

Lemma 3.2. *Let $V$ be a vector space of signature $(p,q)$, where $p < q$. Let $V^+$ be a maximal spacelike subspace of $V$ and let $S^{q-1} := S^+(V^+)$. Let $A : v \to A_v$ be a continuous map from $S^{q-1}$ to $\mathcal{A}(V)$ with $A_{-v} = -A_v$ and with $A_v v = 0$ for all $v \in S^{q-1}$. Assume that $A$ has constant Jordan normal form. Then:*

(1) *The operator $A_v$ is Jordan simple for any $v \in S^{q-1}$;*
(2) *If $p < q - \nu(q)$, then $\operatorname{rank}\{A)\} \le 2 \cdot \nu(q)$.*
(3) *If $q$ is odd, then $A = 0$.*

Proof. By Lemma 2.4, $\operatorname{Spec}\{A\} \subset \sqrt{-1}\,\mathbb{R}$. We use equation (1.1) to define the operators $A_{\lambda,v}$ for $\lambda \in \operatorname{Spec}\{A_v\}$. Since $\lambda$ is purely imaginary,

$$A_{\lambda,v} = \begin{cases} A_v^2 + |\lambda|^2 & \text{if } \lambda \neq 0, \\ A_v & \text{if } \lambda = 0. \end{cases} \tag{3.1}$$

Since $A_v v = 0$, $A_v$ preserves $v^\perp$. We therefore introduce the *reduced generalized eigenspaces*

$$\tilde{E}_{v,\lambda} := \ker\{A_{\lambda,v}^m\} \cap v^\perp.$$

To prove assertion (1), we suppose, to the contrary, that $A_{\lambda,v}$ is not Jordan simple. Since $A_v v = 0$, this means that $A_{\lambda,v} \neq 0$ on the generalized eigenspaces $\tilde{E}_{v,\lambda}$. Choose $k \geq 1$ maximal so $A^k_{\lambda,v} \neq 0$ on $\tilde{E}_{\lambda,v}$. Then $A^{k+1}_{\lambda,v} = 0$. Let

$$U_{\lambda,v} := A^k_{\lambda,v}\tilde{E}_{v,\lambda}.$$

Because $A_v$ has constant Jordan normal form, the vector spaces $U_{\lambda,v}$ glue together smoothly to define a smooth vector bundle $U_\lambda$ over $S^{q-1}$. This bundle descends to a smooth bundle $W_\lambda$ over $\mathbb{RP}^{q-1} = S^{q-1}/\mathbb{Z}_2$ because, by equation (3.1)

$$A_{\lambda,v} = \pm A_{\lambda,-v}.$$

Let $w_i = A^k_{\lambda,v}v_i$ for $v_i \in \tilde{E}_{v,\lambda}$. Since $k \geq 1$, $2k > k$ so:

$$0 = (A^{2k}_{\lambda,v}v_1, v_2) = (A^k_{\lambda,v}v_1, A^k_{\lambda,v}v_2) = (w_1, w_2). \tag{3.2}$$

By equation (3.2), the bundle $W_\lambda$ is totally isotropic. Since $W_\lambda$ contains no spacelike or timelike vectors, the projections $\pi^\pm$ are non-singular on $W_\lambda$. Set

$$W^\pm_\lambda := \pi^\pm(W_\lambda).$$

If $w \in W_{\lambda,v}$, then $w \perp v$. Consequently, by Lemma 2.1, $\pi^+ w \in V^+ \cap v^\perp$. It is immediate that $\pi^- w \in V^-$. Therefore,

$$W^+_\lambda \subset L^\perp \quad \text{and} \quad W^-_\lambda \subset V^-.$$

As $W^+_\lambda$, $W_\lambda$, and $W^-_\lambda$ are isomorphic, this contradicts Lemma 3.1 and proves Lemma 3.2 (1).

By assertion (1), $A_v = 0$ on $E_{0,v}$. Thus

$$\operatorname{rank}\{A\} = 2\sigma \quad \text{where} \quad \sigma := \textstyle\sum_{\Im(\lambda)>0} p_\lambda = \sum_{\Im(\lambda)>0} q_\lambda. \tag{3.3}$$

By Lemma 2.2, choose a maximal spacelike sub-bundle $\tilde{E}^+_{v,\lambda}$ of $\tilde{E}_{v,\lambda}$ of rank $q_\lambda$. Let

$$\tilde{E}^+ = \oplus_{\Im(\lambda)>0}\tilde{E}^+_{v,\lambda} \tag{3.4}$$

be a sub-bundle of $v^\perp$ of rank $\sigma$. Since elements of $\tilde{E}^+$ are all spacelike, the projection $\pi^+$ is injective and, by Lemma 2.1, $\tilde{E}^+$ is a sub-bundle of $TS^{q-1}$ of rank $\sigma$. By Lemma 3.1 (2), either $\sigma \leq \nu(q)$ or $\sigma \geq q-1-\nu(q)$. Since the first inequality implies assertion (2) of Lemma 3.2, we assume, to the contrary, that

$$\sigma \geq q - 1 - \nu(q). \tag{3.5}$$

Displays (3.3) and (3.5) imply

$$p \geq \sigma \geq q - 1 - \nu(q). \tag{3.6}$$

We assumed that $p < q - \nu(q)$. Therefore, by (3.6), $\sigma = p = q - 1 - \nu(q)$. Let $e \in \tilde{E}^+_{\lambda,v}$. By assertion (1), $A_v$ is Jordan simple. Thus $(A_v^2 + |\lambda|^2)e = 0$ so

$$(A_v e, A_v e) = (A_v^2 e, e) = -|\lambda|^2(e, e). \tag{3.7}$$

The decomposition of equation (3.4) is an orthogonal direct sum. Equation (3.7) shows that $A_v\tilde{E}^+_{\lambda,v}$ is a timelike subspace. Because $\operatorname{rank}\{A_v\tilde{E}^+_v\} = p$, $\pi^-$ is an isomorphism from $A_v\tilde{E}^+$ onto $V^-$. Consequently, $\pi^+\tilde{E}^+$ is isomorphic to a trivial bundle so there exist $p$ linearly independent vector fields on $S^{q-1}$. This shows $\sigma \leq p \leq \nu(q)$ and completes the proof of assertion (2).

Assertion (3) follows directly from assertion (2) since $\nu(q) = 0$ if $q$ is odd. □

The following observation is well known, see, for example, [**7**]:

LEMMA 3.3. *Let $\mathfrak{R}$ be an algebraic covariant derivative curvature tensor. If $\mathfrak{S}$ vanishes identically, then $\mathfrak{R} = 0$.*

PROOF OF THEOREM 1.4. Let $\mathfrak{R}$ be a spacelike Jordan Szabó tensor on a vector space $V$ of signature $(p, q)$, where $p < q$. By equation (2.2), we may apply Lemma 3.2. Assertions (1) and (2) of Theorem 1.4 now follow directly from assertions (1) and (2) of Lemma 3.2. If $q$ is odd, we apply Lemma 3.2 (3) to see $\mathfrak{S} = 0$. Assertion (3) now follows from Lemma 3.3. □

## 4. Timelike and spacelike Jordan Szabó Tensors

Let $\mathfrak{R}$ be both timelike and spacelike Jordan Szabó. Let $r_\pm := \operatorname{rank}\{\mathfrak{S}(\cdot)\}$ on $S^\pm(V)$. To establish Theorem 1.5 (1), we must show that $r_+ = r_-$. Fix a vector $v^+ \in S^+(V)$. Choose vectors $\{v_1, ..., v_{r_+}\}$ in $V$ and choose dual elements $\{v_1^*, ..., v_{r_+}^*\}$ in the dual vector space $V^*$ so

$$v_i^* \mathfrak{S}(v^+) v_j = \delta_{ij}.$$

We adopt the notation established in the proof of Lemma 2.2 and complexify. If $v \in V_\mathbb{C}$, then we consider the matrix $A_{ij}(v) := v_i^* \mathfrak{S}(v) v_j$. We define a holomorphic function on $V_\mathbb{C}$ by setting:

$$f(v) := \det\{v_i^* \mathfrak{S}(v) v_j\}.$$

Since $f(v^+) = 1$, $f$ does not vanish identically on $V_\mathbb{C}$. Thus by the identity theorem, there must exist a real timelike element $v^-$ of $V$ so $f(v^-) \neq 0$. This implies that the elements $\{\mathfrak{S}(v^-)v_1, ..., \mathfrak{S}(v^-)v_{r_+}\}$ are linearly independent so

$$r_- = \operatorname{rank}\{\mathfrak{S}(v^-)\} \geq r_+.$$

Similarly we can show $r_+ \geq r_-$. Thus $r_+ = r_-$.

To prove the second assertion, we may suppose without loss of generality that $p < q$. If $v \in V_\mathbb{C}$, then we shall let $E_\lambda^c \subset V_\mathbb{C}$ be the generalized complex eigenspaces of $\mathfrak{S}(v)$. Let $v_\pm \in S^\pm(V)$. As, by Theorem 1.4, $\mathfrak{S}(v^+)$ is Jordan simple,

$$r_+ = \textstyle\sum_{\lambda \neq 0} \dim_\mathbb{C}\{E_\lambda^c\}.$$

In Section 2, we showed that the characteristic polynomials of the two operators $\mathfrak{S}(v^+)$ and $\sqrt{-1}\,\mathfrak{S}(v^-)$ were the same. Thus

$$r_+ = \textstyle\sum_{\lambda \neq 0} \dim_\mathbb{C}\{E_\lambda^c(\mathfrak{S}(v^-))\}.$$

Since $r_+ = r_-$, we conclude that $\mathfrak{S}(v^-)$ must have rank 0 on $E_0^c(\mathfrak{S}(v^-))$ and consequently 0 is a Jordan simple eigenvalue for $\mathfrak{S}(v^-)$. By Theorem 1.3, the eigenvalues of $\mathfrak{S}(v^-)$ are real. By Lemma 2.4, the generalized eigenspaces for $\lambda \in \mathbb{R} - \{0\}$ are spacelike. Since $\mathfrak{S}(v^-)$ is self-adjoint with respect to a definite metric on $E_\lambda$, $\mathfrak{S}(v^-)$ is diagonalizable on $E_\lambda$. This shows $\mathfrak{S}(v^-)$ is Jordan simple and establishes assertion (2) and completes the proof of Theorem 1.5.

## 5. Jordan Szabó Tensors

PROOF OF THEOREM 1.6 (1). Let $\mathfrak{R}$ be a non-trivial Jordan Szabó tensor whose rank on the null cone is $r_0$. We must show that $r_0 < r_+$. Let $v_0 \in \mathcal{N}$. As in the proof of Theorem 1.5, choose vectors $\{v_1, ..., v_{r_0}\}$ in $V$ and $\{v_1^*, ..., v_{r_0}^*\}$ in $V^*$ so $v_i^* \mathfrak{S}(v_0) v_j = \delta_{ij}$. Let $f(v) := \det\{v_i^* \mathfrak{S}(v) v_j\}$. Since $f(v_0) \neq 0$, the holomorphic function $f$ does not vanish identically and thus there exists a timelike vector $v^-$ in $V$ so $f(v^-) \neq 0$. This proves $r_- \geq r_0$.

To show that $r_0 < r_-$, we suppose to the contrary that $r_0 = r_+ = r_-$ and argue for a contradiction; this implies $\mathfrak{S}(v)$ has constant rank $r$ on $V - \{0\}$. Let $V_+$ be a maximal space like subspace of $V$ and let $V_- := V_+^\perp$ be the complementary maximal timelike subspace. If $v \in V$, then we may decompose $v = v^+ + v^-$ for $v^\pm \in V^\pm$. We define a self-adjoint map linear map $\psi$ of $V$ and a positive definite inner product on $V$ by setting:

$$\psi(v) := v^+ - v^- \quad \text{and} \quad (v,w)_e := (\psi v, w) = -(v_-, w_-) + (v_+, w_+). \tag{5.1}$$

Define $A(v) := \mathfrak{S}(\psi v)\psi$. We show that $A$ is self-adjoint with respect $(\cdot,\cdot)_e$:

$$\begin{aligned} (A(v)v_1, v_2)_e &= (\mathfrak{S}(\psi v)\psi v_1, v_2)_e = (\psi\mathfrak{S}(\psi v)\psi v_1, v_2) \\ &= (v_1, \psi\mathfrak{S}(\psi v)\psi v_2) = (v_1, A(v)v_2)_e. \end{aligned}$$

Let $S^{p+q-1} := \{v \in V : (v,v)_e = 1\}$ be the associated unit sphere. Because we have that $A(v)v = \mathfrak{S}(\psi v)\psi v = 0$, we have an orthogonal direct sum decomposition

$$T_v(S^{p+q-1}) = E_+(v) \oplus E_0(v) \oplus E_-(v),$$

where $E_+(v)$, $E_-(v)$, and $E_0(v)$ are the span of the eigenvectors of $A(v)$ corresponding to positive eigenvalues, the zero eigenvalue, and negative eigenvalues, respectively. Since $A(-v) = -A(v)$, $E_\pm(v) = E_\mp(-v)$. As $A$ has constant rank, $\dim(E_\pm)$ is constant so these spaces fit together to define smooth bundles on $S^{p+q-1}$.

Since $\mathfrak{S} \neq 0$, $E_\pm(v) \neq 0$. We choose a section $s_+$ to $E_+$ only vanishing at the north pole. Then $s_-(v) := s_+(-v)$ is a section which only vanishes at the south pole and which satisfies $s_+ \perp s_-$. Setting $s := s_+ + s_-$ then constructs a nowhere vanishing section to $T(S^{p+q-1})$ with $s(-v) = s(v)$; this contradicts Lemma 2.3 and completes the proof of Theorem 1.6 (1). □

The following is a useful technical Lemma.

LEMMA 5.1. *Let $V$ be a vector space of signature $(p,q)$. Let $X$ be a connected topological space on which $\mathbb{Z}_2$ acts. Let $f$ be a continuous map from $X$ to the space of self-adjoint linear transformations of $V$. Assume that $r = \operatorname{rank}(f)$ is constant and that $f(-x) = -f(x)$. Then $r$ is even.*

PROOF. We adopt the notation of equation (5.1) to define $(\cdot,\cdot)_e$ and $\psi$. Let $\tilde{f}(x) := f(x)\psi$. Decompose the trivial bundle $X \times V = E_- \oplus E_0 \oplus E_+$ as the span of the eigenvectors of $\tilde{f}$ corresponding to positive eigenvalues, the zero eigenvalue, and negative eigenvalues, respectively. Since $f$ has constant rank, these define vector bundles over $X$. Since $E_-(x) = E_+(-x)$ and as $X$ is connected, $\dim E_- = \dim E_+$. Since $\tilde{f}$ is self-adjoint with respect to a positive definite inner product, $\tilde{f}$ vanishes on $E_0$. Thus $r = \operatorname{rank}(f) = \operatorname{rank}(\tilde{f}) = \dim E_- \oplus \dim E_+ = 2\dim E_-$ is even. □

PROOF OF THEOREM 1.6 (2). Let $V$ be a vector space of signature $(p,q)$ where $q \equiv 2 \bmod 4$ and where $p < q-1$. By Theorem 1.2, we may assume $p \geq 2$. Let $\mathfrak{R}$ be a Jordan Szabó tensor on $V$. We must show $\mathfrak{R} = 0$. By equation (1.4), $\nu(q) = 1$. Because $p < q - 1 = q - \nu(q)$, Theorem 1.4 (2) shows $r_+ \leq 2$. Thus, by Theorem 1.6 (1), $r_0 < 2$. Since by Lemma 5.1 we have $r_0$ is even, we conclude therefore that $r_0 = 0$. Choose a basis $e_i$ for $V$ and expand $\mathfrak{S}(x)v_i = \sum_j S_{ij}(x)v_j$. The functions $S_{ij}$ are cubic polynomials in $x$ which vanish identically on the real null-cone. Complexification then extends this relationship to the complex null cone.

Since the polynomial $(x, x)$ is irreducible as $\dim(V) \geq 3$, we may use the Hilbert Nullstellungsatz to see there exists a linear function $f(x)$ so that:

$$\mathfrak{S}(x) = (x, x) f(x).$$

Since $\mathfrak{S}(x)x = 0$, we have $f(x)x = 0$ for $(x, x) \neq 0$ and hence $f(x)x = 0$ for all $x$ by continuity. We polarize this relation to see $f(x)y + f(y)x = 0$ for all $x, y \in V$. Dotting this relationship with $y$ then yields

$$0 = (f(x)y, y) + (f(y)x, y) = (f(x)y, y) + (x, f(y)y) = (f(x)y, y)$$

for all $x, y \in V$. Polarization then yields $(f(x)y, z) + (f(x)z, y) = 0$ for all $x, y, z$ in $V$ and hence, as $f$ is symmetric, $2(f(x)y, z) = 0$ for all $x, y, z$ in $V$. Since the metric on $V$ is non-degenerate, we may conclude that $f(x)$ and hence $\mathfrak{S}(x)$ vanishes identically. Theorem 1.6 (2) now follows from Lemma 3.3. □

## References

[1] J. Adams, *Vector fields on spheres*, Annals of Math., **75**, (1962), 603–632.

[2] B. Fiedler and P. Gilkey, *Nilpotent Szabó, Osserman and Ivanov-Petrova pseudo-Riemannian manifolds*, math.DG/0211080.

[3] E. Garciá-Rió, D. Kupeli, and R. Vázquez-Lorenzo, **Osserman Manifolds in Semi-Riemannian Geometry**, Lecture notes in Mathematics, 1777, Springer Verlag (2002), ISBN 3-540-43144-6.

[4] P. Gilkey, **Geometric Properties of Natural Operators Defined by the Riemann Curvature Tensor**, World Scientific Press (2001), ISBN 981-02-04752-4.

[5] P. Gilkey and R. Ivanova, *Spacelike Jordan Osserman algebraic curvature tensors in the higher signature setting*, to appear Proc. Valencia Math Conf., math.DG/0205072.

[6] P. Gilkey, R. Ivanova, and T. Zhang, *Szabó Osserman IP Pseudo-Riemannian manifolds*, to appear Publ. Math. Debrecen, math.DG/0205085.

[7] P. Gilkey and I. Stavrov, *Curvature tensors whose Jacobi or Szabó operator is nilpotent on null vectors*, to appear Bull. London Math. Soc., math.DG/0205074.

[8] P. Gilkey and T. Zhang, *Algebraic curvature tensors for indefinite metrics whose skew-symmetric curvature operator has constant Jordan normal form*, Houston Math J. **28** (2002), 311–328.

[9] S. Ivanov and I. Petrova, *Riemannian manifold in which the skew-symmetric curvature operator has pointwise constant eigenvalues*, Geom. Dedicata, **70**, (1998), 269–282.

[10] R. Ivanova and G. Stanilov, *A skew-symmetric curvature operator in Riemannian geometry*, in **Sympos. Gaussiana, Conf A**, ed. Behara, Fritsch, and Lintz (1995), 391–395.

[11] R. Osserman, Curvature in the eighties, *Amer. Math. Monthly*, **97**, (1990), 731–756.

[12] E. Spanier, **Algebraic Topology**, McGraw-Hill (1966); corrected reprint Springer-Verlag (1981) ISBN: 0-387-90646-0.

[13] I. Stavrov, Ph.D. Thesis, University of Oregon (2003).

[14] Z. I. Szabó, *A short topological proof for the symmetry of* 2 *point homogeneous spaces*, Invent. Math., **106**, (1991), 61–64.

[15] J. Wolf, **Spaces of constant curvature,** 5th ed., Publish or Perish Inc. (1984), ISBN 0-914098-07-1

PG: Mathematics Department, University of Oregon, Eugene OR 97403 USA
*E-mail address*: gilkey@darkwing.uoregon.edu

RI: Mathematics Department, University of Hawaii - Hilo, 200 W. Kawili St., Hilo, HI 96720 USA
*E-mail address*: rivanova@hawaii.edu

IS: Mathematics Department, University of Oregon, Eugene OR 97403 USA
*E-mail address*: stavrov@hopf.uoregon.edu

Contemporary Mathematics
Volume **337**, 2003

# Differential Topology, Differential Geometry, and Hyperbolic Operators

Adam D. Helfer

ABSTRACT. Some of the most important results in mathematical general relativity depend not on fine control of the analytic structure, but on rather general properties of causality and on broad differential inequalities. I indicate here how one can generalize these results, replacing the space–time metric with an arbitrary strongly hyperbolic symbol. Given such a symbol on a manifold, there is a well-defined causal structure and measure of proper time. The variational properties of the proper time are related to the causal structure in much the same way as in the general–relativistic case. One can accordingly establish singularity theorems analogous to those of general relativity.

## 1. Introduction

We know that Lorentzian metrics on manifolds lead to very beautiful causal structures, with deep physical interpretations. And remarkably, some of the most important and profound results in the theory — the singularity theorems and the area theorem — depend essentially on this causal structure and on certain positivity-of-energy assumptions, but not (as one might have expected) on fine control of the joint evolution of the gravitational and matter fields.

Thus some of the seminal results in mathematical general relativity seem to depend only on rather general considerations. It is natural therefore to ask whether one can obtain similar results for other classes of evolution problems, classes in which one can formulate similar notions of causal structure and of positivity of energy.

There is a natural class of evolution problems with qualitatively similar causal structure to Lorentzian metrics: the evolutions given by *hyperbolic partial differential operators*. Indeed, historically, the essential elements of the causal structures

1991 *Mathematics Subject Classification.* 35L25, 53C80, 58J45, 58K99.

*Key words and phrases.* hyperbolic operators, differential topology, singularity theorems, variational principles.

This paper is based on two talks, one given at the 1999 AMS-SMM joint meeting in Denton, and one at the 2003 AMS meeting in Baltimore. I thank the organizers of the special sessions at those meetings: Professors Krishnan Duggal, Philip Parker, Lila del Riego and Ramesh Sharma.

(domains of dependence, futures and pasts) were investigated for general hyperbolic operators before Penrose and others embarked on their deep studies of the Lorentzian case.

In this paper, I shall outline how it is possible to develop a theory for general hyperbolic operators parallel to that of the Lorentzian case. The aim of this paper is only to convey the main ideas. Only very brief indications of some of the proofs are given; a full development of the theory requires a number of lengthy technicalities and will be presented elsewhere.

An accessible discussion of hyperbolic partial differential equations will be found in Courant and Hilbert (1989). A deeper treatment is given in Leray's (1955) classic, which contains the germs of many of the ideas discussed here. For the differential topology of space–time, Penrose's (1972) book gives a rapid, clear treatment.

*Terminology.* There are some differences between the terminologies used in the differential equations literature from those in relativity which, while in principle reconcilable, are in practice confusing. For example, a covector $t_a$ normal to a spacelike surface is called "spacelike" by Courant and Hilbert, whereas a relativist would call it "timelike." In this paper, I shall adopt a terminology which is based on the standard usage in relativity. The reader wishing to compare this work with the differential equations literature should thus be careful to check precisely what the authors' usages are.

*Conventions.* Conventions for differential geometry are those of Penrose (1972). Most computations will be done in a standard index notation in local coordinates. Unless otherwise specified, we work in the smooth category. The base manifold, denoted $N$, will be $n$-dimensional and paracompact. We use $u$ for a parameter along curves.

## 2. Hyperbolicity

I shall begin by explaining the motivations behind the definition of hyperbolicity of a differential operator. The actual definition will be seen to be a pointwise restriction on the highest-order coefficients of the operator, that is, a restriction on the *symbol* of the operator, a symmetric tensor $H^{a_1a_2\cdots a_m}$ of rank $m$. I shall then discuss some important geometric properties of these tensors.

### 2.1. The Idea of Hyperbolicity.

The causal structure of space–time is really a consequence of the fact that disturbances cannot propagate faster than the speed of light. We may thus expect that at least loosely analogous structures will arise whenever we study waves with finite velocities of propagation. This is indeed the case. Such waves are described by *hyperbolic* partial differential equations. (In fact, as we shall see, not all of the differential operator is needed in order to specify its causal structure — only a tensor encoding its highest-order part, the *symbol*, is required.)

Now it is a remarkable fact that the property of finite speed of propagation follows from well-posedness, in an appropriate sense, of the Cauchy problem. This is a basic result in the theory of partial differential equations and will not be proved here (roughly speaking, well-posedness forces certain analyticity properties on the Fourier transform, which translate into finite speed of propagation in direct space); however, it is mentioned because hyperbolicity is defined in terms of well-posedness of the Cauchy problem; finite speed of propagation is a consequence.

We consider a real linear $m^{\text{th}}$ order partial differential operator on scalars, with $m > 1$, given in local coordinates $x^a$ (on a space of $n$ dimensions) by

$$H^{a_1 a_2 \ldots a_m} \partial_{a_1} \partial_{a_2} \cdots \partial_{a_m} \phi + \text{lower-order terms}\,, \tag{2.1}$$

where $\partial_a = \partial / \partial x^a$. The coefficients $H^{a_1 a_2 \ldots a_m}$ are allowed to depend on the coordinates. In the theory of differential equations, there is a well–established general outline for analyzing such operators. One begins by studying the highest–order terms (the ones shown explicitly above), and in fact by "freezing" the coefficients $H^{a_1 a_2 \ldots a_m}$ at each point. One thus studies a linear operator with constant coefficients at each point, and tries to piece together the behaviors of these as the point varies.

Of course, it is easy to find solutions to the homogeneous equation

$$H^{a_1 a_2 \ldots a_m} \partial_{a_1} \partial_{a_2} \cdots \partial_{a_m} \phi = 0\,. \tag{2.2}$$

One can simply Fourier transform it to obtain the algebraic equation

$$H(\zeta_a) \tilde{\phi} = 0\,, \tag{2.3}$$

where

$$H(\zeta_a) = H^{a_1 a_2 \ldots a_m} \zeta_{a_1} \zeta_{a_2} \cdots \zeta_{a_m}\,. \tag{2.4}$$

However, this is rather formal until we make more precise the classes of functions we wish to consider. For this, we investigate the initial–value problem.

The idea of a pointwise *hyperbolic* operator is one for which, at each point, for the linearized and "frozen" operator there is an initial–data hypersurface for which the Cauchy problem is well–posed with oscillatory solutions. Recall that "well–posed" means that the solutions should exist, be unique, and vary continuously with the data. (This is not a precise definition, of course, until one specifies the classes of functions that are allowed. However, we are interested in the concept as motivation, and at this level it will be enough to say that we are interested in "oscillatory" solutions.)

Let us suppose the initial surface is given by $t_a x^a = \text{const}$ for some fixed covector $t_a$. Let us also fix, in the space of covectors, a hyperplane $X$ complementary to $t_a$. Then any covector can be written uniquely as

$$\zeta_a = \omega t_a + \xi_a\,, \tag{2.5}$$

where $\xi_a \in X$. We may regard $\xi_a$ as a Fourier transform variable associated with the hyperplane $t_a x^a = \text{const}$, and $\omega$ as the Fourier transform variable associated with the time $t_a x^a$.

Then the exponential solutions we seek correspond to solutions of the algebraic equation

$$H^{a_1 a_2 \ldots a_m} (\omega t_{a_1} + \xi_{a_1})(\omega t_{a_2} + \xi_{a_2}) \cdots (\omega t_{a_m} + \xi_{a_m}) = 0\,. \tag{2.6}$$

For each fixed $\xi_a$, this is a real polynomial of degree $\leq m$ in $\omega$. In order that it have $m$ independent solutions, none of the roots can be at infinity (and thus the degree must be $m$). In order that the solutions not grow exponentially, all of the roots must be real. Finally, in order that the solutions not grow algebraically, all of the roots must be distinct. This motivates the

DEFINITION 2.1. A symmetric tensor $H^{a_1 a_2 \ldots a_m}$ is said to be *strongly hyperbolic* with respect to $t_a$ if for every covector $\xi_a$ not parallel to $t_a$, the equation

$$H^{a_1 a_2 \ldots a_m} (\omega t_{a_1} + \xi_{a_1})(\omega t_{a_2} + \xi_{a_2}) \cdots (\omega t_{a_m} + \xi_{a_m}) = 0$$

has $m$ distinct real roots.

Note that strong hyperbolicity is an *open* condition: any sufficiently small perturbation of a strongly hyperbolic tensor is strongly hyperbolic. Also note that the absence of a root at infinity implies

$$H^{a_1 a_2 \ldots a_m} t_{a_1} t_{a_2} \ldots t_{a_m} \neq 0 . \tag{2.7}$$

We will usually find it convenient to adjust the sign of $t_a$ or $H^{a_1 a_2 \ldots a_m}$ so that this quantity is positive.

One sometimes says that $t_a$ is a *hyperbolic codirection* for $H^{a_1 a_2 \ldots a_m}$. We shall sometimes say that $t_a$ is *timelike* with respect to $H^{a_1 a_2 \ldots a_m}$. It can be shown that the set of hyperbolic codirections consists of distinct pairs of oppositely–directed convex cones. For $n$ (the dimension of the space) greater than two, there is exactly one such pair. In what follows, we shall almost always be concerned with just one of these components, which will be called the set of *future–hyperbolic codirections.* We assume that a continuous choice of this component, as the point in the base manifold $N$ varies, is possible and has been made.

**2.2. Some Geometric Properties of Hyperbolic Tensors.** In what follows, it will be necessary to have some understanding of the geometry of a hyperbolic tensor $H^{a_1 a_2 \ldots a_m}$ (at a fixed point). See Courant and Hilbert (1989) for proofs of material in this subsection.

Recall that such a tensor is hyperbolic at certain codirections $t_a$; the set of hyperbolic codirections is a union of pairs of opposite open convex cones. (In this paper, a *cone* in a vector space is a set closed under multiplication by positive numbers.) If $n$ (the dimension of space–time) is greater than two, there is necessarily only one pair of such cones. For simplicity of phraseology, I shall restrict to the case $n > 2$.

One should think of the hyperbolic covectors as the appropriate generalization (to the case of general $m$) of the timelike covectors (in the case $m = 2$). Recall that one component of the set of hyperbolic vectors has been distinguished as *future-directed.*

We have seen that the set of future-directed hyperbolic covectors is convex, hence connected, and that the function $H(\zeta_a) = H^{a_1 a_2 \ldots a_m} \zeta_{a_1} \zeta_{a_2} \cdots \zeta_{a_m}$ is nowhere-vanishing on this set. Therefore $H(\zeta_a)$ must have constant sign on the set of future-directed hyperbolic covectors. We may (and shall) assume that the sign of $H(\zeta_a)$ has been chosen positive on this set.

The real variety $H(\zeta_a) = H^{a_1 a_2 \ldots a_m} \zeta_{a_1} \zeta_{a_2} \cdots \zeta_{a_m} = 0$ in the space of cotangent vectors consists (by the definition of hyperbolicity) of $m$ distinct sheets. It can be shown that the boundary of the set of future-directed hyperbolic covectors is one of these sheets. This we shall refer to as the set of *future-directed null covectors*; its negative would be the set of *past-directed null covectors.* The zero covector will also be considered a null covector.

Note that, with this terminology, *not every covector* $\zeta_a$ *satisfying* $H(\zeta_a) = 0$ *is considered* null. Only those on the "inner" sheets of the variety, one sheet to the future and one to the past, are.

**2.3. The Local Metric.** It is a remarkable fact that hyperbolicity localizes, in the sense that at any hyperbolic covector $t_a$ the quantity $H^{ab}_{\text{loc}} = \partial^a \partial^b H = m(m-1) H^{abc_3 c_4 \ldots c_m} t_{c_3} t_{c_4} \cdots t_{c_m}$ is Lorentzian. We shall call this the *inverse local* *metric*, and make use of it both as a technical convenience and also more essentially in establishing some of the variational principles, below.

Let us see why $H^{ab}_{\text{loc}}$ is Lorentzian. First note that $X_t = \{\xi_a \mid \xi_a\partial^a H(t_a) = 0\}$ is a hyperplane complementary to $t_a$ (since $t_a\partial^a H(t_a) = mH(t_a) > 0$). We may thus decompose the space of covectors as a direct sum $\mathbb{R}\, t_a \oplus X_t$. We shall show that this decomposition is orthogonal with respect to $H^{ab}$, with $H^{ab}$ positive–definite on $\mathbb{R}\, t_a$ and negative–definite on $X_t$.

We have $t_a t_b H^{ab}_{\text{loc}} = m(m-1)H(t_a) > 0$ and $t_a\zeta_b H^{ab}_{\text{loc}} = (m-1)\xi_a\partial^a H(t_a) = 0$. We have, then, only to show that $H^{ab}$ is negative–definite on $X_t$. Putting accordingly $\omega t_a + \xi_a$ for an arbitrary covector; that $H$ be strongly hyperbolic at $t_a$ requires $H(\omega t_a + \xi_a)$ to have $m$ distinct real zeroes for $\xi_a \in X_t - \{0\}$. Now we have

$$
\begin{aligned}
&H(\omega t_a + \xi_a) \\
&\quad = \omega^m H(t_a) + \omega^{m-1}\xi_a\partial^a H(t_a) + \omega^{m-2}\xi_a\xi_b\partial^a\partial^b H(t_a) + \cdots \\
&\quad = \omega^m H(t_a) + \omega^{m-2}\xi_a\xi_b H^{ab}_{\text{loc}} + \cdots .
\end{aligned}
\tag{2.8}
$$

In order for this to have $m$ distinct zeroes, the term $\xi_a\xi_b H^{ab}$ must have the opposite sign to $H(t_a)$.

This result can be extended in two important ways. First, a refinement of the computation just given shows that the null cone of $H^{ab}_{\text{loc}}$ lies *outside of* that of $H(\omega t_a + \xi_a)$. (Recall that the null cone of $H(\omega t_a + \xi_a)$ is the boundary of the set of hyperbolic covectors; there will be other sheets of the variety $H(\omega t_a + \xi_a) = 0$, but these will not be called null covectors.) Second, a somewhat deeper argument shows that $H^{ab}_{\text{loc}}(\zeta_a)$ remains Lorentzian even when $\zeta_a \neq 0$ is taken on the boundary of the set of hyperbolic covectors.

## 3. Trajectories and the Length Functional

In the usual development of Lorentzian geometry, one starts with a manifold and a metric $g_{ab}$. One is led almost immediately to consider the length functional, and thence to the geodesic equation.

Our present point of view is somewhat different. Even in the Lorentzian case ($m = 2$), we would begin with the *inverse* metric $g^{ab}$, as it appears in a differential operator

$$
g^{ab}\partial_a\partial_b + \text{lower-order terms}\,. \tag{3.1}
$$

The metric $g_{ab}$ and the equations of the geodesics do indeed arise naturally in the study of this operator, but there are a number of steps involved in getting to them. And while in the Lorentzian case the effect of these steps is to derive an object $g_{ab} = (g^{ab})^{-1}$ from $g^{ab}$, in the general ($m > 2$) case, we must expect something else, for there is no general sense in which $H^{a_1a_2\ldots a_m}$ will have an inverse.

The analogs of the causal geodesics in the general case will be called *causal trajectories*. We shall show how these can be defined, and viewed as the local stationary points of a suitable length (or proper time) functional.

**3.1. Trajectories.** We may regard $H(\zeta_a) = H^{a_1a_2\ldots a_m}\zeta_{a_1}\zeta_{a_2}\cdots\zeta_{a_m}$ as a Hamiltonian function on the cotangent bundle (equipped with its canonical symplectic structure). Then the associated *Hamiltonian trajectories* are the solutions of the system

$$
\dot{x}^a = \partial^a H\,, \qquad \dot{\zeta}_a = -\partial_a H \tag{3.2}
$$

in local coordinates, where $\partial^a = \partial/\partial\zeta_a$ and $\partial_a = \partial/\partial x^a$.

Note that $H(\zeta_a)$ must be constant along any trajectory. Thus a trajectory which at one point has $\zeta_a$ future-directed hyperbolic will have this property at all its points. Similarly one with a future-directed null covector at one point will have the same at all points. (It is not quite obvious, from the theory described so far, that $\zeta_a$ could not pass through zero on a null trajectory. However, this will be a consequence of results below.) We may thus speak of future-directed hyperbolic or null trajectories.

By a slight abuse of terminology, we shall also call the projections of these Hamiltonian trajectories in $T^*N$ to the base space $N$ *trajectories*, hyperbolic or null as the case may be. Through any point in the base space, the set of future-directed hyperbolic or null covectors form initial data for Hamiltonian's equations, and so specify the set of future-directed trajectories passing through the point.

This specification of trajectories in terms of data in the base space is less direct than it might be, however, since the tangent vectors to the trajectories are described only implicitly; it is the "momentum" covector $\zeta_a$ which is given as data. To describe the tangent vectors, we must use a Legendre transformation, as we discuss shortly.

Finally, it should be remarked that most of the literature on partial differential equations deals only with trajectories on which $H(\zeta_a) = 0$ (also known as *bicharacteristic strips*). However we wish to consider causal trajectories.

**3.2. The Legendre Transformation.** In classical mechanics, it is well-known that one can pass from a Hamiltonian description (such as the one just given, in terms of $x$ and $\zeta_a$) to a Lagrange description (in terms of $x$ and $\dot{x}^a$) by means of a *Legendre transformation.*

At any point $x$, this transformation is defined by the first of Hamilton's equations,

$$\dot{x}^a = \partial^a H = mH^{aa_2a_3\ldots a_m}\zeta_{a_2}\zeta_{a_3}\cdots\zeta_{a_m}\,. \tag{3.3}$$

This equation gives $\dot{x}^a$ explicitly in terms of $\xi_a$. For $m = 2$, it is a linear relation

$$\dot{x}^a = 2g^{ab}\zeta_b\,, \tag{3.4}$$

but for larger $m$ it is a more complicated, nonlinear relation.

The Jacobian of the Legendre transformation is $\partial^a\partial^b H = H^{ab}_{\mathrm{loc}}$, which, as discussed above, is a Lorentzian form on the set of future-causal covectors. In particular, it is non-degenerate on this set and must be a local diffeomorphism onto its image. On the other hand, the set of future-causal covectors can be shown to be a convex set, which is contractible. Thus

PROPOSITION 3.1. *The Legendre transformation provides a diffeomorphism from the set of future-causal covectors to a subset of the tangent space.*

We call vectors future-causal or future-hyperbolic if they are the images of the corresponding covectors. It should be remarked that in general the Legendre transformation is *not* onto the entire tangent space, but onto some convex cone. (Again, here a "cone" in a vector space is a set closed under multiplication by positive reals. The cone is not, for $m > 2$, given by a quadratic form, but is a more complicated algebraic variety.) Roughly speaking, one would expect the corresponding set of trajectories to form the causal future of the point in question, at least near the point (and neglecting possible complications from global structure).

Finally, it should be remarked that because of the duality between vectors and covectors, the vectors which are causal with respect to the local metric $(H^{ab})^{-1}$ are causal with respect to the hyperbolic structure: the null cone of the local metric lies inside of the null cone of the hyperbolic structure.

**3.3. The Proper Time Functional.** In classical mechanics, there is a well-defined method for converting a Hamiltonian system (here, the equations for the trajectories as given by data $(x, \zeta_a)$ on the cotangent bundle) to a Lagrangian system (here, the equations given in terms of data $(x, \dot{x}^a)$ on the tangent bundle). We introduce the *Lagrangian*

$$\zeta_a \dot{x}^a - H \tag{3.5}$$

regarded as a function of $(x, \dot{x}^a)$ using the Legendre transformation. Since in our case the Hamiltonian is homogeneous of degree $m$ in $\xi_a$, we have

$$\zeta_a \dot{x}^a - H = \zeta_a \partial^a H - H = (m-1)H\,. \tag{3.6}$$

We shall drop the factor $(m-1)$, so the Lagrangian is just the Hamiltonian for us, regarded however as a function of $(x, \dot{x}^a)$. (In the Lorentzian case, this gives $H = (1/4) g_{ab} \dot{x}^a \dot{x}^b$ as a Lagrangian.)

It is easy to see (and follows from classical mechanics) that the trajectories are then given precisely as the stationary points of the functional

$$\gamma \mapsto \int_\gamma H \, du\,, \tag{3.7}$$

with respect to variations fixing the end-points. Actually, there is some freedom in the choice of functional. We may choose as the integrand a power of $H$, rather than $H$ itself. This is well known in the Lorentzian (or Riemannian) cases: one can work either with the *action functional*

$$\int_\gamma g_{ab} \dot{x}^a \dot{x}^b \, du \tag{3.8}$$

or the *proper time functional*

$$\int_\gamma [g_{ab} \dot{x}^a \dot{x}^b]^{1/2} \, du\,. \tag{3.9}$$

The stationary points of the action functional are the affinely parameterized geodesics; those of the proper time functional (which is reparameterization-invariant) are the arbitrarily parameterized geodesics.

We may do something similar in the general case. Since $H$ is homogenous of degree $m$ in $\zeta_a$ and $\dot{x}^a = \partial^a H$, it follows that $\dot{x}^a$ is homogeneous of degree $m-1$ in $\zeta_a$. Inversely, then, the covector $\zeta_a$ is homogeneous of degree $(m-1)^{-1}$ in $\dot{x}^a$, and so the Lagrangian $H$ is homogeneous of degree $m/(m-1)$ in $\dot{x}^a$. Thus the functional which is reparameterization-invariant, the *proper time functional* is

$$T(\gamma) = \int_\gamma [H(x, \dot{x}^a)]^{(m-1)/m} \, du\,. \tag{3.10}$$

The stationary points of this are the causal trajectories, modulo their parameterization.

It will be convenient to understand how to factor out the reparameterization freedom. At the infinitesimal level, this corresponds to $\delta\gamma^a$'s of the form $\phi\dot{\gamma}^a$, where $\phi$ is a smooth function vanishing at the end-points. Along any strictly future-hyperbolic curve we have $\zeta_a\dot\gamma^a = mH > 0$, and thus we may use the reparameterization freedom to consider only perturbations satisfying $\zeta_a\delta\gamma^a = 0$. We shall not discuss curves which are in part null here.

**3.4. The Exponential Map.** As discussed above, the stationary points of the proper time (or action) are the causal trajectories (modulo reparameterizations, for the proper time). The associated variational equations are non-linear second-order ordinary differential equations, Lagrange's equations.

One could have equivalently used the Legendre transformation to regard $\zeta_a$ as a function of $\dot x^a$ and $x$; then the second of Hamilton's equations ($\dot\zeta_a = -\partial_a H$) becomes the second-order differential equation for the trajectory; it has the form

$$\ddot x^a + \text{lower-order nonlinear terms} = 0\,. \tag{3.11}$$

(A somewhat more explicit form of this will be given in the next subsection, but we do not need the details here.) This will be the analog, for the general case, of the geodesic equation in the case $m = 2$.

There is an important difference, however, between the cases $m = 2$ and $m > 2$. For $m > 2$, not all initial data $(x, \dot x^a)$ are allowed for the equation of the trajectory. This is because the Legendre transformation is only known to map onto the causal vectors, so general non-causal vectors are not admissible as initial data. (In the form (3.11), this shows up in the lower-order terms. They contain inverses of the local metric $H^{ab}_{\text{loc}}(\zeta_a)$; these inverses need not exist for non-causal vectors.)

Let us write $X(x, \dot x^a, u)$ for the trajectory with initial data $(x, \dot x^a)$. From the theory of ordinary differential equations, this is a smooth function if $\dot x^a$ is restricted to be causal and $u$ to be sufficiently small. On the other hand, it is not hard to show, using the homogeneity discussed in the previous subsection, that we have

$$X(x, k\dot x^a, u) = X(x, \dot x^a, ku) \tag{3.12}$$

where defined, for positive scalars $k$. It follows that $X(x, \dot x^a, 1)$ is defined, for each $x$, for causal vectors in a sufficiently small neighborhood of the origin. We may thus define an *exponential map*

$$\exp_x \dot x^a = X(x, \dot x^a, 1) \tag{3.13}$$

from neighborhood of the zero in the bundle of causal vectors, into the base space.

**3.5. Congruences.** It is often useful, both conceptually and technically, to consider a *congruence* of hyperbolic trajectories in a given open set in the base space, that is, a smooth $(n-1)$-dimensional family of trajectories such that each point in the open set lies on precisely one trajectory in the family. Then we may regard $\dot x^a$ and $\zeta_a$ as hyperbolic vector and covector fields over the region in question. Very conveniently, we get a well-defined metric tensor, whose inverse is the local metric $H^{ab}_{\text{loc}} = m(m-1)H^{abc_3c_4\cdots c_m}\zeta_{c_3}\zeta_{c_4}\cdots\zeta_{c_m}$.

We may then usefully rewrite Hamilton's equations in terms of the congruence. We have $\dot x^a = (m-1)^{-1}H^{ab}_{\text{loc}}\zeta_b$.

Rewriting Hamilton's second equation is not hard, but requires some care, since we will shift the sense in which we interpret the partial derivatives. The equation is given canonically in the form $\dot\zeta_a = -\partial_a H$, where the derivative $\partial_a = \partial/\partial x^a$ is taken holding $\zeta_a$ constant. We may equivalently write, then, $\dot\zeta_a =$

$-(\partial_a H^{a_1 a_2 \dots a_m})\zeta_{a_1}\zeta_{a_2}\cdots\zeta_{a_m}$. In this form, we may adapt it to a congruence by writing

$$\begin{aligned}\dot{\zeta}_a &= -(\partial_a H^{a_1 a_2 \dots a_m})\zeta_{a_1}\zeta_{a_2}\cdots\zeta_{a_m} \\ &= -\partial_a(H^{a_1 a_2 \dots a_m}\zeta_{a_1}\zeta_{a_2}\cdots\zeta_{a_m}) + mH^{a_1 a_2 \dots a_m}(\partial_a\zeta_{a_1})\zeta_{a_2}\cdots\zeta_{a_m} \\ &= -\partial_a H + (m-1)^{-1}H^{bc}_{\mathrm{loc}}(\partial_a\zeta_b)\zeta_c \\ &= -\partial_a H + \dot{x}^b\partial_a\zeta_b\,,\end{aligned}\tag{3.14}$$

where now $\zeta_a$ is interpreted as a covector field and $\partial_a$ may act on it. Now, we may also rewrite the left-hand side:

$$\dot{\zeta}_a = \dot{x}^b\partial_b\zeta_a\,,\tag{3.15}$$

and so the second equation becomes

$$\dot{x}^b\partial_b\zeta_a = -\partial_a H + \dot{x}^b\partial_a\zeta_b\,.\tag{3.16}$$

There are various useful ways of rewriting this. If we let $\nabla_a$ be the covariant derivative got from $H^{ab}_{\mathrm{loc}}$, then we have, in covariant form,

$$\dot{x}^b\nabla_b\zeta_a - \dot{x}^b\nabla_a\zeta_b = -\nabla_a H\tag{3.17}$$

(since this is independent of the connection, as long as the connection is symmetric). Perhaps of more present interest is a form we can get using $\dot{x}^a = (m-1)^{-1}H^{ab}_{\mathrm{loc}}\zeta_b$. We find, after a short calculation,

$$\dot{x}^b\nabla_b\dot{x}_a = (m/2)(m-2)\nabla_a H\,.\tag{3.18}$$

Thus the trajectories of the hyperbolic symbol will be geodesics with respect to the local metric if $H = (1-1/m)H_{ab\,\mathrm{loc}}\dot{x}^a\dot{x}^b$ is constant for the congruence. (It will be constant along each trajectory, so if it is constant over one transverse section of the congruence, it will be constant along the congruence. For a hyperbolic congruence, this can always be arranged.)

## 4. The Second Variation, Conjugate Points, and Convergence Conditions

There exists a theory of Jacobi fields, conjugate points, focal points and the second variation of the proper time functional which parallels the Lorentzian case. I shall sketch it here.

**4.1. The Second Variation.** A straightforward computation of the second variation of the length functional, at a trajectory $\gamma$ (and using the reparameterization freedom to require $\zeta_a\delta\gamma^a = 0$) yields

$$\delta^2 T(\gamma) = (\mathrm{const})\int_\gamma \left[P_{ab}\delta\dot{\gamma}^a\delta\dot{\gamma}^b + 2Q_{ab}\dot{\gamma}^a\delta\gamma^b + R_{ab}\delta\gamma^a\delta\gamma^b\right]\,du\tag{4.1}$$

where we shall only need the form of $P_{ab}$ explicitly:

$$P_{ab} = \partial_{\dot{x}^a}\partial_{\dot{x}^b}H - (mH)^{-1}\partial_{\dot{x}^a}H\partial_{\dot{x}^b}H\,.\tag{4.2}$$

A direct computation shows that $P_{ab}$ is precisely the local metric $(H^{ab}_{\mathrm{loc}})^{-1}$ restricted to the spacelike hyperplane orthogonal to $\zeta_a$, and hence negative-definite.

The study of the second variation is really a study of an ordinary differential equation. We may regard the form defining a linear ordinary differential operator $L$ with Dirichlet boundary conditions by setting

$$\delta^2 T = (\text{const}) \int_\gamma P_{ab}\delta\gamma^a (L\delta\gamma)^b \, du \,. \tag{4.3}$$

Then $L$ is a self–adjoint operator with discrete spectrum bounded above (since $P_{ab}$ is negative-definite).

The usual Morse Index Theorem applies to this system (with minor changes because of the sign of $P_{ab}$). Thus the number of positive eigenvalues (counted with multiplicity) of $L$ is the number of independent perturbations causing $\delta^2 T$ to increase. Also this number is equal to the number of conjugate points (counted with multiplicity). This forms the basis for a theory of *maximal* trajectories. (We saw above that trajectories were stationary points of the proper time functional. That they may be extrema is a deeper result.)

In particular, it follows from ODE theory that if we fix a trajectory and a point on it, then for small enough intervals around the point the spectrum of $L$ will be strictly negative. (The "kinetic term"in $L$, which is $\delta\ddot{\gamma}^a$, dominates.) This shows that trajectories are, at a local level, maximizing curves for the proper time functional.

**4.2. Convergence Conditions.** We have seen that a Morse theory applies to the proper time functional in the general hyperbolic case much as it does in the Lorentzian case. The key aspect of this theory, from the point of view of singularity theorems, is that the existence of conjugate points signals the non-maximality of the proper time. In the Hawking–Penrose singularity theorems, the existence of conjugate points is either simply taken as a hypothesis or is a consequence of certain restrictions on the curvature, the *convergence conditions.* (With the usual physical interpretation of Einstein's equation, these convergence conditions are equivalent to certain restrictions on the stress-energy, and so they are often called *energy conditions.*)

We may formulate similar convergence conditions here. While a treatment based on the formalism of the previous subsection is possible, it is rather lengthy (one has to implicitly change coordinates), and I shall give another (equivalent) one here.

Recall that in terms of the local metric $H_{ab\,\mathrm{loc}}$ and covariant derivative $\nabla_a$ associated with a congruence of hyperbolic trajectories, the trajectory equation is

$$\dot{x}^b \nabla_b \dot{x}_a = (m/2)(m-2)\nabla_a H \,. \tag{4.4}$$

If $w^a$ is a connecting field for this congruence (that is, if $[w, \dot{x}] = 0$), then a short calculation shows that

$$(\dot{x}^b \nabla_b)^2 w^a = \dot{x}^r \dot{x}^q R_{pqr}{}^a w^q + (m/2)(m-2) w^b \nabla_b \nabla_a H \,. \tag{4.5}$$

For congruences with $H = \text{const}$, this is the same as the usual geodesic deviation equation. Thus we may take the Lorentzian convergence condition over to the present case:

DEFINITION 4.1. A strongly hyperbolic symbol obeys the *timelike convergence condition* if, for any hyperbolic congruence with $H$ constant, the Ricci tensor derived from the associated local metric satisfies $R_{ab}\dot{x}^a \dot{x}^b \leq 0$.

Since the Jacobi equation in this case is precisely the same as the usual Lorentzian one, the timelike convergence condition here will enforce the development of conjugate points just as in the Lorentzian case: if the timelike convergence condition holds, then any timelike trajectory which is complete (in the sense that a proper-time parameter along it takes all real values) will necessarily contain conjugate points. (In fact, one only needs the trajectory to be half-complete — to extend to either arbitrarily positive or arbitrarily negative proper time values.)

It should be emphasized that here the Ricci tensor depends, in the case $m > 2$, on the congruence, since the local metric does. Explicit expressions for the timelike convergence condition in terms of the hyperbolic tensor $H^{a_1 a_2 \cdots a_m}$ are complicated.

## 5. Differential Topology of Causal Curves

It is possible to develop a theory of the differential topology of causal curves much as in the Lorentzian case. This is perhaps not surprising, but it is not quite automatic, either. The chief reason is that in the standard development of the classical theory, the proofs of a number of important technical results use the existence of *simply convex* neighborhoods — sets in which any two points are joined by a unique geodesic. Since we have no general analog of spacelike geodesics, we cannot have simply convex neighborhoods. We must therefore devise a substitute, and verify that this can be used to get a counterpart of the Lorentzian theory.

I shall sketch here the main ideas; a full technical treatment will be given elsewhere.

We may define the basic elements of the causal structure by straightforward adaptation of the definitions in Penrose (1972). I shall give a rapid treatment of these; see Penrose's book for technical details.

A *trip* is a curve which is piecewise a future-directed hyperbolic trajectory; a *causal trip* is a curve which is piecewise a future-directed causal trajectory. We write $p \ll q$ to mean that there is a trip from $p$ to $q$ and $p \prec q$ to mean that there is a causal trip from $p$ to $q$. If $A$ is a given set, we write $p \ll_A q$ or $p \prec_A q$ to mean there is a trip or causal trip in $A$ from $p$ to $q$.

The timelike and causal futures and pasts may be defined in terms of trips in the usual way. Thus for example the *timelike future* of $p$ as $I^+(p) = \{q \mid p \ll q\}$ and the *causal future* of $p$ as $J^+(p) = \{q \mid p \prec q\}$.

As mentioned above, we cannot define an analog of simply convex sets (sets any two of whose points can be connected by a unique geodesic segment in the set) since we have no general notion of "spacelike trajectories." However, there is a slightly weaker concept which is adequate for much of the theory.

DEFINITION 5.1. An open set $O$ will be called *causally simple* if, whenever $p \prec_O q$, there is a unique causal trajectory (modulo reparameterization) from $p$ to $q$ in $O$. A *causally simple region* is a causally simple set $O$ whose closure is compact and is contained in a causally simple open set.

LEMMA 5.2. *Arbitrarily small causally simple neighborhoods of a point exist.*

PROOF. (Sketch.) Let $b$ be the point in question. Fix a neighborhood $O_1$ of $b$ such that on any causal trajectory in the closure of $O_1$ there is no pair of conjugate points. (This is always possible, from the theory of ordinary differential equations.) It follows that no two points in $O_1$ can be connected by more than one causal trajectory in $O_1$. (The proof is the same as in the Lorentzian case.)

Now let $\gamma$ be a causal trajectory in $O_1$ through $b$, and choose $a$ and $c$ on $\gamma$ with $a \ll b \ll c$, the trips being provided by segments of $\gamma$. We will let $O$ be the set of points in $O_1$ such that there is a hyperbolic trajectory from $a$ to $p$ and a hyperbolic trajectory from $p$ to $c$. Using the absence of conjugate points in $O_1$, it is not hard to see that $O$ is open.

The verification that $p \prec_O q$ implies there exists a unique causal trajectory in $O$ from $p$ to $q$ is by a connectivity argument. If there are causal trajectories $pp_1$ and $p_1p_2$ in $O$, the set of points along $p_1p_2$ which can be reached by causal trajectories from $p$ must be open (since the Jacobian of the exponential map from $p$ cannot be degenerate). On the other hand, by using the fact that the points in $O$ can be reached by unique hyperbolic trajectories from $a$ and to $c$, one can show that this set is closed. □

With this result at hand, the development of the theory follows that of the Lorentzian case. In particular, one can study the space of causal curves and establish its compactness properties. One can use these to prove the existence of maximal trajectories (trajectories with maximal proper time) connecting pairs of causally separated points, or connecting a point to a hypersurface, in suitable circumstances.

## 6. Singularity Theorems

With the concepts and techniques above, one can formulate and prove various singularity theorems. I shall here give only one a "cosmological" one.

THEOREM 6.1. *Let $N$ be a manifold and $H(\zeta_a)$ strongly hyperbolic on $N$, with strong causality holding in $N$.*[1] *Suppose: (a) there exists a compact hypersurface $\Sigma$ in $N$ which every endless causal trip meets exactly once; (b) the normals to $\Sigma$ are diverging (this can be defined in terms of the local metric, cf. section 2.3); (c) the timelike convergence condition holds.*

*Then there must exist a trajectory which is not complete towards the past.*

PROOF. (Sketch.) This is the same as in the Lorentzian case. Were the trajectories all complete towards the past, each would contain, among its points to the past of $\Sigma$, a futuremost focal point to $\Sigma$. The proper time along each trajectory from these focal points to $\Sigma$ would vary continuously, and so would assume a maximal value. However, this would then have to be on the interval of a trajectory from a focal point to $\Sigma$, which is impossible. □

I will close by remarking that it seems likely that one will be able to formulate "trapped surface" singularity theorems for strongly hyperbolic operators (cf. Penrose 1965). These are potentially very interesting, as the convergence conditions that would be expected to be required for them would be weaker than the timelike one used above.

## References

[1] R. Courant and D. Hilbert, *Methods of mathematical physics*, volume II. Wiley, New York et al., 1989.
[2] J. Leray, *Hyperbolic differential equations.* Institute for Advanced Study, Princeton, NJ, 1955.
[3] R. Penrose, Phys. Rev. Lett. **14** (1965) 57.

[1]Strong causality is defined as in the Lorentzian case. The assumption of strong causality here may be redundant.

[4] R. Penrose, *Techniques of differential topology in relativity.* SIAM, Philadelphia, PA, 1972.

Department of Mathematics, University of Missouri–Columbia, Columbia, Missouri
*E-mail address*: adam@math.missouri.edu

Contemporary Mathematics
Volume **337**, 2003

# Examples of Conjugate Loci of Pseudoriemannian 2-step Nilpotent Lie Groups with Nondegenerate Center

Changrim Jang and Phillip E. Parker

ABSTRACT. We present examples of geodesics, geodesic surfaces, and conjugate loci in the 3-dimensional Heisenberg group with signatures $+-+$ and $--+$.

## 1. Introduction

By an *inner product* on a vector space $V$ we shall mean a nondegenerate, symmetric bilinear form on $V$, generally denoted by $\langle\,,\rangle$. Our convention is that $v \in V$ is timelike if $\langle v, v\rangle > 0$, null if $\langle v, v\rangle = 0$, and spacelike if $\langle v, v\rangle < 0$.

Throughout, $N$ will denote a connected, 2-step nilpotent Lie group with Lie algebra $\mathfrak{n}$ having center $\mathfrak{z}$. We shall use $\langle\,,\rangle$ to denote either an inner product on $\mathfrak{n}$ or the induced left-invariant pseudoriemannian (indefinite) metric tensor on $N$.

The only general result on conjugate points is Proposition 4.18 from [**CP1**].

PROPOSITION 1.1. *Let $N$ be a simply connected, 2-step nilpotent Lie group with left-invariant metric tensor $\langle\,,\rangle$, and let $\gamma$ be a geodesic with $\dot\gamma(0) = a \in \mathfrak{z}$. If $\mathrm{ad}^{\dagger}_{\bullet}\, a = 0$, then there are no conjugate points along $\gamma$.*

In [**JP**] we obtained results on conjugate points under the additional assumption that the center $\mathfrak{z}$ of $\mathfrak{n}$ is nondegenerate. Here we give explicit examples of these in the 3-dimensional Heisenberg group $H_3$ where their application can be visualized effectively. Some pictures for the Riemannian (positive-definite) case may be found in [**M**].

---

2000 *Mathematics Subject Classification.* Primary 53C50; Secondary 22E25, 53B30, 53C30.

*Key words and phrases.* Conjugate locus, 2-step nilpotent Lie group, geodesics.

The first author was on sabbatical leave at WSU in 2002–03.

We denote the adjoint with respect to $\langle\,,\rangle$ of the adjoint representation of the Lie algebra $\mathfrak{n}$ on itself by $\mathrm{ad}^\dagger$. In the case of a nondegenerate center, the involution $\iota$ of [**CP1**] is merely given by $\iota(z_\alpha) = \varepsilon_\alpha\, z_\alpha$ and $\iota(e_a) = \bar{\varepsilon}_a\, e_a$ where $\langle z_\alpha, z_\alpha\rangle = \varepsilon_\alpha$ and $\langle e_a, e_a\rangle = \bar{\varepsilon}_a$ on an orthonormal basis of $\mathfrak{n}$. Then the operator $j : \mathfrak{z} \to \mathrm{End}\,(\mathfrak{v})$ is given by $j(z)x = \iota\,\mathrm{ad}_x^\dagger \iota z$. We refer to [**CP1**] and [**O**] for other notations and results.

## 2. Recollections

First, we specialize Proposition 4.8 of [**CP1**] to our present setting. As there, $L_n$ denotes left translation in $N$ by $n \in N$.

PROPOSITION 2.1. *Let $N$ be simply connected, $\gamma$ a geodesic with $\gamma(0) = 1 \in N$, and $\dot\gamma(0) = z_0 + x_0 \in \mathfrak{n}$. Then*

$$\dot\gamma(t) = L_{\gamma(t)*}\left(z_0 + e^{tJ}x_0\right)$$

*where $J = \mathrm{ad}_\bullet^\dagger\, z_0$.*

As in [**JP**], we shall identify all tangent spaces with $\mathfrak{n} = T_1N$. Thus we regard

$$\dot\gamma(t) = z_0 + e^{tJ}x_0$$

as the geodesic equation.

For the convenience of the reader, we recall the major results from [**JP**]. Let $\gamma$ be a geodesic with $\gamma(0) = 1$ and $\dot\gamma(0) = z_0 + x_0 \in \mathfrak{z} \oplus \mathfrak{v}$, respectively, and let $J = \mathrm{ad}_\bullet^\dagger\, z_0$. If $\gamma$ is not null, we may assume $\gamma$ is normalized so that $\langle\dot\gamma, \dot\gamma\rangle = \pm 1$. As usual, $\mathbb{Z}^*$ denotes the set of all integers with 0 removed.

THEOREM 2.2. *Under the preceding assumptions:*

(1) *if $J = 0$ and $x_0 = 0$, then there are no conjugate points along $\gamma$;*
(2) *if $J = 0$ and $x_0 \neq 0$, then $\gamma(t)$ is conjugate to $\gamma(0)$ along $\gamma$ if and only if $-12/t^2$ is an eigenvalue of the linear operator*

$$A : \mathfrak{z} \to \mathfrak{z} : z \mapsto \left[x_0, \mathrm{ad}_{x_0}^\dagger z\right];$$

(3) *if $J \neq 0$ and $x_0 = 0$, then $\gamma(t)$ is conjugate to $\gamma(0)$ along $\gamma$ if and only if*

$$t \in \bigcup_{k=1}^{n} \frac{2\pi}{\lambda_k}\mathbb{Z}^*$$

*where the $-\lambda_k^2$ are the negative eigenvalues of $J^2$.*

If there are no negative eigenvalues in Part 3, then there are of course no conjugate points.

THEOREM 2.3. *In addition, also assume that* $\dim \mathfrak{z} = 1$. *Then:*

(1) *if* $z_0 \neq 0$ *and* $x_0 = 0$, *then* $\gamma(t)$ *is conjugate to* $\gamma(0)$ *along* $\gamma$ *if and only if*

$$t \in \bigcup_{k=1}^{n} \frac{2\pi}{\lambda_k}\mathbb{Z}^*$$

*where the* $-\lambda_k^2$ *are the negative eigenvalues of* $J^2$;

(2) *if* $z_0 = 0$ *and* $x_0 \neq 0$, *then* $\gamma(t)$ *is conjugate to* $\gamma(0)$ *along* $\gamma$ *if and only if*

$$-\frac{12}{t^2} = \varepsilon\langle \mathrm{ad}_{x_0}^{\dagger} z, \mathrm{ad}_{x_0}^{\dagger} z\rangle,$$

*where* $z \in \mathfrak{z}$ *is a unit vector with* $\varepsilon = \langle z, z\rangle$;

(3) *if* $z_0 \neq 0$ *and* $x_0 \neq 0$, *then* $\gamma(t)$ *is conjugate to* $\gamma(0)$ *along* $\gamma$ *if and only if either*

$$t \in \bigcup_{k=1}^{p} \frac{2\pi}{\lambda_k}\mathbb{Z}^*$$

*where the* $-\lambda_k^2$ *are the negative eigenvalues of* $J^2$, *or* $t$ *is a solution of*

$$t\langle Jx_0, \left(e^{-tJ} - I\right)^{-1} x_0\rangle = \langle \dot\gamma, \dot\gamma\rangle.$$

In the next result, we use the orthogonal direct sum decomposition $\bigoplus_{j=1}^{m} \mathfrak{w}_j$ where $J$ leaves each $\mathfrak{w}_j$ invariant and each $\mathfrak{w}_j$ is an eigenspace of $J^2$. Further, $J^2$ is negative definite for $1 \leq j \leq p$ and positive definite for $p+1 \leq j \leq m$.

COROLLARY 2.4. *If* $J^2$ *is moreover diagonalizable, then Parts 2 and 3 can be made completely explicit.*

2. *if* $z_0 = 0$ *and* $x_0 \neq 0$, *then* $\gamma(t)$ *is conjugate to* $\gamma(0)$ *along* $\gamma$ *if and only if*

$$\frac{12}{t^2} = \sum_{l=1}^{m-p} \lambda_{p+l}^2 \varepsilon\langle B_{p+l}, B_{p+l}\rangle - \sum_{k=1}^{p} \lambda_k^2 \varepsilon\langle A_k, A_k\rangle,$$

*where* $x_0 = \sum_k A_k + \sum_l B_{p+l}$, $A_k \in \mathfrak{w}_k$, $B_{p+l} \in \mathfrak{w}_{p+l}$, $-\lambda_k^2$ *and* $\lambda_{p+l}^2$ *are the eigenvalues of* $J^2$ *with* $J = \mathrm{ad}_{\bullet}^{\dagger} z$ *for* $z \in \mathfrak{z}$ *a unit vector, and* $\varepsilon = \langle z, z\rangle$;

3. *if* $z_0 \neq 0$ *and* $x_0 \neq 0$, *then* $\gamma(t)$ *is conjugate to* $\gamma(0)$ *along* $\gamma$ *if and only if either*

$$t \in \bigcup_{k=1}^{p} \frac{2\pi}{\lambda_k}\mathbb{Z}^*$$

*or* $t$ *is a solution of*

$$\sum_{k=1}^{p}\langle A_k, A_k\rangle \frac{\lambda_k t}{2} \cot\frac{\lambda_k t}{2} + \sum_{l=1}^{m-p}\langle B_{p+l}, B_{p+l}\rangle \frac{\lambda_{p+l} t}{2} \coth\frac{\lambda_{p+l} t}{2} = \langle \dot\gamma, \dot\gamma\rangle$$

*where the $A_h$, $B_i$ are as in the preceding part and $\pm\lambda_j^2$ are the eigenvalues of $J^2$ with $J = \mathrm{ad}^\dagger_\bullet z_0$ now.*

## 3. The Examples

We consider the Heisenberg group with nondegenerate center as in Example 2.4 of [**CP1**]. We denote the 3-dimensional Heisenberg algebra by $\mathfrak{h}_3$. On an orthonormal basis $\{e_1, e_2, z\}$ the structure equation is

$$[e_1, e_2] = z\,.$$

**3.1. Signature $+-+$.** As one example, we take the nontrivial inner products as

$$1 = \langle z, z\rangle, \quad 1 = \langle e_1, e_1\rangle, \quad -1 = \langle e_2, e_2\rangle$$

so the nontrivial adjoint maps are

$$\mathrm{ad}^\dagger_{e_1} z = -e_2\,, \quad \mathrm{ad}^\dagger_{e_2} z = -e_1$$

and the rest vanish.

First let $\gamma$ be a unit-speed (non-null) geodesic emanating from the identity, so $\gamma(0) = 1 \in N = H_3$, with $\dot\gamma(0) = z_0 + x_0$ for some $z_0 \in \mathfrak{z}$ and $x_0 \in \mathfrak{v} = \mathfrak{z}^\perp$. Note that we have $\langle z_0 + x_0, z_0 + x_0\rangle = \langle z_0, z_0\rangle + \langle x_0, x_0\rangle = \varepsilon = \pm 1$. We take $\dot\gamma(t) = z_0 + e^{tJ}x_0$ as after Proposition 2.1 above. Setting $z_0 = az$ for some real $a$, we have $\langle z_0, z_0\rangle = a^2$ so $\langle x_0, x_0\rangle = \varepsilon - a^2$,

$$J(e_1) = -a\,e_2 \quad \text{and} \quad J(e_2) = -a\,e_1\,, \tag{3.1}$$

whence $J^2 = a^2 I_2$; *cf.* [**CP1**, p. 30].

When $z_0 \neq 0 \neq x_0$, we are in the second case of Part 3 of Corollary 2.4. If $\gamma(t_0)$ is to be a conjugate point to $\gamma(0)$ along $\gamma$, then for timelike geodesics we must have

$$(1 - a^2)\frac{at_0}{2}\coth\frac{at_0}{2} = 1 \tag{3.2}$$

since the only interesting solutions are nonzero $t_0$. Note that for each $0 < a < 1$ there are exactly two solutions, negatives of each other. Thus all these timelike geodesics $\gamma$ have two conjugate points to $\gamma(0) = 1$, one forward and one backward, so to say. For spacelike and null geodesics, it similarly follows that there are no conjugate points along them.

Next we integrate the geodesic equation. From (3.1),

$$J = \begin{bmatrix} 0 & -a \\ -a & 0 \end{bmatrix} \quad \text{so} \quad e^{tJ} = \begin{bmatrix} \cosh(-at) & \sinh(-at) \\ \sinh(-at) & \cosh(-at) \end{bmatrix}$$

and we obtain

$$\begin{aligned} \dot\gamma_1 &= \cosh(-at)x_{01} + \sinh(-at)x_{02}\,, \\ \dot\gamma_2 &= \sinh(-at)x_{01} + \cosh(-at)x_{02}\,, \\ \dot\gamma_3 &= a \end{aligned}$$

with obvious notation for the components of $x_0$. Integrating, simplifying, and using the initial condition $\gamma(0) = [0,0,0]$ in exponential coordinates, we get

$$\begin{aligned}
\gamma_1(t) &= \frac{1}{a}\left[\sinh(at)x_{01} - \cosh(at)x_{02} + x_{02}\right], \\
\gamma_2(t) &= \frac{1}{a}\left[-\cosh(at)x_{01} + \sinh(at)x_{02} + x_{01}\right], \\
\gamma_3(t) &= a\,t.
\end{aligned} \tag{3.3}$$

To draw pictures of geodesic surfaces and the conjugate locus, we chose a parametrization based on rotating the initial velocity vector about the $e_2$ axis and arranging the axes so that usual $xyz$-pictures correspond to $e_1e_2z$-pictures. Thus we have

$$x_{01} = a\sin\theta\, e_1, \qquad x_{02} = b\, e_2, \qquad z_0 = a\cos\theta\, z$$

where

$$b = \begin{cases} \pm\sqrt{a^2-1} & \text{for } \dot\gamma(0) \text{ timelike,} \\ \pm\sqrt{a^2+1} & \text{for } \dot\gamma(0) \text{ spacelike,} \\ \pm a & \text{for } \dot\gamma(0) \text{ null,} \end{cases} \tag{3.4}$$

since we want $\gamma$ to be unit speed if nonnull. Thus $a^2 \geq 1$ for timelike geodesics while $a$ is any real number for spacelike or null ones. Substituting in the integrated geodesic equations (3.3), we obtain

$$\begin{aligned}
\gamma_1(t) &= \frac{b}{a}\sec\theta\,(1-\cosh(t\,a\cos\theta)) + \tan\theta\sinh(t\,a\cos\theta) \\
\gamma_2(t) &= \tan\theta\,(1-\cosh(t\,a\cos\theta)) + \frac{b}{a}\sec\theta\sinh(t\,a\cos\theta) \\
\gamma_3(t) &= t\,a\cos\theta
\end{aligned} \tag{3.5}$$

in exponential coordinates.

In Figure 1, we show some geodesic surfaces. Starting from the front of the box, we see two spacelike conical surfaces, a null conical surface, three timelike surfaces, and the other half of the null conical surface. The middle timelike surface has initial velocities in the $e_1z$-plane, the other two timelike surfaces have theirs around cones at angles of $\pi/6$ in front of and behind the $e_1z$-plane, the null surfaces start at angles of $\pi/4$ of course, and the outer spacelike cone is $\pi/3$ in front of the $e_1z$-plane. Curves emanating outward from the origin are geodesics and those transverse to them are curves of constant $t$.

Rewriting (3.2) in our exponential coordinates, conjugate points are located *via* the solutions $t$ of the following equation, obtained from the second case in Part 3 of Corollary 2.4. (The first case does not apply here since $J^2$

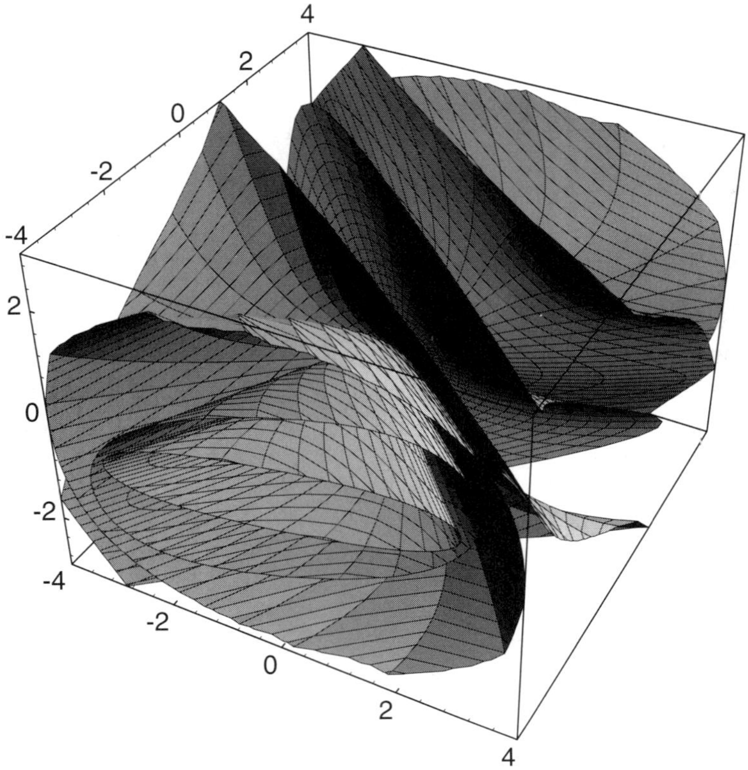


FIGURE 1. Geodesic surfaces in $H_3$ using exponential coordinates with axes $e_1$, $e_2$, $z$ and signature $+-+$.

has only a positive eigenvalue.)

$$(1 - a^2\cos^2\theta)\frac{at\cos\theta}{2}\coth\frac{at\cos\theta}{2} = 1 \tag{3.6}$$

Figures 2 and 3 show the conjugate locus for this geometry as so computed. For $a > 1$ we get $\cos^{-1}(-1/a) > \theta > \cos^{-1}(1/a)$ for $\theta$ near $\pi/2$. There is a similar condition for $a < -1$, and we also solved the equation near $3\pi/2$ to get both "halves" of the conjugate locus. The "horizontal" bands, some with ragged ends, reflect the choices we made for limits on $\theta$ and some limitations we encountered in the numerical runs. It would be at least amusing, probably interesting, to see an explicit formulation for the (more or less) conical surface to which the conjugate locus is asymptotic.

As seems apparent from the figures, it is indeed easy to verify that the locations of conjugate points do go off to infinity as $\theta$ approaches $\cos^{-1}(\pm 1/a)$ appropriately.

When $z_0 = 0$ and $x_0 \neq 0$, we obtain $\gamma(t) = x_0 t = (ae_1 + be_2)t$ as the geodesics in the $e_1e_2$-plane. Note that $J^2 = I_2$ in this case, since $J = \operatorname{ad}^{\dagger}_{\bullet} z$ here. Now part 2 of Corollary 2.4 applies and we obtain conjugate points

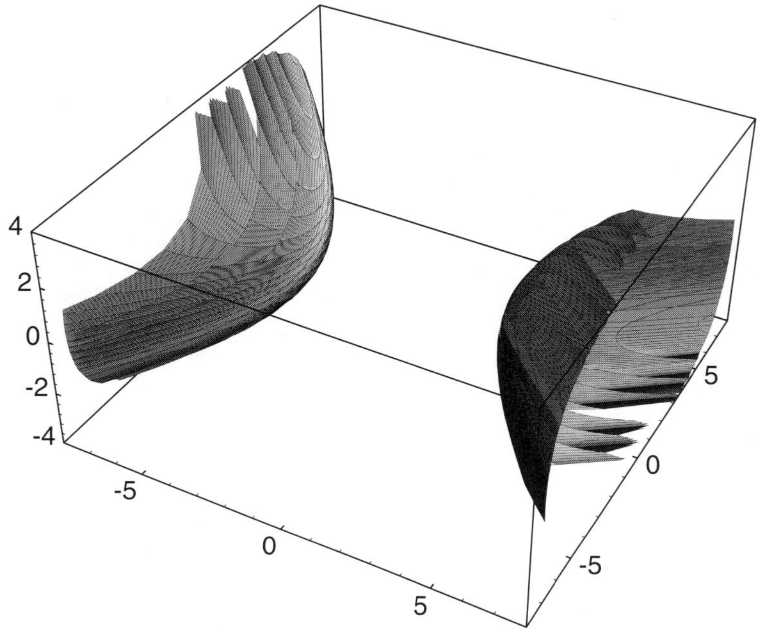


FIGURE 2. Conjugate locus with lines of constant $a$ and $\theta$. The ragged edges are numerical artifacts.

at $t = \pm 2\sqrt{3}$. This is in fact correctly represented in Figures 2 and 3 as the intersection of the surfaces depicted there with the $z = 0$ plane.

The obvious way to see this would be through the Implicit Function Theorem, but unfortunately we happen to be in a degenerate case where it does not apply. So to verify the apparent continuity, we proceed as follows.

LEMMA 3.1. *If* $0 < a^2 < 1$, *then the equation* $(1 - a^2)at = \tanh at$ *has only one solution in* $(0, \infty)$.

PROOF. Consider the function $f(t) = (1-a^2)at - \tanh at$. Differentiating with respect to $t$, we find

$$f'(t) = a(\tanh at - a)(\tanh at + a).$$

Thus $f'(t) = 0$ has exactly two roots $t_1, t_2$ such that $t_1 < 0 < t_2$ and $\tanh at_1 + a = \tanh at_2 - a = 0$. Assuming $a > 0$, we see that $f(t)$ is increasing in $(-\infty, t_1]$ and $[t_2, \infty)$ and decreasing in $[t_1, t_2]$. This implies that $f(t_1) > f(0) = 0 > f(t_2)$. It follows that $f(t) = 0$ has exactly three roots $t_3, 0, t_4$ such that $t_3 < 0 < t_4$. The case $a < 0$ is done similarly. $\square$

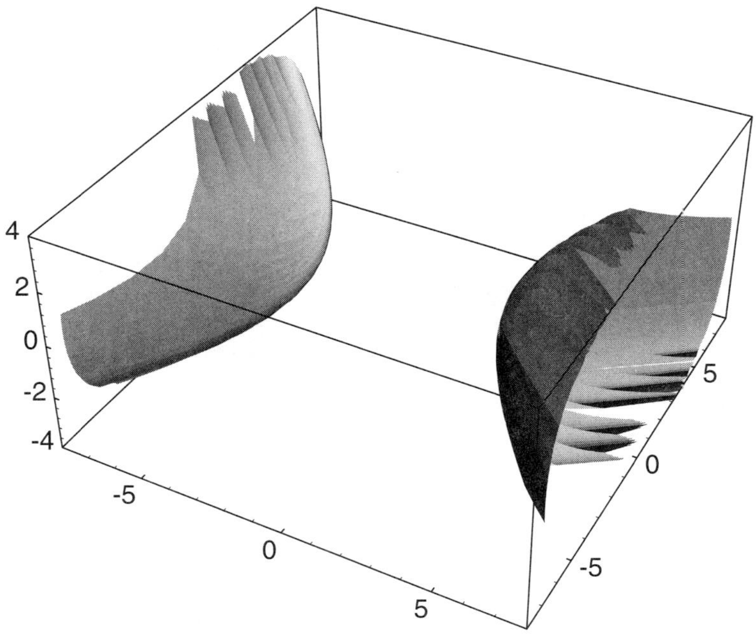


FIGURE 3. Conjugate locus without those lines.

By this lemma we can consider a continuous fuction $t = t(a) > 0$ such that

$$(1 - a^2)at(a) = \tanh at(a)$$

for $-1 < a < 0$ and $0 < a < 1$.

LEMMA 3.2. *For every constant $k > \sqrt{3}$, the following inequality holds for sufficiently small $|a|$.*

$$\sqrt{3} < t(a) < \frac{k}{1 - a^2}$$

PROOF. Since $t(a) = t(-a)$, we may as well assume $a > 0$. Consider

$$g(a) = (1 - a^2)a\sqrt{3} - \tanh \sqrt{3}a\,.$$

Differentiating with respect to $a$, we get

$$g'(a) = \sqrt{3}(\tanh^2 \sqrt{3}a - 3a^2) < 0\,,$$

so we find

$$g(a) < g(0) = 0\,. \tag{3.7}$$

Next consider

$$h(a) = (1-a^2)a\frac{k}{1-a^2} - \tanh\frac{ka}{1-a^2} = ka - \tanh\frac{ka}{1-a^2}$$

for $k > \sqrt{3}$. Differentiating, we have

$$h'(a) = k\frac{(1+a^2)\tanh^2\frac{ka}{1-a^2} + a^4 - 3a^2}{(1-a^2)^2}.$$

Since $\tanh^2 ka - 3a^2 > 0$ for sufficiently small $a$,

$$h'(a) > 0$$

for sufficiently small $a$. This implies that

$$h(a) > h(0) = 0 \tag{3.8}$$

for sufficiently small $a$. Using (3.7) and (3.8), we conclude that

$$\sqrt{3} < t(a) < \frac{k}{1-a^2}.$$

□

PROPOSITION 3.3.

$$\lim_{a\to 0} t(a) = \sqrt{3}.$$

PROOF. Lemma 3.2 implies that

$$\sqrt{3} \le \lim_{a\to 0} t(a) \le k$$

for every $k > \sqrt{3}$. Thus $\lim_{a\to 0} t(a) = \sqrt{3}$. □

Replacing $t$ by $t/2$ yields the desired result: the conjugate locus is indeed a continuous surface crossing the $e_1e_2$-plane.

**3.2. Signature $--+$.** For the other example we take the nontrivial inner products to be

$$1 = \langle z, z\rangle, \quad -1 = \langle e_1, e_1\rangle, \quad -1 = \langle e_2, e_2\rangle$$

whence the nontrivial adjoint maps are

$$\mathrm{ad}^{\dagger}_{e_1} z = -e_2\,, \quad \mathrm{ad}^{\dagger}_{e_2} z = e_1\,.$$

This time, we chose a parametrization based on rotation about the $z$-axis. In the $e_1e_2$-plane, where $z_0 = 0$ and $x_0 \neq 0$, we have

$$J = \begin{bmatrix} 0 & 1 \\ -1 & 0 \end{bmatrix} \quad \text{so} \quad e^{tJ} = \begin{bmatrix} \cos t & \sin t \\ -\sin t & \cos t \end{bmatrix}$$

and for the integrated geodesic equations we obtain

$$\begin{aligned} \gamma_1(t) &= \cos\theta\sin t + \sin\theta(1-\cos t) \\ \gamma_2(t) &= \cos\theta(\cos t - 1) + \sin\theta\sin t \\ \gamma_3(t) &= 0 \end{aligned} \tag{3.9}$$

in the same exponential coordinates as before. Since $J^2$ has the single eigenvalue $-1$, conjugate points are again found at $t = \pm 2\sqrt{3}$. Note that all these geodesics are spacelike.

Figures 4 and 5 show some of these geodesics, with the conjugate locus in red (gray) in the first. As seems apparent in Figure 5, these geodesics in

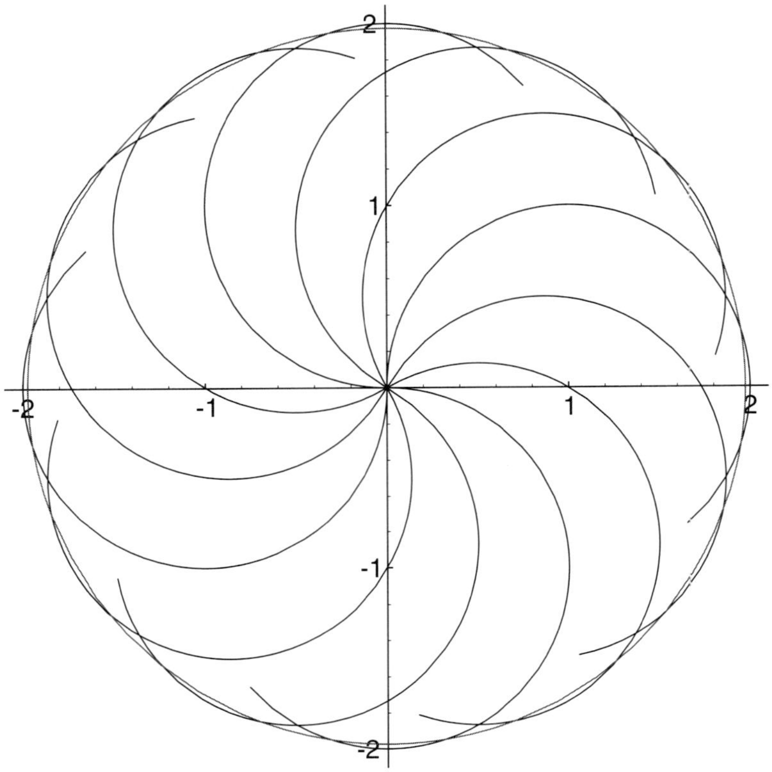


FIGURE 4. Geodesics on the $e_1e_2$-plane in $H_3$ with signature $--+$, $0 \le t \le 4$, and the conjugate locus in red (gray).

the $e_1e_2$-plane are indeed closed, as is easily verified from equations (3.9).

Next, consider $z_0 = az$ and $x_0 = b(\cos\theta e_1 + \sin\theta e_2)$ where $a \neq 0$ (but we do allow $b = 0$). Now we have all three types of geodesics.

$$\begin{array}{rllll} \text{timelike:} & a^2 - b^2 = 1 & \text{so} & b = \pm\sqrt{a^2 - 1} & \text{with} \quad |a| \ge 1 \\ \text{spacelike:} & a^2 - b^2 = -1 & \text{so} & b = \pm\sqrt{a^2 + 1} & \text{with} \quad a \in \mathbb{R} \\ \text{null:} & a^2 - b^2 = 0 & \text{so} & b = \pm a & \text{with} \quad a \in \mathbb{R} \end{array} \tag{3.10}$$

In any case we have

$$J = \begin{bmatrix} 0 & a \\ -a & 0 \end{bmatrix} \quad \text{so} \quad e^{tJ} = \begin{bmatrix} \cos at & \sin at \\ -\sin at & \cos at \end{bmatrix}$$

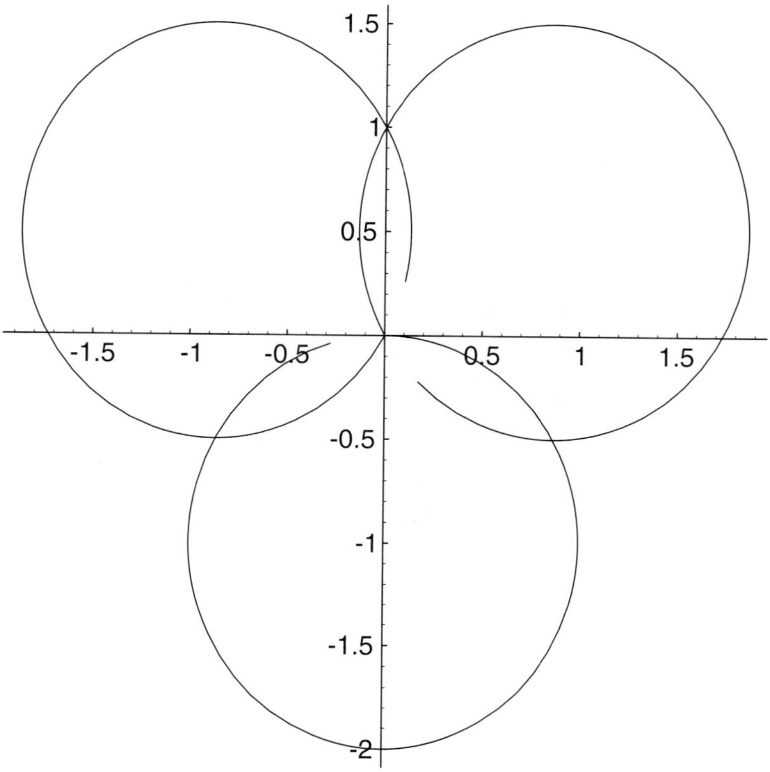


FIGURE 5. Geodesics on the $e_1e_2$-plane in $H_3$ with signature $--+$ and $0 \leq t \leq 6$.

and the integrated geodesic equations follow.

$$
\begin{aligned}
\gamma_1(t) &= \frac{b}{a}\left[\cos\theta \sin at + \sin\theta(1 - \cos at)\right] \\
\gamma_2(t) &= \frac{b}{a}\left[\cos\theta(\cos at - 1) + \sin\theta \sin at\right] \\
\gamma_3(t) &= a\,t
\end{aligned}
\tag{3.11}
$$

Figure 6 shows some of each type of helical geodesic. Lines emanating outward from the origin are geodesics and lines transverse to them are curves of constant geodesic parameter $t$. It is readily apparent that all these geodesics should have conjugate points.

Since $J^2$ has the single eigenvalue $-a^2$, all of Part 3 of Corollary 2.4 applies. For any $b$, we have conjugate points at

$$t \in \frac{2\pi}{a}\mathbb{Z}^*$$

and for $b \neq 0$ also at solutions $t$ of

$$(1 \mp a^2)\frac{at}{2}\cot\frac{at}{2} = 1$$

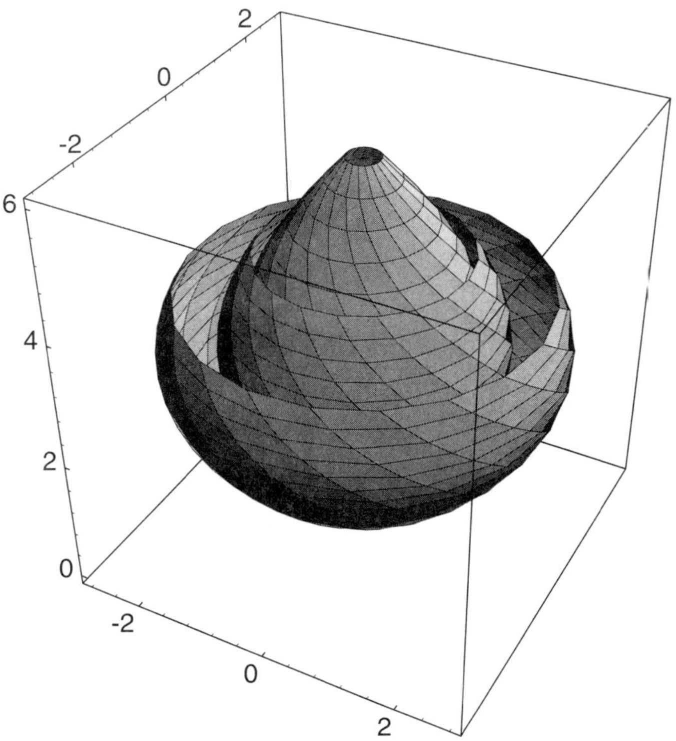


FIGURE 6. Geodesic surfaces in $H_3$ with signature $--+$, spacelike, null, and timelike, respectively, from outside in.

along timelike and spacelike geodesics, respectively, and of

$$\frac{at}{2}\cot\frac{at}{2} = 0$$

along null geodesics.

Substituting in (3.11), we see that the three types of geodesic surfaces appear as "beads on a string" along the $z$-axis. All the "inter-bead" nodes occur at the conjugate points $[0, 0, 2\pi\mathbb{Z}^*]$ where all three types of geodesic cross the $z$-axis at the same place, although not generally at the same "time" (more precisely, geodesic parameter) $t$.

Figure 7 shows the first two timelike "beads" up the $z$-axis for $a = 2$. The first two conjugate point nodes at $z = 2\pi$ and $z = 4\pi$ are clearly visible. The heavier black circles around the middles of the two "beads" show the second kind of timelike conjugate points.

The second kind of conjugate points are somewhat peculiar. They occur in infinitely many discrete levels parallel to the $e_1e_2$-plane, both above and below it. Figure 8 shows part of a typical level. The timelike conjugate points go off to infinity "up the spike" as $a \downarrow 1$. (Below the plane they go "down the spike" as $a \uparrow -1$.)

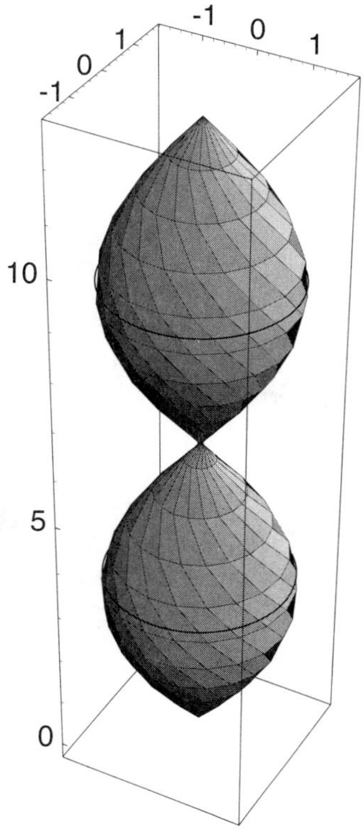


FIGURE 7. Timelike geodesic surface in $H_3$ with signature $--+$, $a = 2$, $0 \leq t \leq 2\pi$, the first two conjugate point nodes, and the first two levels of the second kind of timelike conjugate points as the heavier black "waist band" circles.

Figure 9 shows how one level of the spacelike conjugate locus is asymptotic to a plane parallel to the $e_1e_2$-plane, going off to infinity as $a \downarrow 0$. The vertical range is almost impossible to read; it extends from 9.05 through 9.2 in steps of 0.05.

Figures 10 and 11 show the level curves of the functions locating the second kind of timelike, respectively spacelike, conjugate points; to wit, of

$$f(a,t) = (1 \mp a^2)\frac{at}{2}\cot\frac{at}{2} = 1\,,$$

respectively. The vertical line $a = 1$ is an asymptote for all the level curves shown in the first figure. In the second one, however, the vertical line $a = 0$ is *not* an asymptote for the lowest level curve shown, although it is one for the rest. One may readily verify these facts by computing the appropriate limits. Pictures of the entire $at$-plane may be obtained in each case by symmetry. That the conjugate locus continues smoothly through the $e_1e_2$-plane is readily apparent in Figure 12. A proof that this happens in general is found in [**JP**].

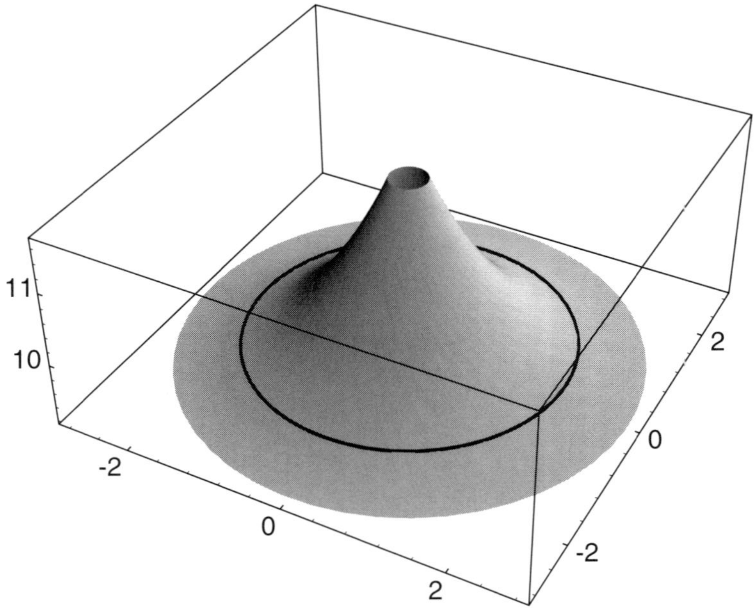


FIGURE 8. One level of conjugate loci in $H_3$ with signature $--+$, spacelike, null (black circle), and timelike, respectively, from outside in.

According to [**CP, CP1**], there is only one other indefinite geometry on $H_3$, that with a degenerate center. But it is flat so there are no conjugate points at all there.

## References


[CP] L. A. Cordero and P. E. Parker, *Left-invariant Lorentzian metrics on 3-dimensional Lie groups*, Rend. Mat. Appl. **17** (1997), 129–155.

[CP1] L. A. Cordero and P. E. Parker, *Pseudoriemannian 2-step nilpotent Lie groups*, Santiago-Wichita preprint DGS/CP4, 1999. arXiv: math.DG/9905188

[JP] C. Jang and P. E. Parker, Conjugate loci of pseudoriemannian 2-step nilpotent Lie groups with nondegenerate center, Wichita-Ulsan preprint DGS/JP1, 2003. arXiv: math.DG/0302029

[M] A. Marenich, *Computational geometry in Heisenberg group Heis*$^3$, preprint 2002. arXiv: math.DG/0204105

[O] B. O'Neill, *Semi-Riemannian Geometry.* New York: Academic Press, 1983.

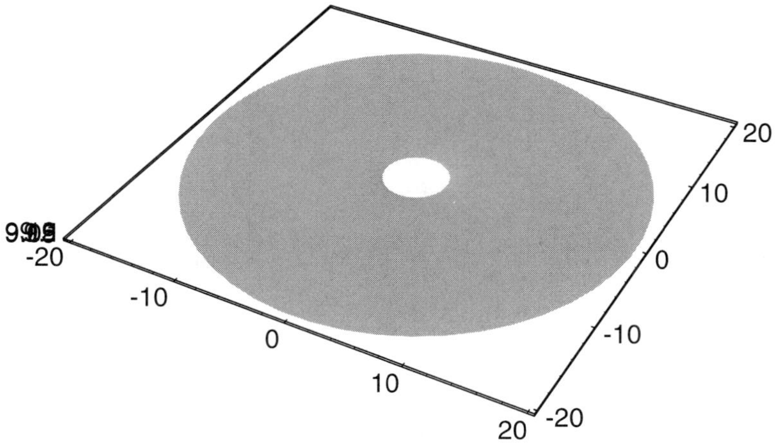


FIGURE 9. Spacelike conjugate locus in $H_3$ with signature $--+$, $0.1 \le a \le 1$, where $a$ increases from outside in. The vertical range is almost impossible to read; it extends from 9.05 through 9.2 in steps of 0.05.

MATHEMATICS DEPARTMENT, WICHITA STATE UNIVERSITY, WICHITA KS 67260-0033, USA

*Current address*: School of Mathematics and Applied Physics, College of Natural Sciences, University of Ulsan, Ulsan 680-749, Republic of Korea

*E-mail address*: crjang@mail.ulsan.ac.kr

MATHEMATICS DEPARTMENT, WICHITA STATE UNIVERSITY, WICHITA KS 67260-0033, USA

*E-mail address*: phil@math.twsu.edu

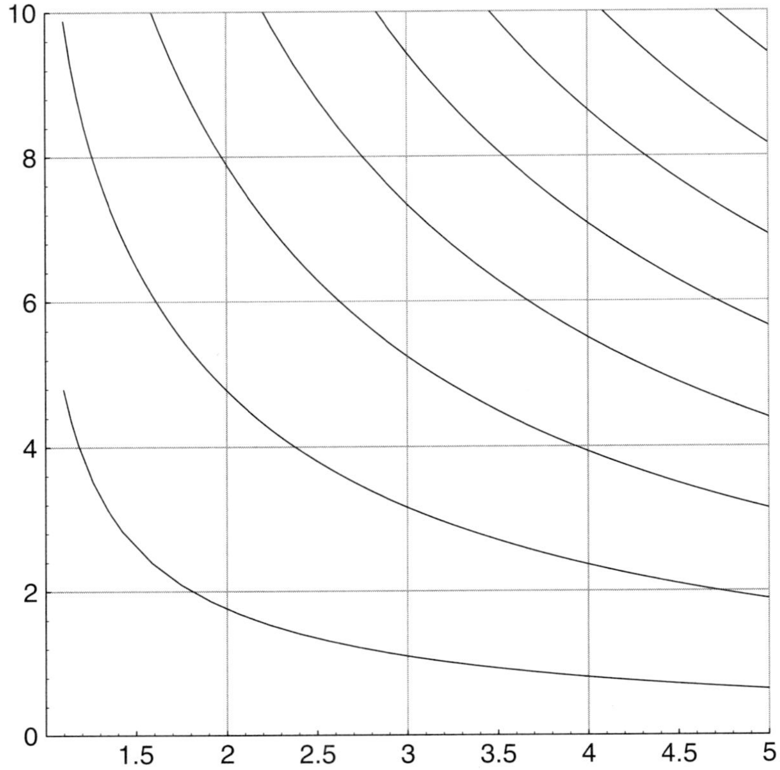


FIGURE 10. Level curves $c = 1$ for the function locating the second kind of timelike conjugate points.

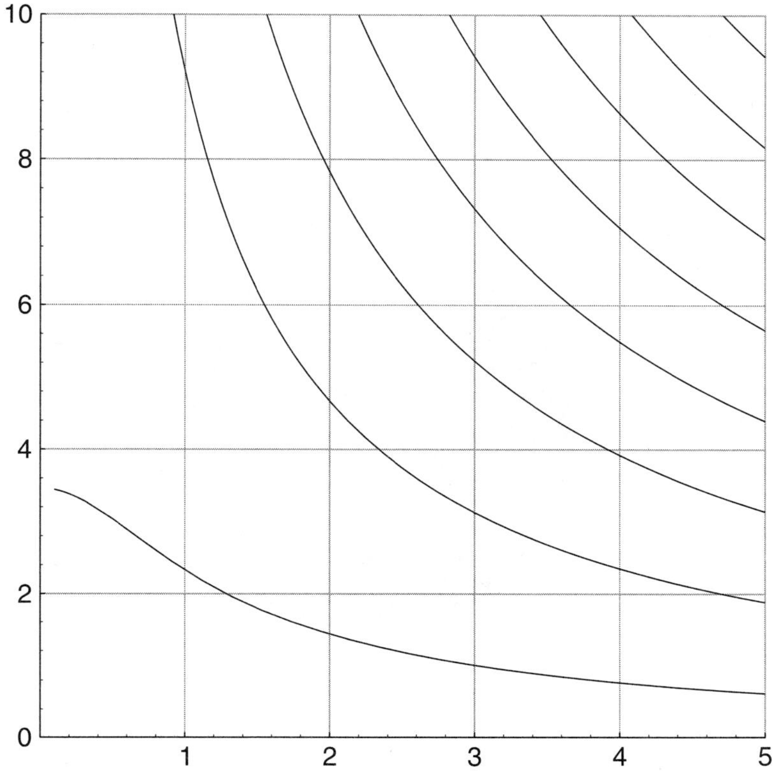


FIGURE 11. Level curves $c = 1$ for the function locating the second kind of spacelike conjugate points.

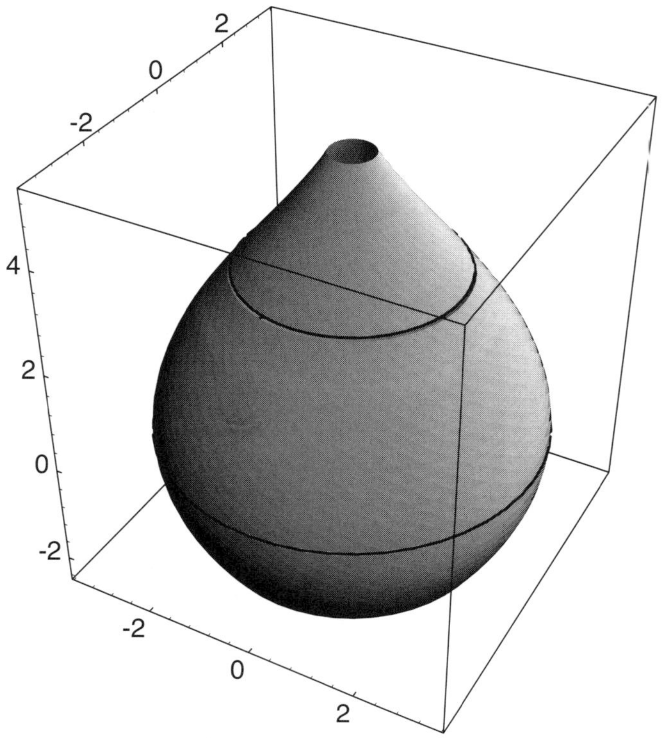


FIGURE 12. Timelike (at the top) and spacelike (the two lower sections) conjugate points spanning the $e_1e_2$-plane. The upper black circle is a null locus, as in Figure 8, and the lower one is the circle from Figure 4.

Contemporary Mathematics
Volume **337**, 2003

# Group invariant classification of orthogonal coordinate webs

Raymond G. McLenaghan, Roman G. Smirnov, and Dennis The

ABSTRACT. We establish a connection between classical invariant theory and the study of Killing tensors in Riemannian geometry. New invariants of the vector spaces of Killing tensors of valence two defined in the Euclidean plane are brought to light and employed to classify orthogonal coordinate webs.

## 1. Introduction

The main purpose of this survey paper is to present some recent results by the authors that incorporate the basic ideas of classical invariant theory into the study of Killing tensors in Riemannian geometry [**MST1, MST2, MST3, MST4, MST5**].

The theory flourished in the 19th century and attained its fame through the celebrated works of David Hilbert [**Hil**]. It is presently undergoing a renaissance due to many new applications in various areas of pure and applied mathematics as well as a recent development of modern computational techniques effectively used in finding new invariants that were not available at the time the theory was conceived (see Olver [**Olv**] for more details and the relevant references therein). As is well known, classical invariant theory involves the study of intrinsic properties of homogeneous polynomials under the action of the general linear group. In the modern mathematical language the main problem of the classical invariant theory can be formulated as follows:

PROBLEM 1.1. Determine the linear action of a group $G$ on a $K$-vector space $V$. Then in the ring of polynomial functions $K[V]$ describe the subring $K[V]^G$ of all polynomial functions on $V$ which are invariant under the action of $G$.

The determination of the invariants for a particular vector space of homogeneous polynomials allows one to solve the intimately related problems of *equivalence* and *finding canonical forms.*

---

1991 *Mathematics Subject Classification.* Primary 53A55, 53A60; Secondary 70G45.

*Key words and phrases.* Theory of invariants, Riemannian geometry, Killing tensors of valence two, problems of equivalence and canonical forms, orthogonal coordinate webs in the Euclidean plane.

The first author was supported in part by NSERC.

The second author was supported in part by the Alexander von Humboldt Foundation.

The third author was supported in part by NSERC.

All of these issues properly re-formulated can be addressed within the framework of the study of vector spaces of Killing tensors defined on Riemannian manifolds of constant curvature under the action of the isometry group. The key idea rests on a *new understanding* of Killing tensors. We view them as elements of a vector space, or algebraic quantities defined by the parameters that determine the dimension of the vector space. Although this idea originally came to light in an application involving Killing tensors of valence two to the Hamilton-Jacobi theory of separation of variables [**MST1**], our experience with the vector spaces of Killing tensors since then can be best described in the words of Sylvester (quoted in [**Olv**]):

> "As all roads lead to Rome so I find in my own case at least that all algebraic inquiries, sooner or later, end at the Capitol of Modern Algebra over whose shining portal is inscribed the Theory of Invariants."

We illustrate our approach by considering from this perspective the vector space of Killing tensors of valence two defined in the Euclidean plane, determine the invariants and solve the corresponding problems of *equivalence* and *canonical forms.* This leads to an invariant classification of orthogonal coordinate webs defined by the eigenvalues and eigenvectors of the Killing tensors.

## 2. Classical invariant theory: A motivating example

We begin by recalling some basic ideas of the classical invariant theory that we extend in the sequel to the study of Killing tensors. Consider a general quadratic form over $\mathbb{R}$:

$$f(x, y) = a_1x^2 + 2a_2xy + a_3y^2. \tag{2.1}$$

The formula (2.1) represents the space of homogeneous polynomials of degree $n = 2$ in $m = 2$ variables. The parameters $a_1, a_2, a_3 \in \mathbb{R}$ determine the dimension $d = 3$ of the space, which can alternatively be computed by making use of the formula

$$d = \binom{n+m-1}{n}\Big|_{n=m=2} = 3.$$

Problem 1.1 for this space requires determination of the sets of *invariants* (polynomial functions of $a_1, a_2$ and $a_3$) and *covariants* (polynomial functions of $a_1, a_2, a_3, x$ and $y$) that remain fixed under the action of the general linear group $GL_2(\mathbb{R})$, which acts on the space by linear substitutions. The key requirement is that the form of (2.1) remain unchanged under the transformation rules induced by $GL_2(\mathbb{R})$. Let

$$T = \begin{bmatrix} \alpha_{11} & \alpha_{12} \\ \alpha_{21} & \alpha_{22} \end{bmatrix} \in GL_2(\mathbb{R}),$$

then

$$\begin{array}{l} x = \alpha_{11}\overline{x} + \alpha_{12}\overline{y}, \\ y = \alpha_{21}\overline{x} + \alpha_{22}\overline{y}, \end{array} \quad , \quad \alpha_{11}\alpha_{22} - \alpha_{12}\alpha_{21} \neq 0$$

and (2.1) yield

$$\overline{f}(\overline{x}, \overline{y}) = \overline{a}_1\overline{x}^2 + 2\overline{a}_2\overline{x}\,\overline{y} + \overline{a}_3\overline{y}^2,$$

where

$$\begin{aligned} \overline{a}_1 = &\ a_1\alpha_{11}^2 + 2a_2\alpha_{11}\alpha_{21} + a_3\alpha_{21}^2, \\ \overline{a}_2 = &\ a_1\alpha_{11}\alpha_{12} + a_2(\alpha_{11}a_{22} + a_{12}\alpha_{21}) \\ &+ a_3\alpha_{21}\alpha_{22}, \\ \overline{a}_3 = &\ a_1\alpha_{12}^2 + 2a_2\alpha_{12}\alpha_{22} + a_3\alpha_{22}^2. \end{aligned}$$

We immediately note that (2.1) is itself a *covariant*, while the quantity $\Delta_f = a_1a_3 - a_2^2$ is an *invariant* up to the factor $\det(T)^2$. Indeed, we have in this case

$$\overline{\Delta}_f = \det(T)^2(a_1a_3 - a_2^2) = det(T)^2\Delta_f.$$

It is well known that the one-element set $\{\Delta_f\}$ constitute the *Hilbert basis* for the invariants of the space determined by (2.1), that is every polynomial invariant of (2.1) is of the form $P(\Delta_f)$, where $P$ denotes a polynomial function over $\mathbb{R}$.

The invariant $\Delta_f$ can be used to solve the problem of equivalence (i.e., to determine when a polynomial of the type (2.1) can be transformed via a transformation $T \in GL_2(\mathbb{R})$ into another homogeneous polynomial belonging to the vector space defined by (2.1)) and the intimately related problem of finding canonical forms (i.e., finding a particular simple form of (2.1) within each of the equivalence classes). Noting that the sign of (2.1) is also an invariant, we sum up the results in Table 1 (see [**Olv**] for more details).

TABLE 1. Equivalent classes and canonical forms of the vector space of homogeneous polynomials in two variables of order two

| Equivalent classes | Canonical forms | Invariants |
|---|---|---|
| $I$ | $x^2 + y^2$ | $\Delta_f > 0, f(x,y) \geq 0$ |
| $II$ | $-x^2 - y^2$ | $\Delta_f > 0, f(x,y) \leq 0$ |
| $III$ | $xy$ (or $x^2 - y^2$) | $\Delta_f < 0$ |
| $IV$ | $x^2$ | $\Delta_f = 0, f(x,y) \geq 0$ |
| $V$ | $-x^2$ | $\Delta_f = 0, f(x,y) \leq 0$ |
| $VI$ | $0$ | $f(x,y) = 0$ |

The invariant $\Delta_f$ may also be used to describe the underlying geometry of the vector space given by (2.1). Indeed, we can classify (centrally symmetric) conic sections in $\mathbb{R}^2$ given by

$$f(x,y) = a_1x^2 + 2a_2xy + a_3y^2 = \text{const} \tag{2.2}$$

as follows

(a) if $\Delta_f$, $a_1$ and $f(x,y)$ are of the same sign, then (2.2) defines an ellipse,
(b) if $\Delta_f a_1 < 0$, then (2.2) defines a hyperbola,
(c) if $\Delta_f = 0$, $a_1 \neq 0$ then (2.2) defines a pair of straight lines.

Our new (invariant) approach to the study of Killing tensors can be viewed as a natural analogue of the classical invariant theory whose basic ideas have been illustrated above. This analogy is elaborated in the next section.

## 3. Vector spaces of Killing tensors

Let $(M, \mathbf{g})$ be a Riemannian manifold of constant curvature, where $\mathbf{g}$ denotes the metric tensor.

DEFINITION 3.1. A *Killing tensor* $\mathbf{K}^p$ *of valence* $p$ *defined in* $(M, \mathbf{g})$ is a symmetric $(p, 0)$ tensor satisfying the Killing tensor equation:

$$[\mathbf{K}^p, \mathbf{g}] = 0, \tag{3.1}$$

where [ , ] denotes the Schouten bracket [**Sch**]. When $p = 1$, $\mathbf{K}^1$ is said to be a *Killing vector (infinitesimal isometry)* and the equation (3.1) reads:

$$\mathcal{L}_{\mathbf{K}^1}\mathbf{g} = 0, \tag{3.2}$$

where $\mathcal{L}$ denotes the Lie derivative operator.

In this case the dimension of the Lie algebra of Killing vectors $i(M)$ is $d = \dim i(M) = \frac{1}{2}n(n+1)$, where $n = \dim M$.

It follows immediately from (3.1) that the Killing tensors of the same valence $p$ constitute a vector space $\mathcal{K}^p(M)$. Moreover, the following properties are valid:

$$[\ ,\ ]\colon\ \mathcal{K}^p(M) \oplus \mathcal{K}^q(M) \to \mathcal{K}^{p+q-1}(M), \tag{3.3}$$

$$[\mathbf{K}^p, \mathbf{K}^q] = -[\mathbf{K}^q, \mathbf{K}^p] \quad \text{(skew-symmetry)}, \tag{3.4}$$

$$[[\mathbf{K}^p, \mathbf{K}^q], \mathbf{K}^r] + \text{(cycle)} = 0 \quad \text{(Jacobi identity)}. \tag{3.5}$$

Thus, we have a graded Lie algebra of Killing tensors defined on $(M, \mathbf{g})$ with respect to the Schouten bracket [ , ]:

$$\mathcal{K}_{alg} = \mathcal{K}^0(M) \oplus \mathcal{K}^1(M) \oplus \mathcal{K}^2(M) \oplus \cdots \oplus \mathcal{K}^p(M) \oplus \cdots, \tag{3.6}$$

where $\mathcal{K}^0(M) = \mathbb{R}$, $\mathcal{K}^1(M) = i(M)$ and $p = 0, 1, 2, \ldots$, denotes the valence of the Killing tensors belonging to the corresponding space $\mathcal{K}^p(M)$. There are two fundamental *structural properties* (referred throughout this paper as SPI and SPII respectively) that determine invariant properties of Killing tensors defined on Riemannian manifolds of constant curvature and which are very useful in applications.

SPI A Killing tensor $\mathbf{K}^p$ defined on $(M, \mathbf{g})$ is a *sum of symmetrized tensor products* of Killing vectors, or infinitesimal generators of the Lie algebra of the corresponding isometry group.

SPII The set of all Killing tensors of valence $p \geq 1$ defined on $(M, \mathbf{g})$, where $\dim(M) = n$ is a *vector space of dimension* $d$ determined by the *Delong-Takeuchi-Thompson (DTT) formula* [**De, Ta, Th**]:

$$d = \frac{1}{n}\binom{n+p}{p+1}\binom{n+p-1}{p}, \quad p \geq 1. \tag{3.7}$$

Accordingly, a Killing tensor of valence $p$ is an *algebraic object* determined by its $d$ parameters, or an element of the corresponding vector space of Killing tensors.

We remark that SPI has been widely used by many authors to study Killing tensors [**Ta, Th**] and in applications. We propose to employ SPII for these purposes. Indeed, the formula (3.7) makes it possible to view the vector spaces of Killing tensors as analogs of vector spaces of homogeneous polynomials in the classical invariant theory. This idea is illustrated in the following example.

REMARK 3.1. Here and below $\mathbf{K}$ denotes the elements of the vector space $\mathcal{K}^2(\mathbb{R}^2)$.

EXAMPLE 3.2. Let $\mathbf{K} \in \mathcal{K}^2(\mathbb{R}^2)$ be the general solution to the Killing tensor equation (3.1) when $p = 2$ and $M = \mathbb{R}^2$ (Euclidean plane). In view of (3.7) $d = \dim\mathcal{K}^2(\mathbb{R}^2) = 6$. According to SPI, $\mathbf{K}$ is given by

$$\mathbf{K} = A\mathbf{X} \odot \mathbf{X} + B\mathbf{X} \odot \mathbf{Y} + C\mathbf{Y} \odot \mathbf{Y} + \alpha\mathbf{X} \odot \mathbf{R} + \beta\mathbf{Y} \odot \mathbf{R} + \gamma\mathbf{R} \odot \mathbf{R}, \tag{3.8}$$

where $\odot$ denotes the symmetric tensor product, **X**, **Y** and **R** are the generators of the corresponding Lie algebra of Killing vectors $i(\mathbb{R}^2)$ and $A, B, C, \alpha, \beta, \gamma \in \mathbb{R}$ are arbitrary constants. Note that the vector fields **X**, **Y** and **R** given in Cartesian coordinates by

$$\mathbf{X} = \frac{\partial}{\partial x}, \quad \mathbf{Y} = \frac{\partial}{\partial y}, \quad \mathbf{R} = x\frac{\partial}{\partial y} - y\frac{\partial}{\partial x}, \tag{3.9}$$

correspond to the infinitesimal translations along $x$, $y$ and a rotation respectively. They satisfy the following commutator relations

$$[\mathbf{X}, \mathbf{Y}] = 0, \quad [\mathbf{X}, \mathbf{R}] = -\mathbf{Y}, \quad [\mathbf{Y}, \mathbf{R}] = \mathbf{X}. \tag{3.10}$$

Alternatively, SPII can be used to interpret $\mathcal{K}^2(\mathbb{R}^2)$ as a six-dimensional vector space, each element of which is given with respect to Cartesian coordinates by the following formula.

$$\mathbf{K} = \begin{pmatrix} A + 2\alpha y + \gamma y^2 & C - \alpha x - \beta y - \gamma xy \\ C - \alpha x - \beta y - \gamma xy & B + 2\beta x + \gamma x^2 \end{pmatrix}. \tag{3.11}$$

The formula (3.11) represents the general solution of the Killing tensor equation (3.1) with respect to Cartesian coordinates with the six parameters $A, B, C, \alpha, \beta, \gamma \in \mathbb{R}$ being the constants of integration. Thus, (3.11) can be looked upon as a natural analogue of the general formula (2.1) for the vector space of quadratic forms over reals. Accordingly, we look at each element $\mathbf{K}_0 \in \mathcal{K}^2(\mathbb{R}^2)$ as an algebraic object *determined by its six parameters* $A_0, B_0, C_0, \alpha_0, \beta_0, \gamma_0$: $\mathbf{K}_0 = \mathbf{K}_0(A_0, B_0, C_0, \alpha_0, \beta_0, \gamma_0)$.

In what follows, we make use of this observation to study the vector spaces Killing tensors defined in Riemannian spaces of constant curvature within the framework of the classical invariant theory.

## 4. Isometry group invariants of Killing tensors

To illustrate our approach we focus our attention on the vector space $\mathcal{K}^2(\mathbb{R}^2)$ described in the previous section. We first observe that the action of the group $GL_2(\mathbb{R})$ does not preserve the general form of (3.11). This fact follows from the application of the appropriate tensor transformation law and the structure of the components of **K**. We thus turn our attention to the isometry group $I(\mathbb{R}^2)$, which is a Lie subgroup of the diffeomorphisms of $\mathbb{R}^2$. Indeed, its Lie algebra $i(\mathbb{R}^2)$ is generated by the basic Killing vectors **X**, **Y** and **R** given by (3.9) and so, in view of the properties (3.5), $I(\mathbb{R}^2)$ is an automorphism of $\mathcal{K}^2(\mathbb{R}^2)$. In the spirit of classical invariant theory we consider therefore the action of $I(\mathbb{R}^2)$ in the vector space $\mathcal{K}^2(\mathbb{R}^2)$, which induces the corresponding action in the space $\Sigma$ spanned by the parameters $A, B, C, \alpha, \beta, \gamma$ of (3.11). Hence we are led to formulate the following analogue of Problem 1.1:

PROBLEM 4.1. Determine the action induced by the isometry group $I(\mathbb{R}^2)$ : $\mathbb{R}^2 \to \mathbb{R}^2$ in the space $\Sigma$ spanned by the parameters $A, B, C, \alpha, \beta, \gamma$ of $\mathcal{K}^2(\mathbb{R}^2)$. Then in the space of functions defined in $\Sigma$ describe the subspace of all functions that remain invariant under the action induced by $I(\mathbb{R}^2)$.

To solve the first part of the problem, namely to determine the induced action of $I(\mathbb{R}^2)$ in $\Sigma$, we note first that the dimension of $\Sigma$ can be reduced by one. Indeed, we remark that the general representation (3.11) of the vector space $\mathcal{K}^2(\mathbb{R}^2)$ contains the metric $\mathbf{g}$ as one of its basic elements. Hence, we can bring the formula (3.11) to the form

$$\mathbf{K} = \mathbf{Q} + B\mathbf{g}, \quad B \in \mathbb{R}, \tag{4.1}$$

where

$$\mathbf{Q} = \begin{pmatrix} A' + 2\alpha y + \gamma y^2 & C - \alpha x - \beta y - \gamma xy \\ C - \alpha x - \beta y - \gamma xy & 2\beta x + \gamma x^2 \end{pmatrix}, A' = A - B. \tag{4.2}$$

We note that $\mathbf{Q}$ represents the five-dimensional vector space $\mathcal{Q}$ of *non-trivial* Killing tensors, none of which is a multiple of $\mathbf{g}$. Accordingly, we consider the reduced parameter space $\Sigma' \simeq \mathbb{R}^5$ spanned by $A', C, \alpha, \beta, \gamma$, where $A' = A - B$. In view of (4.1), the action of the isometry group $I(\mathbb{R}^2)$ in $\Sigma$ is equivalent to its action in $\Sigma'$. Therefore, we consider each element $Q_0$ of the vector space $\mathcal{Q}$ to be a function of its five parameters $Q_0 = Q_0(A'_0, C_0, \alpha_0, \beta_0, \gamma_0)$.

Let $\mathbf{K} \in \mathcal{K}^2(\mathbb{R}^2)$ be the general Killing tensor (3.11) determined by arbitrary parameters $A, B, C, \alpha, \beta, \gamma$. Denote by Diff$\mathcal{K}^2(\mathbb{R}^2)$ all Killing tensors, whose parameters are smooth functions of $A, B, C, \alpha, \beta, \gamma$ that define the general $\mathbf{K}$ given by (3.11). In essense, Diff$\mathcal{K}^2(\mathbb{R}^2)$ is generated by Diff$\Sigma$ - the group of all diffeomorphisms of $\Sigma \simeq \mathbb{R}^6$. To determine the action of $I(\mathbb{R}^2)$ in $\Sigma'$, we introduce the *projection map* $\pi$ : Diff$\mathcal{K}^2(\mathbb{R}^2) \to \mathcal{X}(\Sigma')$ defined by

$$\pi(\mathbf{K}_0) := (A_0 - B_0)\frac{\partial}{\partial(A-B)} + C_0\frac{\partial}{\partial C} + \alpha_0\frac{\partial}{\partial \alpha} + \beta_0\frac{\partial}{\partial \beta} + \gamma_0\frac{\partial}{\partial \gamma}, \tag{4.3}$$

where $\mathcal{X}(\Sigma')$ is the space of all vector fields in $\Sigma'$, $\mathbf{K}_0 \in$ Diff$\mathcal{K}^2(\mathbb{R}^2)$ is determined by six smooth functions $\mathbb{R}^6 \to \mathbb{R}$: $A_0 = A_0(A, B, C, \alpha, \beta, \gamma), \ldots, \gamma_0 = \gamma_0(A, B, C, \alpha, \beta, \gamma)$. Next, we make use of the *composition map* $\pi \circ \mathcal{L}$, where $\mathcal{L}$ is the Lie derivative operator, as follows. Define the vector fields $\mathbf{V}_i \in \mathcal{X}(\Sigma'), i = 1, 2, 3$ by

$$\mathbf{V_1} = \pi(\mathcal{L}_\mathbf{X}(\mathbf{K})), \quad \mathbf{V_2} = \pi(\mathcal{L}_\mathbf{Y}(\mathbf{K})), \quad \mathbf{V_3} = \pi(\mathcal{L}_\mathbf{R}(\mathbf{K})),$$

where $\mathbf{K}$ is the general Killing tensor given by (3.11). Note that the tensors $\mathcal{L}_\mathbf{X}(\mathbf{K})$, $\mathcal{L}_\mathbf{Y}(\mathbf{K})$ and $\mathcal{L}_\mathbf{X}(\mathbf{K}) \in$ Diff$\mathcal{K}^2(\mathbb{R}^2)$. Therefore the composition map $\pi \circ \mathcal{L}$ is well-defined. The vector fields $\mathbf{V}_i \in \mathcal{X}(\Sigma'), i = 1, 2, 3$ are found to be

$$\begin{aligned} \mathbf{V_1} &= -2\beta\frac{\partial}{\partial A'} - \alpha\frac{\partial}{\partial C} + \gamma\frac{\partial}{\partial \beta}, \\ \mathbf{V_2} &= 2\alpha\frac{\partial}{\partial A'} - \beta\frac{\partial}{\partial C} + \gamma\frac{\partial}{\partial \alpha}, \\ \mathbf{V_3} &= -4C\frac{\partial}{\partial A'} + A'\frac{\partial}{\partial C} + \beta\frac{\partial}{\partial \alpha} - \alpha\frac{\partial}{\partial \beta}. \end{aligned} \tag{4.4}$$

Our next observation is that the vector fields $\mathbf{V}_i \in \mathcal{X}(\Sigma'), i = 1, 2, 3$ are closed under the Lie commutator. Moreover, they satisfy the same commutator relations as the generators of the Lie algebra $i(\mathbb{R}^2)$ $\mathbf{X}$, $\mathbf{Y}$ and $\mathbf{R}$ in $\mathcal{X}(\mathbb{R}^2)$ (compare with (3.10)):

$$[\mathbf{V_1}, \mathbf{V_2}] = 0, \quad [\mathbf{V_1}, \mathbf{V_3}] = -\mathbf{V_2}, \quad [\mathbf{V_2}, \mathbf{V_3}] = \mathbf{V_1}. \tag{4.5}$$

Therefore $\mathbf{V}_i \in \mathcal{X}(\Sigma'), i = 1, 2, 3$ form a basis of a Lie algebra $i_{\Sigma'}(\mathbb{R}^2)$ in $\mathcal{X}(\Sigma')$ and the following result holds true.

PROPOSITION 4.1. *The vector space $i_{\Sigma'}(\mathbb{R}^2) \subset \mathcal{X}(\Sigma')$ is a Lie subalgebra of $\mathcal{X}(\Sigma')$ isomorphic to the Lie algebra of Killing vectors $i(\mathbb{R}^2)$.*

This proposition which we have proved directly from (4.4) is a particular case of a more general theorem [**MST4**] that can be proven without the explicit knowledge of the vector fields $\mathbf{V}_i \in \mathcal{X}(\Sigma'), i = 1, 2, 3$. However, we need them in order to compute the invariants of the isometry group of the vector space $\mathcal{K}^2(\mathbb{R}^2)$.

DEFINITION 4.1. A smooth function $F : \Sigma' \to \mathbb{R}$ is said to be an $I(\mathbb{R}^2)$-*invariant of* $\mathcal{Q} \subset \mathcal{K}^2(\mathbb{R}^2)$ iff

$$\mathbf{V}_i(F) = 0, \quad i = 1, 2, 3, \tag{4.6}$$

where $\mathbf{V}_i \in \mathcal{X}(\mathbb{R}^2)$, $i = 1, 2, 3$ are the generators of the Lie algebra $i_{\Sigma'}(\mathbb{R}^2)$ isomorphic to the Lie algebra $i(\mathbb{R}^2)$ of Killing vectors defined on $\mathbb{R}^2$.

In view of Definition 4.1 and Proposition 4.1 in order to find the $I(\mathbb{R}^2)$-invariants of $\mathcal{K}^2(\mathbb{R}^2)$, we have to solve the system of PDE's (4.6). Applying the method of characteristics, we arrive at the following result:

THEOREM 4.1. Let $\mathcal{Q} \subset \mathcal{K}^2(\mathbb{R}^2)$ be the vector subspace of non-trivial Killing tensors of valence two defined in the Euclidean plane $\mathbb{R}^2$. Then every smooth $I(\mathbb{R}^2)$-invariant of $\mathcal{Q}$ enjoys the following form:

$$F(\Delta^1_{\mathbb{R}^2}, \Delta^2_{\mathbb{R}^2}), \tag{4.7}$$

where $F : \mathbb{R}^2 \to \mathbb{R}$ is a smooth function,

$$\Delta^1_{\mathbb{R}^2} = \gamma, \quad \Delta^2_{\mathbb{R}^2} = (\alpha^2 - \beta^2 + \gamma A')^2 + 4(\gamma C + \alpha\beta)^2, \tag{4.8}$$

and $A', C, \alpha, \beta, \gamma$ are the parameters defined in (4.2).

DEFINITION 4.2. The polynomials $\Delta^i_{\mathbb{R}^2}, i = 1, 2$ defined by (4.8) are called the *fundamental* $I(\mathbb{R}^2)$*-invariants of* $\mathcal{K}^2(\mathbb{R}^2)$.

Note that the fundamental $I(\mathbb{R}^2)$-invariants are homogeneous polynomials in $A', C, \alpha, \beta, \gamma$.

Having found the fundamental $I(\mathbb{R}^2)$-invariants of the vector space of non-trivial Killing tensors defined in the Euclidean plane, we can now solve, in complete analogy with the classical invariant theory of homogeneous polynomials, the corresponding problems of equivalence and canonical forms.

DEFINITION 4.3 ([**Olv**]). Given a transformation group $G$ acting on $X$. Two points $x, y \in X$ are called *equivalent* iff they lie in the same orbit. The *equivalence problem* for $G$ is to find necessary and sufficient conditions for this to hold.

A *canonical form* of an element $x \in X$ just means a distinguished, "simple" representative $x_0 \in \mathcal{O}_x$ of the orbit containing $x$.

Indeed, we can classify invariantly the non-trivial Killing tensors of $\mathcal{K}^2(\mathbb{R}^2)$ according to whether the fundamental invariants $\Delta^i_{\mathbb{R}^2}, i = 1, 2$ vanish or not (which is an invariant property). Note $\Delta^2_{\mathbb{R}^2} \geq 0$. This leads to the following *four equivalence classes (EC)*:

$$\begin{aligned} I: &\quad \Delta^1_{\mathbb{R}^2} = 0, \quad \Delta^2_{\mathbb{R}^2} = 0, \\ II: &\quad \Delta^1_{\mathbb{R}^2} \neq 0, \quad \Delta^2_{\mathbb{R}^2} = 0, \\ III: &\quad \Delta^1_{\mathbb{R}^2} = 0, \quad \Delta^2_{\mathbb{R}^2} \neq 0, \\ IV: &\quad \Delta^1_{\mathbb{R}^2} \neq 0, \quad \Delta^2_{\mathbb{R}^2} \neq 0. \end{aligned} \tag{4.9}$$

Within each of the four classes of equivalence we choose canonical forms as follows:

$$
\begin{aligned}
I: \quad & K_{ij}^{(I)} := \begin{pmatrix} 1 & 0 \\ 0 & 0 \end{pmatrix}, \\
II: \quad & K_{ij}^{(II)} := \begin{pmatrix} y^2 & -xy \\ -xy & x^2 \end{pmatrix}, \\
III: \quad & K_{ij}^{(III)} := \begin{pmatrix} 0 & y \\ y & -2x \end{pmatrix} \text{ or } \begin{pmatrix} -2y & x \\ x & 0 \end{pmatrix}, \\
IV: \quad & K_{ij}^{(IV)} := \begin{pmatrix} k^2+y^2 & -xy \\ -xy & x^2 \end{pmatrix},
\end{aligned} \tag{4.10}
$$

where $K_{ij}^{(I)}, K_{ij}^{(II)}$, $K_{ij}^{(III)}$ and $K_{ij}^{(IV)}$ denote the components of the corresponding *canonical* Killing tensors given in Cartesian coordinates $(x, y)$. Note the (five) Killing tensors given by (4.10) span the vector space of non-trivial Killing tensors determined by the general form (4.2).

Every continuous transformation of the Lie group $I(\mathbb{R}^2) : (x, y) \to (\overline{x}, \overline{y})$ can be parametrized as follows (see (3.9)):

$$
\begin{pmatrix} \overline{x} \\ \overline{y} \end{pmatrix} = \begin{pmatrix} \cos\theta & -\sin\theta \\ \sin\theta & \cos\theta \end{pmatrix} \begin{pmatrix} x \\ y \end{pmatrix} + \begin{pmatrix} a \\ b \end{pmatrix} \tag{4.11}
$$

for some fixed parameters $\theta, a, b \in \mathbb{R}$. In turn, the coordinate transformation (4.11) induces the following transformation of the five parameters that appear in (4.2):

$$
\begin{aligned}
\overline{A}' &= A'(\cos^2\theta - \sin^2\theta) - 4C\sin\theta\cos\theta + 2(a\beta - b\alpha)\cos\theta \\
&\quad -2(b\beta + a\alpha)\sin\theta + \gamma(b^2 - a^2), \\
\overline{C} &= A'\sin\theta\cos\theta + C(\cos^2\theta - \sin^2\theta) + (a\alpha + b\beta)\cos\theta \\
&\quad +(a\beta - b\alpha)\sin\theta - \gamma ab, \\
\overline{\alpha} &= \alpha\cos\theta + \beta\sin\theta - \gamma b, \\
\overline{\beta} &= \beta\cos\theta - \alpha\sin\theta - \gamma a, \\
\overline{\gamma} &= \gamma,
\end{aligned} \tag{4.12}
$$

where the quantities with a bar refer to the components of the general non-trivial Killing tensor (4.2) in the $(\overline{x}, \overline{y})$ coordinates system. Note the formulas (4.12) reveal immediately that the parameter $\gamma$ is an $I(\mathbb{R}^2)$-invariant.

Employing the formula (4.11) and tensor transformation laws, we find the components of the canonical Killing tensors (4.10) with respect to $(\overline{x}, \overline{y})$ (we selected only one canonical Killing tensor in EC III):

$$\overline{K}_{ij}^{(I)} := \begin{pmatrix} \cos^2\theta & \cos\theta\sin\theta \\ \cos\theta\sin\theta & \sin^2\theta \end{pmatrix},$$

$$\overline{K}_{ij}^{(II)} := \begin{pmatrix} (\overline{y}-b)^2 & -(\overline{x}-a)(\overline{y}-b) \\ -(\overline{x}-a)(\overline{y}-b) & (\overline{x}-a)^2 \end{pmatrix},$$

$$\overline{K}_{ij}^{(III)} := \begin{pmatrix} 2\sin\theta(b-\overline{y}) & (\overline{y}-b)\cos\theta+(\overline{x}-a)\sin\theta \\ (\overline{y}-b)\cos\theta+(\overline{x}-a)\sin\theta & 2\cos\theta(a-\overline{x}) \end{pmatrix},$$

$$\overline{K}_{ij}^{(IV)} := \begin{pmatrix} k^2\cos^2\theta+(\overline{y}-b)^2 & k^2\cos\theta\sin\theta-(\overline{x}-a)(\overline{y}-b) \\ k^2\cos\theta\sin\theta-(\overline{x}-a)(\overline{y}-b) & k^2\sin^2\theta+(\overline{x}-a)^2 \end{pmatrix}.$$

The formulas above give the most general forms of Killing tensors within each equivalence class characterized by the fundamental $I(\mathbb{R}^2)$-invariants $\Delta^i_{\mathbb{R}^2}, i = 1, 2$ according to (4.9). Using the transformations induced by the isometry group, we can always bring any non-trivial Killing tensor to the canonical form of the equivalence class it belongs to.

We conclude therefore that the problems of equivalence and canonical forms have been solved for the vector space of the Killing tensors of valence two defined in the Euclidean plane.

## 5. Invariant classification of orthogonal coordinate webs

Now we turn our attention to the underlying geometry of the equivalent classes of Killing tensors presented in the previous section.

Recall that the eigenvalues and eigenvectors of the non-trivial Killing tensors of the vector space $\mathcal{K}^2(\mathbb{R}^2)$ define families of *confocal conics* in $\mathbb{R}^2$, which are also called *orthogonal coordinate webs* in $\mathbb{R}^2$ [**Be**]. They are the integral curves of the eigenvectors. The eigenvectors are necessarily integrable for the non-trivial elements of $\mathcal{K}^2(\mathbb{R}^2)$. The orthogonal coordinate webs may also be defined by the eigenvalues $\lambda_i, i = 1, 2$ of a given Killing tensor: $\lambda_i = \text{const}, i = 1, 2$, provided they are distinct and nowhere constant. Otherwise, one also has to consider the eigenvectors and solve the corresponding differential equations (see [**Be**] for more details).

There are exactly four types of orthogonal coordinate webs, which to no surprise, correspond to the four equivalent classes of non-trivial Killing tensors classified above. More specifically, they are the well-known Cartesian, polar, parabolic and elliptic-hyperbolic coordinate systems extensively used in many applications. In summary:

*Each non-trivial Killing tensor of $\mathcal{K}^2(\mathbb{R}^2)$ determines an orthogonal coordinate web.*

The canonical forms given by (4.10) with respect to the Cartesian coordinates $(x, y)$ can be transformed by making use of the standard coordinate transformations to diagonal forms. Below we list the corresponding metrics, coordinate transformations and associated diagonalized canonical Killing tensors:

$$I: \quad \text{Cartesian:} \quad \begin{cases} ds^2 = du^2 + dv^2 \\ x = u, \; y = v \\ K_{ij}^{(I)} = \mathrm{diag}(1, 0) \end{cases}$$

$$II: \quad \text{Polar:} \quad \begin{cases} ds^2 = du^2 + u^2 dv^2 \\ x = u\cos v, \; y = u \sin v \\ K_{ij}^{(II)} = \mathrm{diag}(0, u^4) \end{cases}$$

$$III: \quad \text{Parabolic:} \quad \begin{cases} ds^2 = (u^2 + v^2)(du^2 + dv^2) \\ x = \frac{1}{2}(u^2 - v^2), \; y = uv \\ K_{ij}^{(III)} = (u^2 + v^2)\,\mathrm{diag}(v^2, -u^2) \end{cases}$$

$$IV: \quad \text{Elliptic-Hyperbolic:} \quad \begin{cases} ds^2 = k^2(\cosh^2 u - \cos^2 v)(du^2 + dv^2) \\ x = k\cosh u \cos v, \; y = k \sinh u \sin v \\ K_{ij}^{(IV)} = k^4(\cosh^2 u - \cos^2 v)\,\mathrm{diag}(\cos^2 v, \cosh^2 u) \end{cases}$$

Note that the orthogonal coordinate webs generated by the canonical forms (4.10) are properly aligned with respect to the Cartesian coordinates $(x, y)$ of (4.10), which means that there is no action of the isometry group involved. In a more general case, an orthogonal coordinate web can be rotated and/or translated with respect to the original Cartesian coordinates $(x, y)$. Ultimately, this changes the form of the corresponding Killing tensor, but not the underlying geometry generated by it. The latter being an invariant property is characterized by the fundamental $I(\mathbb{R}^2)$-invariants $\Delta^i_{\mathbb{R}^2}, i = 1, 2$. In summary:

> *The non-trivial Killing tensors within each of the equivalence classes I-IV (4.9) give rise to a particular type of an orthogonal coordinate web.*

The invariant analysis of the geometry of orthogonal coordinate webs exhibited in this paper would be incomplete without considering one more *invariant* which is associated to EC IV whose elements generate elliptic-hyperbolic coordinate systems. We are referring to the parameter $k$ which represent (half) the distance between the foci (see the last formula in (4.10)). The elements of the other three equivalent classes appear to be degenerate cases of those constituting EC IV. Indeed, the polar system (EC II) is a limiting case of the elliptic-hyperbolic system when the foci coincide, i.e., $k = 0$. The parabolic system is also a limiting case of the elliptic-hyperbolic system for which one of the two foci is at infinity. Finally, for the Cartesian system both foci are at infinity (for more details and a different classification see [**BR, MST1**]). Therefore we conclude that the parameter $k$ is also an $I(\mathbb{R}^2)$-invariant and as such, in view of Theorem 4.1, must be related to the fundamental $I(\mathbb{R}^2)$-invariants $\Delta^i_{\mathbb{R}^2}, i = 1, 2$ when both of them do not vanish (i.e., for the elements of EC IV). Indeed, such a formula has been found [**MST1**]; it determines $k$ as a function of $\Delta^1_{\mathbb{R}^2}$ and $\Delta^2_{\mathbb{R}^2}$ as follows:

$$k^2(\Delta^1_{\mathbb{R}^2}, \Delta^2_{\mathbb{R}^2}) = \frac{\sqrt{\Delta^2_{\mathbb{R}^2}}}{(\Delta^1_{\mathbb{R}^2})^2}. \tag{5.1}$$

Now for the elements of EC II and III generating polar and parabolic coordinate systems we have:

$$k_{II}^2 = \lim_{\Delta^2_{\mathbb{R}^2}\to 0} \sqrt{\Delta^2_{\mathbb{R}^2}/(\Delta^1_{\mathbb{R}^2})^2} = 0$$

and

$$k_{III}^2 = \lim_{\Delta^1_{\mathbb{R}^2}\to 0} \sqrt{\Delta^2_{\mathbb{R}^2}/(\Delta^1_{\mathbb{R}^2})^2} = \infty$$

respectively, that shows that they are indeed limiting cases of the elements of the most general EC IV. This observation completes the classification.

## 6. Concluding remarks

We have presented a new approach to the study of Killing tensors defined in Riemannian spaces of constant curvature by incorporating into the study the basic ideas of the classical invariant theory. The elucidating example of the vector space $\mathcal{K}^2(\mathbb{R}^2)$ employed in this article to illustrate the approach appears to be an analogue of the space of quadratic forms we used in Section 2 as a basic example to present the main concepts of the classical invariant theory. The general theory can be applied to other (more complicated) vector spaces of Killing tensors defined in pseudo-Riemannian spaces of constant curvature (see, for instance [**MST4, MST5**]). The main idea that has been used here is to associate a vector space $\mathcal{K}^p(M)$ of Killing tensors of valence $p$ defined on a Riemannian space $(M, \mathbf{g})$ of constant curvature with its parameter space $\Sigma$ of dimension $d$ given by (3.7). The next step is to determine the action of the isometry group $I(M)$ (or another transformation group $G : \mathcal{K}^p(M) \to \mathcal{K}^p(M)$) and compute the $I(M)$-invariants, which are polynomial functions of the parameters that span $\Sigma$. The procedure can be summarized in the following scheme:

$$I(M) \to \mathcal{K}^p(M) \simeq \Sigma \to \text{Invariants of } \mathcal{K}^p(M) \text{ as functions in } \Sigma.$$

It should be mentioned that the computation of the $I(M)$-invariants of vector spaces of Killing tensors becomes more difficult as the dimension of $M$ and/or valence $p$ are increased. In the case of higher dimensions (valences) this technical problem requires the use of computer algebra packages, such as MAPLE, MuPAD, etc.

We finally remark that the $I(M)$-invariants of vector spaces of Killing tensors defined on Riemannian spaces of constant curvature can be effectively used in various problems arising in such areas of mathematical physics as classical mechanics, field theory and general relativity.

The authors wish to thank Sergio Benenti, Claudia Chanu and Giovanni Rastelli for many illuminating discussions pertinent to the subject of the paper.

The second author (RGS) expresses his gratitude to Krishan L. Duggal and Ramesh Sharma for the opportunity to present and discuss these results at the AMS Special Session on Recent Advances in Riemannian and Lorentzian Geometries (Joint Mathematics Meetings, Baltimore, January 15-18, 2003); he also acknowledges the hospitality of Mathematics Department, University of Turin where a part of this work has been written.

## References

[MST1] R. G. McLenaghan, R. G. Smirnov and D. The, *Group invariant classification of separable Hamiltonian systems in the Euclidean plane and the O(4)-symmetric Yang-Mills theories of Yatsun.* J. Math. Phys. **43** (2002) 1422–1440.

[MST2] R. G. McLenaghan, R. G. Smirnov and D. The, *The 1881 problem of Morera revisited,* Diff. Geom. Appl., Proceedings Conf. Opava (Czech Republic), August 27-31, 2001, Silesian University, Opava, 2001, 333–341. `http://8icdga.math.slu.cz/proceedings.html`

[MST3] R. G. McLenaghan, R. G. Smirnov and D. The, *Group invariants of Killing tensors in the Minkowski plane.* Proceedings of "Symmetry and Perturbation Theory - SPT2002", the conference held in Cala Gonone, 19-26 May 2002, S. Abenda, G. Gaeta and S. Walcher eds., World Scientific, 2003, 153–162.

[MST4] R. G. McLenaghan, R. G. Smirnov and D. The, *An extension of the classical theory of algebraic invariants to pseudo-Riemannian geometry and Hamiltonian mechanics,* preprint, 62 pages, 2002.

[MST5] R. G. McLenaghan, R. G. Smirnov and D. The, *An invariant classification of cubic integrals of motion,* preprint, 19 pages, 2003.

[Hil] D. Hilbert, *Theory of algebraic invariants.* Cambridge University Press, 1993.

[Olv] P. J. Olver, *Classical Invariant Theory.* London Mathematical Society, Student Texts **44**, Cambridge University Press, 1999.

[Sch] J. A. Schouten, *Uber Differentalkomitanten zweier kontravarianter Grösser.* Proc. Kon. Ned. Akad. Amsterdam **43**, (1940) 449–452.

[De] R. P. Delong, Jr., *Killing tensors and Hamilton-Jacobi equation.* PhD thesis, University of Minnesota, 1982.

[Ta] M. Takeuchi, *Killing tensor fields on spaces of constant curvature.* Tsukuba J. Math. **7** (1983) 233–255.

[Th] G. Thompson, *Killing tensors in spaces of constant curvature.* J. Math. Phys. **27** (1986) 2693–2699.

[Be] S. Benenti, *Intrinsic characterization of the variable separation in the Hamilton-Jacobi equation.* J. Math. Phys. **38** (1997) 6578–6602.

[BR] S. Benenti and G. Rastelli, *Sistemi di coordinate separabili nel piano euclideo.* Mem. Acad. Sci. Torino **15** (1991), 1–21.

Department of Applied Mathematics, University of Waterloo, Waterloo, Ontario, Canada, N2L 3G1

*E-mail address*: `rgmclena@sirius.math.uwaterloo.ca`

Universität Paderborn, Fachbereich 17 Mathematik, Pohlweg 98, D-33098 Paderborn, Germany (Current address: Department of Mathematics and Statistics, Dalhousie University, Halifax, Nova Scotia, Canada, B3H 4H8)

*E-mail address*: `rsmirnov@sirius.math.uwaterloo.ca`

Department of Mathematics, University of British Columbia, 121-1984 Mathematics Rd., Vancouver, British Columbia, Canada V6T 1Z2 (Current address: Department of Mathematics and Statistics, McGill University, Montreal, Quebec, Canada, H3A 2K6)

*E-mail address*: `dthe@math.mcgill.ca`

Contemporary Mathematics
Volume **337**, 2003

# Spectral Geometry and the Kaehler Condition for Hermitian Manifolds with Boundary

JeongHyeong Park

ABSTRACT. Let $(M, g, J)$ be a compact Hermitian manifold with a smooth boundary. Let $\Delta_{p,\mathcal{B}}$ and $\square_{p,\mathcal{B}}$ be the realizations of the real and complex Laplacians on $p$ forms with either Dirichlet or Neumann boundary conditions. We generalize previous results in the closed setting to show that $(M, g, J)$ is Kaehler if and only if $\text{Spec}(\Delta_{p,\mathcal{B}}) = \text{Spec}(2\,\square_{p,\mathcal{B}})$ for $p = 0, 1$. We also give a characterization of manifolds with constant sectional curvature or constant Ricci tensor (in the real setting) and manifolds of constant holomorphic sectional curvature (in the complex setting) in terms of spectral geometry.

## 1. Introduction

The relationship between the spectrum of certain natural operators of Laplace type and the underlying geometry of a Riemannian manifold has been studied by many authors. Let $(M, g)$ be a compact Riemannian manifold with smooth boundary $\partial M$. Let $V$ be a smooth Hermitian vector bundle over $M$ and let $D$ be a formally self-adjoint operator of Laplace type acting on the space of smooth sections $C^\infty(V)$. Let $D_\mathcal{B}$ denote the realization of $D$ with respect to either the Dirichlet $(\mathcal{B} = \mathcal{B}_D)$ or the Neumann $(\mathcal{B} = \mathcal{B}_N)$ boundary operators. Then $D_\mathcal{B}$ is self-adjoint and has a complete discrete spectral resolution $\mathcal{S}(D_\mathcal{B}) = \{(\phi_\nu, \lambda_\nu)\}$. The $\phi_\nu \in C^\infty(V)$ form a complete orthonormal basis for $L^2(V)$ such that

$$D\phi_\nu = \lambda_\nu \phi_\nu \quad \text{and} \quad \mathcal{B}\phi_\nu = 0.$$

We let the spectrum $\text{Spec}(D_\mathcal{B}) = \{\lambda_\nu\}$ be the collection of eigenvalues. We repeat the eigenvalues according to multiplicity and order the eigenvalues so $\lambda_1 \leq \lambda_2 \dots$.

---

2000 *Mathematics Subject Classification.* 58J50.

*Key words and phrases.* Dirichlet boundary conditions, Neumann boundary conditions, Kaehler, Hermitian manifold, spectral geometry, constant sectional curvature, constant holomorphic sectional curvature, Einstein manifold.

This work was supported by Korea Research Foundation Grant (KRF-2000-015-DS0003).

For example, if $M = [0,\pi]$ and if $D = -\partial_x^2$ on $C^\infty(M)$, then:

$$\begin{aligned}
&\mathcal{S}(D_{\mathcal{B}_D}) = \left\{\left(\sqrt{\tfrac{2}{\pi}}\sin(nx), n^2\right)\right\}_{n=1}^{\infty}\\
&\operatorname{Spec}(D_{\mathcal{B}_D}) = \{1,4,9,...\}\\
&\mathcal{S}(D_{\mathcal{B}_N}) = \left\{\left(1,\sqrt{\tfrac{1}{\pi}}\right)\right\}\cup\left\{\left(\sqrt{\tfrac{2}{\pi}}\cos(nx), n^2\right)\right\}_{n=1}^{\infty}\\
&\operatorname{Spec}(D_{\mathcal{B}_N}) = \{0,1,4,9,...\}.
\end{aligned}$$

Let $(M,g,J)$ be a Hermitian manifold of complex dimension $\bar{m}$ and corresponding real dimension $m = 2\bar{m}$; here $J$ is an integrable almost complex structure which is unitary with respect to the Riemannian metric $g$. Let $\Lambda^n M$ be the bundle of complex $n$ forms on $M$. Let

$$\Delta_n := d^*d + dd^* \quad \text{and} \quad \Box_n := \bar\partial^*\bar\partial + \bar\partial\bar\partial^* \quad \text{on} \quad C^\infty(\Lambda^n M)$$

be real and complex form valued Laplacians. We further decompose

$$\Box_n = \oplus_{p+q=n}\Box_{(p,q)} \quad \text{on} \quad C^\infty(\Lambda^{(p,q)}M).$$

We introduce the associated *Kaehler form* $\Omega(X,Y) := g(X,JY)$. Extend the metric $g$ to be Hermitian on the complexified tangent bundle. Let $\nabla$ be the Levi-Civita connection. The following notions are equivalent and any defines the notion of a *Kaehler manifold*:

(1) For every $P$ in $M$, there exist local holomorphic coordinates so $dg(P) = 0$.
(2) We have $d\Omega = 0$.
(3) We have $\nabla J = 0$.

Let $\mathcal{I} := \operatorname{int}(\Omega)$. Let $\delta'$ be the formal adjoint of $\partial$ and $\delta''$ be the formal adjoint of $\bar\partial$. For a Kaehler manifold, one has thefollowing relationships:

(1) $\bar\partial\mathcal{I} - \mathcal{I}\bar\partial = \sqrt{-1}\,\delta'$.
(2) $\bar\partial\delta' + \delta'\bar\partial = 0$ and $\partial\delta'' + \delta''\partial = 0$.
(3) $d^*d + dd^* = \partial\delta' + \delta'\partial + \bar\partial\delta'' + \delta''\bar\partial$.
(4) $\partial\bar\partial\mathcal{I} - \partial\mathcal{I}\bar\partial = \sqrt{-1}\,\partial\delta'$ and $\bar\partial\mathcal{I}\partial - \mathcal{I}\bar\partial\partial = \sqrt{-1}\,\delta'\partial$.
(5) $\sqrt{-1}\,(\partial\delta' + \delta'\partial) = \partial\bar\partial\mathcal{I} - \partial\mathcal{I}\bar\partial + \bar\partial\mathcal{I}\partial - \mathcal{I}\bar\partial\partial$.
(6) $\partial\delta' + \delta'\partial = \bar\partial\delta'' + \delta''\bar\partial$.

The following well known result is now immediate.

THEOREM 1.1. *Let $(M,g,J)$ be a compact Kaehler manifold without boundary of complex dimension $\bar{m}$. Then $\Delta = 2\,\Box$ and so* $\operatorname{Spec}(\Delta_p) = \operatorname{Spec}(2\,\Box_p)$ *for all $p$.*

Conversely, one has the following result to T. Tsujishita (reported by Gilkey [**8**]):

THEOREM 1.2. *Let $(M,g,J)$ be a compact Hermitian manifold without boundary. If* $\operatorname{Spec}(\Delta_0) = \operatorname{Spec}(2\Box_0)$ *and if* $\operatorname{Spec}(\Delta_1) = \operatorname{Spec}(2\Box_1)$, *then $(M,g,J)$ is Kaehler.*

Donnelly [**2**] established a similar characterization of the Kaehler property using the reduced complex Laplacian. Pak [**12**] extended these results to the context of almost isospectral manifolds.

Theorem 1.2 is sharp. We refer to Gilkey [**7**] for the proof of the following result:

THEOREM 1.3. *Let* $ds^2 = dz_1 \circ d\bar{z}_1 + e^{\psi(z_1)} dz_2 \circ d\bar{z}_2 + e^{-\psi(z_1)} dz_3 \circ d\bar{z}_3$ *be a Hermitian metric on the torus* $M_3$ *where* $\psi(z_1)$ *is an arbitrary smooth real valued function. Then* $\Delta_0 = 2\,\Box_0$ *but the metric is not Kaehler.*

A Riemannian manifold of constant sectional curvature $c$ is said to be a *space form*; a Kaehler manifold of constant holomorphic sectional curvature $c$ is said to be a *complex space form*. Modulo rescaling, any space form is locally isometric to the unit sphere, to flat space, or to hyperbolic space. Similarly, modulo rescaling, any complex space form is locally isometric to complex projective space, to flat space, or to the negative curvature dual. Thus the geometries are very rigid in this context.

Patodi [**11**] established the following spectral characterization of space forms:

THEOREM 1.4. *Let* $(M_i, g_i)$ *be compact Riemannian manifolds without boundary. Assume that* $\operatorname{Spec}(\Delta_p, M_1) = \operatorname{Spec}(\Delta_p, M_2)$ *for* $0 \le p \le 2$. *Then:*

(1) *The manifold* $M_1$ *has constant scalar curvature* $c$ *if and only if the manifold* $M_2$ *has constant scalar curvature* $c$.
(2) *The manifold* $M_1$ *is Einstein if and only if the manifold* $M_2$ *is Einstein.*
(3) *The manifold* $M_1$ *has constant sectional curvature* $c$ *if and only if the manifold* $M_2$ *has constant sectional curvature* $c$.

Donnelly [**2**] and Gilkey and Sacks [**9**] extended Theorem 1.4 to the complex setting – see also related work by Friedland [**5, 6**], C.C. Hsiung et. al. [**10**], and Pak [**12**].

THEOREM 1.5. *Let* $(M_i, g_i, J_i)$ *be compact Kaehler manifolds without boundary. Assume that* $\operatorname{Spec}(\Box_{p,q}, M_1) = \operatorname{Spec}(\Box_{p,q}, M_2)$ *for* $0 \le p \le 2$ *and* $0 \le q \le 2$. *Then the manifold* $M_1$ *has constant holomorphic sectional curvature* $c$ *if and only if the manifold* $M_2$ *has constant holomorphic sectional curvature* $c$.

We can extend Theorem 1.2 to manifolds with boundary:

THEOREM 1.6. *Let* $\mathcal{B}$ *denote either Dirichlet or Neumann boundary conditions. Let* $(M, g, J)$ *be a compact Hermitian manifold with smooth boundary* $\partial M$. *Assume that* $\operatorname{Spec}(\Delta_{0,\mathcal{B}}) = \operatorname{Spec}(2\,\Box_{0,\mathcal{B}})$ *and that* $\operatorname{Spec}(\Delta_{1,\mathcal{B}}) = \operatorname{Spec}(2\,\Box_{1,\mathcal{B}})$. *Then* $(M, g, J)$ *is Kaehler.*

We can also generalize Theorems 1.4 and 1.5 to the category of manifolds with boundary under the additional technical hypothesis that the manifolds in question have constant scalar curvature.

THEOREM 1.7. *Let* $\mathcal{B}$ *denote either Dirichlet or Neumann boundary conditions. Let* $(M_i, g_i)$ *be compact Riemannian manifolds with smooth boundaries* $\partial M_i$ *which have constant scalar curvatures* $\tau_i$. *Assume that* $\operatorname{Spec}(\Delta_{p,\mathcal{B}}, M_1) = \operatorname{Spec}(\Delta_{p,\mathcal{B}}, M_2)$ *for* $0 \le p \le 2$. *Then:*

(1) $\tau_1 = \tau_2$.
(2) *The manifold* $M_1$ *is Einstein if and only if the manifold* $M_2$ *is Einstein.*
(3) *The manifold* $M_1$ *has constant sectional curvature* $c$ *if and only if the manifold* $M_2$ *has constant sectional curvature* $c$.

THEOREM 1.8. *Let* $\mathcal{B}$ *denote either Dirichlet or Neumann boundary conditions. Let* $(M_i, g_i, J_i)$ *be compact Kaehler manifolds with smooth boundaries* $\partial M_i$ *which have constant scalar curvatures* $\tau_i$. *Assume that* $\operatorname{Spec}(\Box_{p,q,\mathcal{B}}, M_1) = \operatorname{Spec}(\Box_{p,q,\mathcal{B}}, M_2)$

*for $0 \le p \le 2, 0 \le q \le 2$. Then the manifold $M_1$ has constant holomorphic sectional curvature $c$ if and only if the manifold $M_2$ has constant holomorphic sectional curvature $c$.*

Here is a brief outline to the remainder of this paper. In Section 2, we review some previous results concerning the heat trace asymptotics. In Section 3, we complete the proof of Theorem 1.6 and in Section 4, we complete the proof of Theorems 1.7 and 1.8. In Remark 4.1, we extend Theorems 1.7 and 1.8 to more general boundary conditions of Robin type where the auxiliary endomorphism $S$ is 'universal' in a certain sense.

It is a pleasant task to thank the referee for helpful suggestions concerning the manuscript.

## 2. Heat trace asymptotics

Let $M$ be a compact Riemannian manifold of real dimension $m$ with smooth boundary $\partial M$ and let $D_{\mathcal{B}}$ be the realization of an operator of Laplace type on $M$ with respect to either Dirichlet or Neumann boundary conditions. Let $e^{-tD_{\mathcal{B}}}$ be the fundamental solution of the heat equation. This operator is of trace class and as $t \downarrow 0$ there is a complete asymptotic expansion with locally computable coefficients in the form:

$$\mathrm{Tr}_{L^2}\, e^{-tD_{\mathcal{B}}} \sim \textstyle\sum_{n\ge 0} t^{(n-m)/2} a_n(D_{\mathcal{B}}).$$

To study the heat trace coefficients $a_n(D_{\mathcal{B}})$, we must introduce a bit of additional notation. There is a canonically defined connection $\nabla = \nabla(D)$ and a canonically defined endomorphism $E = E(D)$ so that

$$D = -(\mathrm{Tr}(\nabla^2) + E).$$

Let indices $i$, $j$, $k$ range from 1 to $m$ and index a local orthonormal frame $\{e_1, ..., e_m\}$ for $TM$. Let indices $a$, $b$, and $c$ range from 1 to $m-1$ and index a local orthonormal frame $\{e_1, ..., e_{m-1}\}$ for $T\partial M$; on $\partial M$, we let $e_m$ be the inward unit normal vector field. Let $\Omega$ be the curvature of $\nabla$, let $\tau := R_{ijji}$ be the normalized scalar curvature, let $\rho_{ij} := R_{ikkj}$ be the Ricci tensor, and let $L_{ab}$ be the second fundamental form. We adopt the Einstein convention and sum over repeated indices. Let ';' denote multiple covariant differentiation. We refer to [**1**] for the proof of the following result:

THEOREM 2.1. *Let $D$ be an operator of Laplace type on the space of sections $C^\infty(V)$ to a vector bundle $V$ over a compact manifold $M$ with smooth boundary $\partial M$. Let $I$ be the identity endomorphism of $V$. With Dirichlet boundary conditions, we have:*

(1) $a_0(D_{\mathcal{B}_D}) = (4\pi)^{-m/2} \int_M \mathrm{Tr}\{I\}$.

(2) $a_1(D_{\mathcal{B}_D}) = -(4\pi)^{-(m-1)/2} \frac{1}{4} \int_{\partial M} \mathrm{Tr}\{I\}$.

(3) $a_2(D_{\mathcal{B}_D}) = (4\pi)^{-m/2} \frac{1}{6} \int_M \mathrm{Tr}\{6E + \tau I\} + (4\pi)^{-m/2} \frac{1}{6} \int_{\partial M} \mathrm{Tr}\{2L_{aa} I\}$.

(4) $a_3(D_{\mathcal{B}_D}) = -\frac{1}{384}(4\pi)^{-(m-1)/2} \int_{\partial M} \mathrm{Tr}\{96E + (16\tau + 8R_{amam} + 7L_{aa}L_{bb} - 10L_{ab}L_{ab})I\}$.

(5) $a_4(D_{\mathcal{B}_D}) = (4\pi)^{-m/2} \frac{1}{360} \int_M \mathrm{Tr}\{60E_{;kk} + 60\tau E + 180E^2 + 30\Omega^2 + (12\tau_{;kk} + 5\tau^2 - 2|\rho|^2 + 2|R|^2)I\} + (4\pi)^{-m/2} \frac{1}{360} \int_{\partial M} \mathrm{Tr}\{-120E_{;m} + 120EL_{aa} + (-18\tau_{;m} + 20\tau L_{aa} + 4R_{amam}L_{bb} - 12R_{ambm}L_{ab} + 4R_{abcb}L_{ac} + \frac{40}{21}L_{aa}L_{bb}L_{cc} - \frac{88}{7}L_{ab}L_{ab}L_{cc} + \frac{320}{21}L_{ab}L_{bc}L_{ac})I\}$.

Let $\nabla_m$ denote the covariant derivative with respect to $e_m$ on $\partial M$. Let $S$ be an auxiliary endomorphism of $V|_{\partial M}$. The Robin boundary operator is then given by:

$$\mathcal{B}_S\phi := (\nabla_m\phi + S\phi)|_{\partial M}\,.$$

We take $S = 0$ to define Neumann boundary conditions. Again, we refer to [**1**] for the proof of the following result:

THEOREM 2.2. *With Robin boundary conditions, we have:*
(1) $a_0(D_{\mathcal{B}_S}) = (4\pi)^{-m/2}\int_M \operatorname{Tr}\{I\}$.
(2) $a_1(D_{\mathcal{B}_S}) = (4\pi)^{(1-m)/2}\frac14\int_{\partial M}\operatorname{Tr}\{I\}$.
(3) $a_2(D_{\mathcal{B}_S}) = (4\pi)^{-m/2}\frac16\int_M \operatorname{Tr}\{6E+\tau I\}$
$+(4\pi)^{-m/2}\frac16\int_{\partial M}\operatorname{Tr}\{2L_{aa}I+12S\}$.
(4) $a_3(D_{\mathcal{B}_S}) = (4\pi)^{(1-m)/2}\frac{1}{384}\int_{\partial M}\operatorname{Tr}\{96E+(16\tau-8R_{amma}+13L_{aa}L_{bb}$
$+2L_{ab}L_{ab})I+96SL_{aa}+192S^2\}$.
(5) $a_4(D_{\mathcal{B}_S}) = (4\pi)^{-m/2}\frac{1}{360}\int_M \operatorname{Tr}\{60E_{;kk}+60\tau E+180E^2+30\Omega^2$
$+(12R_{ijji;kk}+5\tau^2-2|\rho|^2+2|R|^2)I\}+(4\pi)^{-m/2}\frac{1}{360}\int_{\partial M}\operatorname{Tr}\{240E_{;m}$
$+120EL_{aa}+(42R_{ijji;m}+20\tau L_{aa}+4R_{amam}L_{bb}-12R_{ambm}L_{ab}$
$+4R_{abcb}L_{ac}+\frac{40}{3}L_{aa}L_{bb}L_{cc}+8L_{ab}L_{ab}L_{cc}+\frac{32}{3}L_{ab}L_{bc}L_{ac})I+120S\tau$
$+720SE+144SL_{aa}L_{bb}+48SL_{ab}L_{ab}+480S^2L_{aa}+480S^3\}$.

## 3. The proof of Theorem 1.6

Let $\star$ be the Hodge operator. We introduce the following invariants:
(1) Let $K_1 := \star(d\bar\partial\Omega\wedge\Omega^{\hat m-2})$ for $\hat m\geq 2$.
(2) Let $K_2 := \frac12|d\Omega|^2$ for $\hat m\geq 2$.
(3) Let $K_3 := \star(d\Omega\wedge\bar\partial\Omega\wedge\Omega^{\hat m-3})$ for $\hat m\geq 3$.
(4) Let $\kappa := L_{aa}$ be the geodesic curvature of the boundary.

We may then use Theorem 2.1 to extend results of Gilkey [**7**] to see:

$$a_2(2\,\square_{0,\mathcal{B}}) = (4\pi)^{-\hat m}\tfrac16\{\textstyle\int_{\partial M}2\kappa+\int_M(\tau+3K_2+3K_3)\}$$
$$a_2(\Delta_{0,\mathcal{B}}) = (4\pi)^{-\hat m}\tfrac16\{\textstyle\int_{\partial M}2\kappa+\int_M\tau\}$$
$$a_2(2\,\square_{1,0,\mathcal{B}}) = (4\pi)^{-\hat m}\tfrac16\{\textstyle\int_{\partial M}2\hat m\kappa+(\hat m-3)\int_M(\tau+3K_2+3K_3)$$
$$+\textstyle\int_M(-6K_1+6K_2+3K_3)\}$$
$$a_2(2\,\square_{0,1,\mathcal{B}}) = (4\pi)^{-\hat m}\tfrac16\{\textstyle\int_{\partial M}2\hat m\kappa+(\hat m-3)\int_M(\tau+3K_2+3K_3)$$
$$+\textstyle\int_M(6K_1+6K_2+3K_3)\}$$
$$a_2(\Delta_{1,\mathcal{B}}) = (4\pi)^{-\hat m}\tfrac16\{\textstyle\int_{\partial M}4\hat m\kappa+2(\hat m-3)\int_M\tau\}.$$

Since $a_2(\Delta_{0,\mathcal{B}}) = a_2(2\,\square_{0,\mathcal{B}})$, we have

$$\textstyle\int_M(3K_2+3K_3)=0. \tag{3.1}$$

We use this relation and the relation $a_2(\Delta_{1,\mathcal{B}}) = a_2(2\,\square_{1,\mathcal{B}})$ to see

$$\textstyle\int_M(6K_2+3K_3)=0. \tag{3.2}$$

Equations (3.1) and (3.2) then imply $\int_M K_2 = 0$ and hence $M$ is Kaehler.

If $\hat m = 2$, we have the formulae:

$$\begin{aligned} a_2(2\,\square_{0,\mathcal{B}}) &= (4\pi)^{-\hat m}\tfrac16\{\textstyle\int_{\partial M}2\kappa+\int_M(\tau+3K_2)\}\\ a_2(\Delta_{0,\mathcal{B}}) &= (4\pi)^{-\hat m}\tfrac16\{\textstyle\int_{\partial M}2\kappa+\int_M\tau\}.\end{aligned} \tag{3.3}$$

Since $a_2(\Delta_{0,\mathcal{B}}) = a_2(2\,\square_{0,\mathcal{B}})$, $\int_M K_2 = 0$ and $M$ is Kaehler; the condition relating $\Delta_{1,\mathcal{B}}$ and $2\,\square_{1,\mathcal{B}}$ is not necessary in this instance. Finally, if $\hat{m} = 1$, then $M$ is automatically Kaehler. $\square$

## 4. Proof of Theorems 1.7 and 1.8

We set $S = 0$. Let $\varepsilon := \frac{1}{4}$ for Neumann boundary conditions and $\varepsilon := -\frac{1}{4}$ for Dirichlet boundary conditions. By Theorems 2.1 and 2.2,

$$\begin{aligned}&\operatorname{Tr}_{L^2}(e^{-t\Delta_{0,\mathcal{B}}}) = (4\pi t)^{-m/2}\{\operatorname{Vol}(M) + \varepsilon\sqrt{t}\operatorname{Vol}(\partial M) + O(t)\} \quad \text{so}\\ &\operatorname{Vol}(M_1) = \operatorname{Vol}(M_2), \quad \operatorname{Vol}(\partial M_1) = \operatorname{Vol}(\partial M_2), \quad \text{and}\\ &\dim_{\mathbb{R}}(M_1) = \dim_{\mathbb{R}}(M_2).\end{aligned} \tag{4.1}$$

We set $m := \dim_{\mathbb{R}}(M_i)$ to this common value and compute:

$$\begin{aligned}a_2(\Delta_{0,\mathcal{B}}, M_i) &= (4\pi)^{-m/2}\tfrac{1}{6}\{\textstyle\int_{M_i}\tau_i + \int_{\partial M_i} 2\kappa_i\}\\ a_2(\Delta_{1,\mathcal{B}}, M_i) &= (4\pi)^{-m/2}\tfrac{1}{6}\{\textstyle\int_{M_i}(m-6)\tau_i + \int_{\partial M_i} 2m\kappa_i\}.\end{aligned}$$

We may then establish assertion (1) by computing:

$$\begin{aligned}\tau_1 &= (4\pi)^{m/2}\operatorname{Vol}(M_1)^{-1}\{ma_2(\Delta_{0,\mathcal{B}}, M_1) - a_2(\Delta_{1,\mathcal{B}}, M_1)\}\\ &= (4\pi)^{m/2}\operatorname{Vol}(M_2)^{-1}\{ma_2(\Delta_{0,\mathcal{B}}, M_2) - a_2(\Delta_{1,\mathcal{B}}, M_2)\}\\ &= \tau_2\,.\end{aligned}$$

For subsequent use, we compute similarly that:

$$\begin{aligned}\textstyle\int_{\partial M_1}\kappa_1 &= 3(4\pi)^{m/2}a_2(\Delta_{0,\mathcal{B}}, M_1) - \tfrac{1}{2}\textstyle\int_{M_1}\tau_1\\ &= 3(4\pi)^{m/2}a_2(\Delta_{0,\mathcal{B}}, M_2) - \tfrac{1}{2}\textstyle\int_{M_2}\tau_2\\ &= \textstyle\int_{\partial M_2}\kappa_2\,.\end{aligned} \tag{4.2}$$

The interior integrands defining $a_4(\Delta_p)$ have been determined by Patodi [**11**]. Motivated by his work, we introduce constants:

$$\begin{aligned}c^1_{m,p} &= \tfrac{1}{72}\tfrac{m!}{p!(m-p)!} - \tfrac{1}{6}\tfrac{(m-2)!}{(p-1)!(m-p-1)!} + \tfrac{1}{2}\tfrac{(m-4)!}{(p-2)!(m-p-2)!},\\ c^2_{m,p} &= -\tfrac{1}{180}\tfrac{m!}{p!(m-p)!} + \tfrac{1}{2}\tfrac{(m-2)!}{(p-1)!(m-p-1)!} - 2\tfrac{(m-4)!}{(p-2)!(m-p-2)!},\\ c^3_{m,p} &= \tfrac{1}{180}\tfrac{m!}{p!(m-p)!} - \tfrac{1}{12}\tfrac{(m-2)!}{(p-1)!(m-p-1)!} + \tfrac{1}{2}\tfrac{(m-4)!}{(p-2)!(m-p-2)!},\\ c^4_{m,p} &= \tfrac{1}{30}\tfrac{m!}{p!(m-p)!} - \tfrac{1}{6}\tfrac{(m-2)!}{(p-1)!(m-p-1)!}\,.\end{aligned}$$

The work of Patodi then shows if $\partial M$ is empty that:

$$a_4(\Delta_p) = (4\pi)^{-m/2}\textstyle\int_M\{c^1_{m,p}\tau^2 + c^2_{m,p}|\rho|^2 + c^3_{m,p}|R|^2 + c^4_{m,p}\tau_{;ii}\}\,. \tag{4.3}$$

To simplify the notation, we introduce reduced invariants

$$\tilde{a}_n(\Delta_{p,\mathcal{B}}) := a_n(\Delta_{p,\mathcal{B}}) - \tfrac{m!}{p!(m-p)!}a_n(\Delta_{0,\mathcal{B}})\,.$$

The terms

$$\{R_{amma}L_{bb},\ R_{ammb}L_{ab},\ R_{abcb}L_{ac}\ L_{aa}L_{bb}L_{cc},\ L_{ab}L_{ab}L_{cc},\ L_{ab}L_{bc}L_{ac}\}$$

appearing in Theorem 2.1 are all multiplied by $\operatorname{Tr}(I_{\Lambda^p}) = \frac{m!}{p!(m-p)!}$. Thus they do not appear in $\tilde{a}_4(\Delta_{p,\mathcal{B}})$; only the terms involving $E(\Delta_p)$ survive in the boundary contributions. One can use the Weitzenböch formula to see that

$$\operatorname{Tr}(E(\Delta_p)) = -\tfrac{(m-2)!}{(p-1)!(m-p-1)!}\tau\,.$$

Using equation (4.3), we see there exist universal constants so

$$\begin{aligned}\tilde a_4(\Delta_{p,\mathcal{B}}) &= (4\pi)^{-m/2}\int_M\{\tilde c^1_{m,p}\tau^2+\tilde c^2_{m,p}|\rho|^2+\tilde c^3_{m,p}|R|^2+\tilde c^4_{m,p}\tau_{;kk}\}\\ &+ (4\pi)^{-m/2}\int_{\partial M}\{\tilde c^5_{m,p}\tau\kappa+\tilde c^6_{m,p}\tau_{;m}\},\end{aligned}$$

where, by Patodi's result, we have:

$$\begin{aligned}\tilde c^1_{m,p} &:= c^1_{m,p}-\tfrac{m!}{p!(m-p)!}c^1_{m,0} = -\tfrac16\tfrac{(m-2)!}{(p-1)!(m-p-1)!}+\tfrac12\tfrac{(m-4)!}{(p-2)!(m-p-2)!},\\ \tilde c^2_{m,p} &:= c^2_{m,p}-\tfrac{m!}{p!(m-p)!}c^2_{m,0} = \tfrac12\tfrac{(m-2)!}{(p-1)!(m-p-1)!}-2\tfrac{(m-4)!}{(p-2)!(m-p-2)!},\\ \tilde c^3_{m,p} &:= c^3_{m,p}-\tfrac{m!}{p!(m-p)!}c^3_{m,0} = -\tfrac1{12}\tfrac{(m-2)!}{(p-1)!(m-p-1)!}+\tfrac12\tfrac{(m-4)!}{(p-2)!(m-p-2)!},\\ \tilde c^4_{m,p} &:= c^4_{m,p}-\tfrac{m!}{p!(m-p)!}c^4_{m,0} = -\tfrac16\tfrac{(m-2)!}{(p-1)!(m-p-1)!}\,.\end{aligned}\tag{4.4}$$

For $p=1,2$, we have, by assumption, that:

$$\begin{aligned}&\tilde a_4(\Delta_{p,\mathcal{B}},M_1)=a_4(\Delta_{p,\mathcal{B}},M_1)-\tfrac{m!}{p!(m-p)!}a_4(\Delta_{0,\mathcal{B}},M_1)\\ =\;&\tilde a_4(\Delta_{p,\mathcal{B}},M_2)=a_4(\Delta_{p,\mathcal{B}},M_2)-\tfrac{m!}{p!(m-p)!}a_4(\Delta_{0,\mathcal{B}},M_2)\,.\end{aligned}\tag{4.5}$$

By equation (4.2), $\int_{\partial M_1}\kappa_1=\int_{\partial M_2}\kappa_2$. By assertion (1), $\tau_1=\tau_2$. Since the scalar curvature is constant, $\tau_{;m}=0$. Thus since $\mathrm{Vol}(\partial M_1)=\mathrm{Vol}(\partial M_2)$, the boundary integrals are equal. Furthermore, since $\mathrm{Vol}(M_1)=\mathrm{Vol}(M_2)$, the interior integrals of $\tau^2$ are equal. Since $\tau_{;ii}=0$, we have

$$\int_{M_1}(\tilde c^2_{m,p}|\rho_1|^2+\tilde c^3_{m,p}|R_1|^2)=\int_{M_2}(\tilde c^2_{m,p}|\rho_2|^2+\tilde c^3_{m,p}|R_2|^2)\tag{4.6}$$

for $p=1,2$; these two equations are independent since, by display (4.4),

$$\det\begin{pmatrix}\tilde c^2_{m,1} & \tilde c^3_{m,1}\\ \tilde c^2_{m,2} & \tilde c^3_{m,2}\end{pmatrix}=\det\begin{pmatrix}\frac12 & -\frac1{12}\\ \frac{m-2}{2}-2 & -\frac{m-2}{12}+\frac12\end{pmatrix}=\tfrac14-\tfrac16\neq0\,.$$

Consequently

$$\int_{M_1}|\rho_1|^2=\int_{M_2}|\rho_2|^2\quad\text{and}\quad\int_{M_1}|R_1|^2=\int_{M_2}|R_2|^2.\tag{4.7}$$

A manifold $M$ has constant sectional curvature $c$ if and only if

$$0=\int_M|R_{ijkl}-c(\delta_{il}\delta_{jk}-\delta_{ik}\delta_{jl})|^2=\int_M(|R|^2-4c\tau+c^2\varepsilon_m)$$

where $\varepsilon_m:=|\delta_{il}\delta_{jk}-\delta_{ik}\delta_{jl}|^2$ is polynomial in $m$. We use equation (4.1), equation (4.7), and assertion (1) to complete the proof of Theorem 1.7 (3) by computing:

$$\int_{M_1}(|R_1|^2-4c\tau_1+c^2\varepsilon_m)=\int_{M_2}(|R_2|^2-4c\tau_2+c^2\varepsilon_m)\,.$$

Note that $M$ is Einstein if and only if there is a constant $c$ so

$$0=|\rho_{ij}-c\delta_{ij}|^2=|\rho|^2-2c\tau+mc^2\,.$$

Thus Theorem 1.7 (2) can be established by verifying that:

$$\int_{M_1}(|\rho_1|^2-2c\tau_1+mc^2)=\int_{M_2}(|\rho_2|^2-2c\tau_2+mc^2)\,.$$

As a similar argument based on the results of [**3, 9**] establishes Theorem 1.8, we shall omit the details of the proof of Theorem 1.8 in the interests of brevity.

REMARK 4.1. We can generalize Theorems 1.7 and 1.8 to the context of Robin boundary conditions as follows. One could take $S=c_1+c_2\kappa$; the same cancellation argument as that given above to establish equation (4.6) shows the additional boundary terms cancel off for the reduced invariant. What is crucial is that the boundary condition be natural and universal in the context in which we are working.

## References

[1] T. Branson and P. Gilkey, The asymptotics of the Laplacian on a manifold with boundary, *Comm. Partial Differential Equations* **15** (1990), 245–272.

[2] H. Donnelly, A spectral condition determining the Kaehler property, *Proc. Amer. Math. Soc.* **47** (1975), 187–194.

[3] —, Minakshisundaram's coefficients on Kaehler manifolds, *Proc. Sympos. Pure Math.* **27** (1975), 195–203.

[4] J.S. Dowker and D. Kennedy, Finite temperature and boundary effects in static space times, *J. Phys. A* **11** (1978), 895–920.

[5] L. Friedland, Spectral geometry on certain almost Hermitian Einstein manifolds, *Publ. Math. Debrecen* **46** (1995), 63-70.

[6] —, Spectral geometry and almost Hermitian manifolds, *Rev. Roumaine Math. Pures Appl.* **41** (1996), 627–633.

[7] P. Gilkey, Spectral Geometry and the Kaehler condition for complex manifolds, *Invent. math.* **26** (1974), 231–258.

[8] —, Errata, *Inventiones math.* **29** (1975), 81–82.

[9] P. Gilkey and J. Sacks, Spectral geometry and manifolds of constant holomorphic sectional curvature, *Proc. Sympos. Pure Math.*, **27** (1975), 281–285.

[10] C.C. Hsiung, W. Yang, and B. Xiong, The spectral geometry of some almost Hermitian manifolds, *SUT J. Math.* **32** (1996), 163–178.

[11] V. K. Patodi, Curvature and the fundamental solution of the heat operator, *J. Indian Math. Soc.* **34** (1970), 269–285.

[12] H. K. Pak, Spectral geometry for almost isospectral Hermitian manifolds, *Geom. Dedicata* **47** (1993), 15-23.

[13] —, On the spectral rigidity of almost isospectral manifolds, *Bull. Korean Math. Soc.* **29** (1992), 237–243.

[14] R. T. Seeley, Complex powers of an elliptic operator, *Proc. Sympos. Pure Math.* **10** (1968), 288–307.

JHP Dept. of Computer & Applied Mathematics, Honam University, Seobongdong 59, Gwangsanku, Gwangju, 506-714 South Korea. Email:jhpark@honam.ac.kr

Contemporary Mathematics
Volume **337**, 2003

# Energy, volume and deformation of contact metrics

Philippe Rukimbira

ABSTRACT. We introduce a new type of deformation of contact metric structures and determine the conditions under which characteristic vector fields in those deformed structures are minimal and/or harmonic maps into the corresponding unit tangent bundles.

## 1. Introduction

The main tool of this paper is a new type of deformation of contact metric structures, called "contact metric structures of type $l$" where $l$ is a positive function. In contrast with the usual D-homothetic deformations found in Tanno's work [**TAN**], deformations of type $l$ keep the contact form fixed and the deformation parameter is a smooth positive function rather than just a constant. The usual contact metric structures correspond to our contact metric structures of constant type 1.

We derive some useful identities for those deformed structures and prove the following theorem.

**Theorem** *Let* $(M_i, \alpha_i, Z_i, J_i, g_{l_i})$, $i = 1, 2$ *be contact metric manifolds of type* $l_i$, *each satisfying at least one of the following conditions.*

1. $l_i$ *is constant.*
2. $M_i$ *is 3-dimensional and* $dl_i(Z_i) = 0$.

*Then any* $J$*-holomorphic map between the two riemannian manifolds is a harmonic map.*

The above theorem is more interesting in the three dimensional case where the type function could be nonconstant. The case of constant type is merely a slight generalization of the result in [**IAN**].

For contact metric structures of constant type 1, the above result is due to Ianus and Pastore [**IAN**]. As a consequence of this theorem, we show that when the deformation parameter $l$ is properly related to the tensor field $h = \frac{1}{2}L_Z J$ on a

---

1991 *Mathematics Subject Classification.* 53C43.

*Key words and phrases.* Harmonic section, contact metric.

The author acknowledges the hospitality of Penn State's Department of Mathematics during Spring 2002, when this work was carried out. He is very grateful to Professor Augustin Banyaga for his kind invitation and helpful discussions.

contact metric structure of type $l$, then the characteristic vector field $Z$ is a minimal and a harmonic map into the corresponding unit tangent sphere bundle.

Let us point out that other examples of characteristic vector fields which are minimal and harmonic have been found by Boeckx and Vanhecke [**BVA**]. They have shown among other things, that the geodesic flow vector field on the unit tangent sphere bundle of a two-point homogeneous space is both minimal and harmonic, and determines a harmonic map.

The last section provides some applications to the 3-dimensional contact metric geometry, leading to a class of metrics in all of which some characteristic vector field $Z$ is minimal.

## 2. Definitions and properties

Consider $\mathbf{R}^{2(n+1)}$ with coordinates $x_i$, $y_i$, $i = 1, 2, .., n+1$. We denote by $\phi$ the standard complex structure defined by

$$\phi(\frac{\partial}{\partial y_i}) = -\frac{\partial}{\partial x_i},$$

and $\phi^2 = -Id$. The standard symplectic form on $\mathbf{R}^{2(n+1)}$ is

$$\omega = \sum_{i=1}^{n+1} dx_i \wedge dy_i.$$

If we denote by $g$ the euclidean metric on $\mathbf{R}^{2(n+1)}$, then $\omega$, $\phi$ and $g$ satisfy

$$\omega(X, Y) = g(X, \phi Y)$$

for any two vector fields $X$ and $Y$ on $\mathbf{R}^{2(n+1)}$.

Now consider the unit sphere

$$\mathbf{S}^{2n+1} = \{(x_1, y_1, ..., x_{n+1}, y_{n+1}) \in \mathbf{R}^{2(n+1)};\ \sum_{i=1}^{n+1} x_i^2 + y_i^2 = 1\}.$$

The vector field

$$N = \sum_{i=1}^{n+1} x_i \frac{\partial}{\partial x_i} + y_i \frac{\partial}{\partial y_i}$$

is the outward unit normal vector field along $\mathbf{S}^{2n+1}$. The standard contact structure on $\mathbf{S}^{2n+1}$ is given by the 1-form

$$\alpha = \sum_{i=1}^{n+1} x_i dy_i - y_i dx_i$$

whose characteristic vector field (see below) is

$$Z = \sum_{i=1}^{n+1} x_i \frac{\partial}{\partial y_i} - y_i \frac{\partial}{\partial x_i} = \phi N.$$

The restriction of $\phi$ to the tangent subbundle $D = kernel\ \alpha$ (also called the contact distribution) gives an almost complex structure compatible with the metric induced by $g$ on $\mathbf{S}^{2n+1}$. We define a partial almost complex operator field $J$ on $\mathbf{S}^{2n+1}$ by declaring $JZ = 0$ and $JE = \phi E$ for any section $E$ of the subbundle

$D = kernel\ \alpha$. Denoting by $g$ the restriction of $g$ to $\mathbf{S}^{2n+1}$, one easily verifies that the tensors fields $\alpha, Z, J$ and $g$ on $\mathbf{S}^{2n+1}$ satisfy the following identities.

$$J^2X = -X + \alpha(X)Z \tag{1}$$

$$\alpha(X) = g(Z, X) \tag{2}$$

$$d\alpha(X, Y) = 2g(X, JY) \tag{3}$$

$$g(JX, JY) = g(X, Y) - \alpha(X)\alpha(Y). \tag{4}$$

Identity (4) is an immediate consequence of (1), (2) and (3). We call $(\alpha, Z, J, g)$ the standard contact metric structure on $\mathbf{S}^{2n+1}$. In general, a contact form on a $2n+1$ dimensional manifold $m$ is a 1-form $\alpha$ such that $\alpha \wedge (d\alpha)^n$ is everywhere nonzero on $M$. Associated with a contact form is a characteristic vector field $Z$ determined uniquely by $\alpha(Z) = 1$ and $i_Z d\alpha = 0$. There is always a partial almost complex operator field $J$ satisfying

$$J^2X = -X + \alpha(X)Z$$

for any vector field $X$ on $M$ (See [**BLA**]).

DEFINITION 2.1. Given a contact form $\alpha$ on a manifold $M$ with characteristic vector field $Z$ and partial almost complex operator $J$, a riemannian metric $g$ is called a contact metric of type 1 if identities (2), (3) and (4) are satisfied.

Given a contact metric structure $(\alpha, Z, J, g)$ of type 1 on $M$, we introduce deformations $g_l$ of the metric $g$, called contact metrics of type $l$ where $l$ is a positive function on $M$.

$$g_l = lg + (1 - l)\alpha \otimes \alpha.$$

The data $(\alpha, Z, J, g_l)$ will be called a contact metric structure of type $l$. On a contact metric structure of any type $l$, we will consider the tensor field $h = \frac{1}{2}L_Z J$ where $L_Z$ is the usual Lie derivative. We will denote by ${}^l\nabla$, the Levi-Civita covariant derivative operator determined by the metric $g_l$. We have adopted the natural convention about exterior differentiation of forms so that for a 1-form $\omega$ and any vector fields $X$ and $Y$ on $M$, $d\omega$ is defined by

$$d\omega(X, Y) = X\omega(Y) - Y\omega(X) - \omega([X, Y]).$$

PROPOSITION 2.1. *On a contact metric structure $(\alpha, Z, J, g_l)$ of type $l$, the following identities are satisfied.*

$$\alpha(X) = g_l(Z, X) \tag{5}$$

$$d\alpha(X, Y) = \frac{2}{l} g_l(X, JY) \tag{6}$$

$$g_l(JX, JY) = g_l(X, Y) - \alpha(X)\alpha(Y) \tag{7}$$

$${}^l\nabla_X Z = -\frac{1}{l}JX - JhX - \frac{dl(Z)}{2l}J^2X. \tag{8}$$

PROOF.
$$g_l(Z, X) = lg(Z, X) + (1 - l)\alpha(X) = \alpha(X),$$
proving (5).
$$\frac{2}{l} g_l(X, JY) = 2g(X, JY) = d\alpha(X, Y),$$
proving (6).
$$\begin{aligned} g_l(JX, JY) &= \frac{l}{2} d\alpha(JX, Y) = -\frac{l}{2} d\alpha(Y, JX) \\ &= -g_l(Y, J^2X) = -g_l(Y, -X + \alpha(X)Z) \\ &= g_l(X, Y) - \alpha(X)\alpha(Y), \end{aligned}$$
proving (7). The proof of (8) uses the usual invariant definition of the Levi-Civita connection as found for instance in [**POO**].

$$\begin{aligned} 2g_l({}^l\nabla_X Z, Y) = Xg_l(Z, Y) + Zg_l(Y, X) - Yg_l(X, Z) + g_l(Y, [X, Z]) \\ g_l(Z, [Y, X]) - g_l(X, [Z, Y]). \end{aligned} \tag{9}$$

Since $g_l(Z, Z) = 1$, setting $Y = Z$ in (9), we have automatically that
$$g_l({}^l\nabla_X Z, Z) = 0. \tag{10}$$
Now, choosing $Y$ such that $\alpha(Y) = 0$ and setting $X = Z$ in (9), we see that
$$g_l({}^l\nabla_Z Z, Y) = 0 \tag{11}$$
for any $Y$ satisfying $\alpha(Y) = 0$. Combining (10) with (11), we obtain the identity
$$\nabla_Z Z = 0. \tag{12}$$

Next we choose $X$, $Y$ such that $\alpha(X) = 0 = \alpha(Y) = 0$ in (9). Using (7), (6) and $L_Z d\alpha = 0$, we see that
$$\begin{aligned} 2g_l({}^l\nabla_X Z, Y) &= Zg_l(Y, X) + g_l(Y, [X, Z]) + \alpha([Y, X]) - g_l(X, [Z, Y]) \\ &= Zg_l(JY, JX) + g_l(JY, J[X, Z]) + \\ &+ \alpha([Y, X]) - g_l(JX, J[Z, Y]) \\ &= Z(\frac{l}{2} d\alpha(JX, Y)) + \frac{l}{2} d\alpha(J[X, Z], Y) + \alpha([Y, X]) - \\ &\frac{l}{2} d\alpha(JX, [Z, Y]) \\ &= \frac{1}{2} dl(Z) d\alpha(JX, Y) + \frac{l}{2} L_Z d\alpha(JX, Y) + \frac{l}{2} d\alpha(L_Z JX, Y) + \\ &\alpha([Y, X]) \\ &= -\frac{dl(Z)}{l} g_l(J^2X, Y) - g_l(Y, JL_Z JX) - i_Y d\alpha(X) \\ &= -\frac{dl(Z)}{l} g_l(J^2X, Y) - g_l(Y, JL_Z JX) - d\alpha(Y, X) \\ &= -\frac{dl(Z)}{l} g_l(J^2X, Y) - g_l(Y, JL_Z JX) - \frac{2}{l} g_l(Y, JX) \end{aligned}$$
$$g_l(2{}^l\nabla_X Z, Y) = g_l(-\frac{dl(Z)}{l} J^2X - JL_Z JX - \frac{2}{l} JX, Y)$$

Therefore, for any $X$ such that $\alpha(X) = 0$, we have

(13) $$ {}^l\nabla_X Z = -\frac{1}{l}JX - JhX - \frac{1}{2l}dl(Z)J^2X. $$

Combining (12) and (13), we obtain for arbitrary $X$,

$$ \begin{aligned} {}^l\nabla_X Z &= {}^l\nabla_{(X-\alpha(X)Z)}Z + {}^l\nabla_{\alpha(X)Z}Z \\ &= -\frac{1}{l}J(X-\alpha(X)Z) - Jh(X-\alpha(X)) - \frac{1}{2l}dl(Z)J^2(X-\alpha(X)Z) \\ &= -\frac{1}{l}JX - JhX - \frac{1}{2l}dl(Z)J^2X. \end{aligned} $$

This completes the proof of Proposition 2.1. □

The expression of the covariant derivative of $J$ is given by the following lemma.

LEMMA 2.1. *Let $(M, \alpha, Z, J, g_l)$ be a contact metric manifold of type $l$ satisfying at least one of the following conditions.*

1. *$l$ is constant,*
2. *$M$ is 3-dimensional and $dl(Z) = 0$.*

*Then, the identity*

(14) $$ \begin{aligned} 2g_l(({}^l\nabla_X J)Y, W) = g_l(&-[Y,W] + [JY,JW] - J[Y,JW] - J[JY,W], JX) \\ &+ d\alpha(JY,X)\alpha(W) - d\alpha(JW,X)\alpha(Y) \end{aligned} $$

*holds for any vector fields $X$, $Y$ and $W$ on $M$.*

PROOF.

$$\begin{aligned}
2g_l(({}^l\nabla_X J)Y, W) &= 2g_l({}^l\nabla_X JY, W) + 2g_l({}^l\nabla_X Y, JW) \\
&= Xg_l(JY, W) + JYg_l(W, X) - Wg_l(X, JY) \\
&+ g_l(W, [X, JY]) + g_l(JY, [W, X]) - g_l(X, [JY, W]) \\
&+ Xg_l(Y, JW) + Yg_l(JW, X) - JWg_l(X, Y) \\
&+ g_l(JW, [X, Y]) + g_l(Y, [JW, X]) - g_l(X, [Y, JW]) \\
&= X(\frac{l}{2}d\alpha(W, Y)) + JY(\frac{l}{2}d\alpha(JW, X) + \alpha(W)\alpha(X)) \\
&- W(\frac{l}{2}d\alpha(X, Y)) + \frac{l}{2}d\alpha(JW, [X, JY]) + \alpha(W)\alpha([X, JY]) \\
&+ \frac{l}{2}d\alpha([W, X], Y) - \frac{l}{2}d\alpha(JX, [JY, W]) - \alpha(X)\alpha([JY, W]) \\
&+ X(\frac{l}{2}d\alpha(Y, W)) + Y(\frac{l}{2}d\alpha(X, W)) - JW(\frac{l}{2}d\alpha(JX, Y) + \\
&\alpha(X)\alpha(Y)) + \frac{l}{2}d\alpha([X, Y], W) + \frac{l}{2}d\alpha(JY, [JW, X]) + \\
&\alpha(Y)\alpha([JW, X]) - \frac{l}{2}d\alpha(JX, [Y, JW]) - \alpha(X)\alpha([Y, JW]) \\
&= d(\frac{l}{2}d\alpha)(X, W, Y) + d(\frac{l}{2}d\alpha)(JY, JW, X) \\
&+ \frac{l}{2}d\alpha([JY, JW], X) + \frac{l}{2}d\alpha([W, Y], X) - \frac{l}{2}d\alpha(JX, [JY, W]) \\
&- \frac{l}{2}d\alpha(JX, [Y, JW]) + L_{JY}\alpha(W)\alpha(X) + \alpha(W)L_{JY}\alpha(X) \\
&- L_{JW}\alpha(X)\alpha(Y) - \alpha(X)L_{JW}\alpha(Y).
\end{aligned}$$

If $l$ is constant or $M$ is 3-dimensional and $dl(Z) = 0$, then

$$d(\frac{l}{2}d\alpha) = 0.$$

Indeed, on a 3-dimensional contact manifold $(M, \alpha, Z)$, if $dl(Z) = 0$, then

$$d(ld\alpha) = dl \wedge d\alpha = dl(Z)\alpha \wedge d\alpha = 0.$$

Under these conditions, the above calculation is continued as follows.

$$\begin{aligned}
2g_l(({}^l\nabla_X J)Y, W) &= g_l([JY, JW], JX) + g_l([W, Y], JX) - g_l(JX, J[JY, W]) \\
&- g_l(JX, J[Y, JW]) + d\alpha(JY, W)\alpha(X) + \\
&\alpha(W)d\alpha(JY, X) - d\alpha(JW, X)\alpha(Y) - \alpha(X)d\alpha(JW, Y) \\
&= g_l(-[Y, W] + [JY, JW] - J[JY, W] - J[Y, JW], JX) \\
&+ \alpha(W)d\alpha(JY, X) - \alpha(Y)d\alpha(JW, X).
\end{aligned}$$

□

As a consequence of Lemma 2.1, we have the following proposition.

PROPOSITION 2.2. *Let $(M, \alpha, Z, J, g_l)$ be a contact metric manifold of type $l$ satisfying at least one of the following conditions.*
1. *$l$ is constant,*
2. *$M$ is 3-dimensional and $dl(Z) = 0$.*

*Then the identity*

$$({}^l\nabla_X J)Y + ({}^l\nabla_{JX} J)JY = \frac{2}{l} g_l(Y, X)Z - \frac{1}{l}\alpha(Y)(X + lhX + \alpha(X)Z) \tag{15}$$

*holds for any vector fields $X$ and $Y$ on $M$.*

PROOF. Applying identity (14) to LHS $=({}^l\nabla_X J)Y + ({}^l\nabla_{JX} J)JY$, one sees that

$$\begin{aligned}
2g_l(LHS, W) &= 2g_l(({}^l\nabla_X J)Y, W) + 2g_l(({}^l\nabla_{JX} J)JY, W)\\
&= g_l(-[Y, W] + [JY, JW] - J[Y, JW] - J[JY, W], JX)\\
&+ d\alpha(JY, X)\alpha(W) - d\alpha(JW, X)\alpha(Y) + g_l(-[JY, W] + [J^2Y, JW]\\
&- J[JY, JW] - J[J^2Y, W], J^2X) + d\alpha(J^2Y, JX)\alpha(W)\\
&= g_l(-J[Y, W] + J[JY, JW] - J^2[Y, JW] - J^2[JY, W], J^2X)\\
&+ g_l(-[JY, W] + [J^2Y, JW] - J[JY, JW] - J[J^2Y, W], J^2X)\\
&+ d\alpha(JY, X)\alpha(W) - d\alpha(JW, X)\alpha(Y) - d\alpha(Y, JX)\alpha(W)\\
&= g_l(-J[Y, W] + J[JY, JW] + [Y, JW] - \alpha([Y, JW])Z\\
&+ [JY, W] - \alpha([JY, W])Z - [JY, W] - [Y, JW] + [\alpha(Y)Z, JW]\\
&- J[JY, JW] + J[Y, W] - J[\alpha(Y)Z, W], J^2X)\\
&+ \frac{4}{l} g_l(JY, JX) g_l(Z, W) - \frac{2}{l}\alpha(Y)(g_l(X, W) - \alpha(X) g_l(Z, W))\\
&= g_l(-\alpha(Y)[Z, JW] - \alpha(Y)J[Z, W], J^2X)\\
&+ g_l(\frac{4}{l} g_l(JY, JX)Z - \frac{2}{l}\alpha(Y)(X - \alpha(X)Z), W)\\
&= g_l(\alpha(Y)(L_Z J)W, J^2X) + g_l(\frac{4}{l} g_l(JY, JX)Z-\\
&\frac{2}{l}\alpha(Y)(X - \alpha(X)Z), W)\\
&= g_l(\alpha(Y)(L_Z J)J^2X, W) + g_l(\frac{4}{l} g_l(JY, JX)Z-\\
&\frac{2}{l}\alpha(Y)(X - \alpha(X)Z), W).
\end{aligned}$$

Therefore,

$$\begin{aligned}
({}^l\nabla_X J)Y + ({}^l\nabla_{JX} J)JY &= \frac{1}{2}\{\alpha(Y)(L_Z J)(-X + \alpha(X)Z) + \frac{4}{l} g_l(Y, X)Z-\\
&\frac{4}{l}\alpha(Y)\alpha(X)Z - \frac{2}{l}\alpha(Y)(X - \alpha(X)Z\}\\
&= \frac{2}{l} g_l(Y, X)Z - \frac{1}{l}\alpha(Y)(X + lhX + \alpha(X)Z).
\end{aligned}$$

$\square$

## 3. Minimal and harmonic maps

For a smooth immersion $f: M \to (N, g)$ of a manifold $M$ into a riemannian manifold $N$, the second fundamental form $B$ is a normal bundle valued symmetric, bilinear form on $f(M)$ defined for any two vector fields $X$, $Y$ by $B(X,Y) = (\nabla_X Y)^{\perp}$. The notation $X^{\perp}$ is for the component of $X$ normal to $f(M)$ in $N$. The symbol $\nabla$ represents the usual Levi-Civita covariant derivative operator determined by the metric $g$. The mean curvature vector field $H$ of the immersion $f$ is the trace of the second fundamental form $B$, that is, given an orthonormal basis $\{E_i\}_{i=1,\dots,m}$ on $f(M)$,

$$H = \sum_{i=1}^{m} B(E_i, E_i).$$

The mean curvature vector field $H$ appears in the first variation formula for the volume functional $\mathcal{F}$, which is defined for a compact manifold $M$ by (see [**EEL**])

$$\mathcal{F}(f) = \int_M f^*\Omega$$

where $\Omega$ is the riemannian volume on $N$. An immersion $f$ is a critical point of the volume functional, or simply a minimal immersion, if $H = 0$. An immersion of a not necessarily compact manifold $M$ into a riemannian manifold $N$ will also be called a minimal immersion if its mean curvature vector field $H$ vanishes.

For a map $f: (M, g_1) \to (N, g_2)$ between two riemannian manifolds, the energy $\mathcal{E}(f)$ is defined by (see [**GIO**])

$$\mathcal{E}(f) = \frac{1}{2} \int_M \|f_* E_i\|^2 dM$$

where $\{E_i\}_{i=1,\dots,m}$ is any $g_1$ orthonormal frame field on $M$ and $dM$ is the riemannian volume of $(M, g_1)$. The tension field $\tau(f)$ of the map $f$ is given by (see [**EEL**])

$$\tau(f) = \sum_{i=1}^{m} {}^2\nabla_{f_* E_i} f_* E_i - f_*({}^1\nabla_{E_i} E_i).$$

In the above identity, ${}^i\nabla$ represents the Levi-Civita connection of the metric $g_i$. A map $f$ is critical for the energy functional if and only if its tension field $\tau(f)$ vanishes, in which case the map is called a harmonic map [**EEL**]. The vanishing of $\tau$ will be used as the criticality condition for the energy even when $M$ is not compact. It is worth pointing out now that for isometric immersions, minimality and harmonicity are equivalent. Using the notation $f_*$ for the differential of a map $f$, we formulate the following definition.

DEFINITION 3.1. A map $f: (M_1, \alpha_1, Z_1, J_1) \to (M_2, \alpha_2, Z_2, J_2)$ between two contact manifolds is said to be $J$-holomorphic if the identity

$$f_* J_1 X = J_2 f_* X$$

is satisfied for any vector field $X$ on $M_1$.

For $J$-holomorphic maps, one has the following lemma, the proof of which can be found in [**IAN**].

LEMMA 3.1. *Let* $f\colon (M_1, \alpha_1, Z_1, J_1) \to (M_2, \alpha_2, Z_2, J_2)$ *be a* $J$*-holomorphic map between two contact manifolds. Then*

$$f_* Z_1 = c Z_2 \tag{16}$$

*for some constant* $c$.

J-holomorphic maps between contact metric structures have the following remarkable property.

THEOREM 3.1. *Let* $(M_i, \alpha_i, Z_i, J_i, g_{l_i})$, $i = 1, 2$ *be contact metric manifolds of type* $l_i$, *each satisfying at least one of the following conditions.*

1. $l_i$ *is constant.*
2. $M_i$ *is 3-dimensional, and* $dl_i(Z_i) = 0$.

*Then, any* $J$*-holomorphic map*

$$f\colon (M_1, \alpha_1, Z_1, J_1, g_{l_1}) \to (M_2, \alpha_2, Z_2, J_2, g_{l_2})$$

*is a harmonic map.*

PROOF. Let the dimension of $M_1$ be $2n+1$ and $\{Z_1,\ E_i,\ JE_i\}$, $i = 1, ..., n$ be an adapted local orthonormal frame field on $M_1$. In view of Lemma 3.1 and the identity ${}^{l_1}\nabla_{Z_1} Z_1 = 0 = {}^{l_2}\nabla_{Z_2} Z_2$, the tension field $\tau(f)$ can be written as

$$\sum_{i=1}^{n} {}^{l_2}\nabla_{f_* E_i} f_* E_i + {}^{l_2}\nabla_{f_* J_1 E_i} f_* J_1 E_i - f_*({}^{l_1}\nabla_{E_i} E_i + {}^{l_1}\nabla_{J_1 E_i} J_1 E_i).$$

Setting $X = f_* E_i$, $Y = J_2 f_* E_i$ in formula (15), one obtains

$$({}^{l_2}\nabla_{f_* E_i} J_2) J_2 f_* E_i + ({}^{l_2}\nabla_{J_2 f_* E_i} J_2) J_2 f_* E_i = 0. \tag{17}$$

That is,

$$-{}^{l_2}\nabla_{f_* E_i} f_* E_i - J_2 {}^{l_2}\nabla_{f_* E_i} J_2 f_* E_i - {}^{l_2}\nabla_{J_2 f_* E_i} J_2 f_* E_i + J_2 {}^{l_2}\nabla_{J_2 f_* E_i} f_* E_i = 0. \tag{18}$$

Similarily, setting $X = E_i$, $Y = J_1 E_i$ in formula (15), one obtains

$$-{}^{l_1}\nabla_{E_i} E_i - J_1 {}^{l_1}\nabla_{E_i} J_1 E_i - {}^{l_1}\nabla_{J_1 E_i} J_1 E_i + J_1 {}^{l_1}\nabla_{J_1 E_i} E_i = 0. \tag{19}$$

From identity (18), we deduce that

$${}^{l_2}\nabla_{f_* E_i} f_* E_i + {}^{l_2}\nabla_{J_2 f_* E_i} J_2 f_* E_i = J_2 [J_2 f_* E_i, f_* E_i] \tag{20}$$

and from identity (19), we get

$${}^{l_1}\nabla_{E_i} E_i + {}^{l_1}\nabla_{J_1 E_i} J_1 E_i = J_1 [J_1 E_i, E_i]. \tag{21}$$

It follows from identities (20) and (21) that the tension field $\tau(f)$ is given by

$$\begin{aligned}
\tau(f) &= \sum_{i=1}^{n} J_2 [J_2 f_* E_i, f_* E_i] - f_* J_1 [J_1 E_i, E_i] \\
&= \sum_{i=1}^{n} J_2 [J_2 f_* E_i, f_* E_i] - J_2 f_* [J_1 E_i, E_i] \\
&= \sum_{i=1}^{n} J_2 [f_* J_1 E_i, f_* E_i] - J_2 [f_* J_1 E_i, f_* E_i] = 0,
\end{aligned}$$

hence, $f$ is a harmonic map. □

Another key concept in this paper is contained in the following definition.

DEFINITION 3.2. A map $f: M \to (N, \alpha, Z, J)$ from an odd dimensional manifold $M$ into a contact manifold $(N, \alpha, Z, J)$ is said to be contact invariant if $Z$ is everywhere tangent to $f(M)$ and $JX$ is tangent to $f(M)$ whenever $X$ is tangent to $f(M)$.

It is clear from the definition that a contact invariant immersion

$$f: M \to (N, \alpha, Z, J)$$

induces a contact structure $(f^*\alpha, Z', J')$ on $M$ where $Z'$ and $J'$ are determined respectively by the equation $f_* Z'(p) = Z(f(p))$ and $f_* J'X = Jf_*X$. With these new structure tensors,

$$f: (M, f^*\alpha, Z', J') \to (N, \alpha, Z, J)$$

becomes a $J$-holomorphic map.

## 4. The unit tangent sphere bundle

The Liouville 1-form $\beta$ is defined on the cotangent bundle $T^*M$ of a manifold $M$ by

$$\beta_\mu(v) = \mu(\pi_* v)$$

where $v \in T_\mu T^*M$ and

$$\pi: T^*M \to M$$

is the bundle projection map. The 2-form $\Omega = -d\beta$ is a symplectic form on $T^*M$. Given a riemannian metric $g$ on $M$, the 1-form $\beta$ pulls back under the duality map to a 1-form $\beta'$ on the tangent bundle $TM$ of $M$. There also, the two form $\Omega' = -d\beta'$ is a symplectic form (See [**POO**], [**BLA**] or [**TON**]). If we denote by

$$\kappa: TTM \to TM$$

the Levi-Civita connection map and by $\pi$ the tangent bundle projection map

$$\pi: TM \to M,$$

then we have a vector bundle isomorphism along $\pi: TM \to M$,

$$(\pi_*, \kappa): TTM \to TM \oplus TM. \tag{22}$$

Under this identification, one defines an almost complex structure $J_{TM}$ on $TM$ as follows. If $x \in TTM$ and $(\pi_*, \kappa)(x) = (u, v)$, then

$$(\pi_*, \kappa)(J_{TM}x) = (-v, u).$$

The tangent bundle of a riemannian manifold $M$ carries a distinguished vector field $S$, the geodesic spray. For $(p, v) \in TM$, $S(p, v)$ is determined by the equation

$$(\pi_*, \kappa)(S(p, v)) = (v, 0).$$

The Sasaki metric $G$ on $TM$ is defined by by

$$G(X, Y) = g(\pi_* X, \pi_* Y) + g(\kappa X, \kappa Y)$$

for any vector fields $X$, $Y$ on $TM$. With respect to the Sasaki metric $G$, the 1-form $\beta'$ is the dual of $S$, that is, the identity

$$\beta'(X) = G(X, S)$$

is satisfied for any vector field $X$ on $TM$. The symplectic form $\Omega' = -d\beta'$ and the metric $G$ are compatible in the sense that the identity

$$\Omega'(X, Y) = G(X, J_{TM}Y)$$

is satisfied.

Setting

$$j: T^1M \to TM$$

to be the inclusion of the unit tangent sphere bundle as a hypersurface in $TM$, the 1-form $\eta = j^*\beta$ is a contact form with characteristic vector field $S$. The restriction of $J_{TM}$ to the kernel of $\eta$ extends to a partial almost complex operator $J_{T^1M}$ by requiring $J_{T^1M}S = 0$.

According to our definition and remembering our convention that

$$d\eta(X, Y) = X\eta(Y) - Y\eta(X) - \eta([X, Y]),$$

one easily checks that the contact metric structure determined by the tensor fields $(\eta, S, J_{T^1M}, j^*G)$ is of type 2, since $\eta$ and $j^*G$ satisfy $d\eta(X, Y) = j^*G(X, J_{T^1M}Y)$ (see [**BLA**]). We will refer to this structure as the Sasaki contact metric structure on $T^1M$.

Under the identification (22), a tangent vector to $T^1M$ at $(p, v) \in T^1M$ is just a couple $(u, y) \in T_pM \times T_pM$ where $u = 0$ or $y = \nabla_u W$ and $W$ is a local unit vector field on $M$ such that $W(p) = v$. Now, if $Z$ is a unit vector field on $M$, then a tangent vector to $Z(M)$ at $(p, Z)$ is a couple $(u, \nabla_u Z)$, where $u \in T_pM$. From now on, by contact metric structure of type $l$ on $T^1M$ we will mean a contact metric structure in the deformation class of the Sasaki contact metric structure, that is, the deformation on which $l = 2$ corresponds to the Sasaki contact metric structure we have just described. We refer to [**DOM**] for more details on the geometry of the tangent bundles.

THEOREM 4.1. *Let $(M, \alpha, Z, J, g_l)$ be a contact metric manifold of type $l$ and $T^1M$ be endowed with any contact metric structure $(\eta, S, J_{T^1M}, G_s)$ of type $s$ in the deformation class of the Sasaki contact metric structure determined by $g_l$. Then the vector field*

$$Z: M \to (T^1M, \eta, S, J_{T^1M}, G_s)$$

*is a contact invariant embedding if and only if $dl(Z) = 0$ and $h^2X = (\frac{1}{l^2} - 1)X$ for any vector field $X$ on $M$ such that $\alpha(X) = 0$.*

PROOF. The geodesic spray $S$ is clearly tangent to $Z(M)$ since $(\pi_*, \kappa)S(p, v) = (Z, 0) = (Z, \nabla_Z Z)$. Let $(u, {}^l\nabla_u Z)$ be a tangent vector at $(p, Z) \in Z(M)$ such that $\eta((u, {}^l\nabla_u Z)) = 0$, that is $g_l(u, Z) = 0$. By formula (8), we have

$$J_{T^1M}(u, {}^l\nabla_u Z) = (-{}^l\nabla_u Z, u) = (\frac{1}{l}Ju + Jhu + \frac{dl(Z)}{2l}J^2u, u). \tag{23}$$

Therefore we see that $J_{T^1M}(u, {}^l\nabla_u Z)$ will be tangent to $Z(M)$ if and only if

$${}^l\nabla_{\frac{1}{l}Ju + Jhu + \frac{dl(Z)}{2l}J^2u} Z = u.$$

A careful calculation shows that

$${}^l\nabla_{\frac{1}{l}Ju + Jhu + \frac{dl(Z)}{2l}J^2u} Z = (\frac{1}{l^2} - \frac{dl(Z)^2}{4l^2})u + \frac{dl(Z)}{l^2}Ju - h^2u. \tag{24}$$

So, $J_{T^1M}(u, \nabla_u Z)$ will be tangent to $Z(M)$ if and only if

$$(\frac{1}{l^2} - \frac{dl(Z)^2}{4l^2} - 1)u + \frac{dl(Z)}{l^2}Ju - h^2u = 0. \tag{25}$$

Pick $u$ to be an eigenvector of $h$ with eigenvalue $\lambda$, then substituting in (25),

$$(\frac{1}{l^2} - \frac{dl(Z)^2}{4l^2} - 1)u + \frac{dl(Z)}{l^2}Ju - \lambda^2u = 0. \tag{26}$$

Since $Ju$ and $u$ are linearly independent, identity (26) holds if and only if $dl(Z) = 0$ and $\frac{1}{l^2} - 1 - \lambda^2 = 0$. □

The following corollary is an immediate consequence of Theorem 3.1 and Theorem 4.1.

COROLLARY 4.1. *Let $(M, \alpha, Z, J, g_l)$ be a contact metric manifold of type $l$ satisfying at least one of the following conditions.*

1. *$l$ is constant,*
2. *$M$ is 3-dimensional and $dl(Z) = 0$.*

*Let $(\eta, S, J_{T^1M}, G_s)$ denote a contact metric structure of constant type $s$ on $T^1M$ in the deformation class of the Sasaki contact metric structure determined by $g_l$. If*

$$h^2X = (\frac{1}{l^2} - 1)X$$

*for any section $X$ of the contact distribution on $M$, then the vector field*

$$Z: M \to (T^1M, \eta, S, J_{T^1M}, G_s)$$

*is a minimal embedding.*

We will call K-contact structure of type $l$ any contact metric structure of type $l$ whose tensor field $h$ vanishes identitcally. Note that the characteristic vector field of a K-contact structure $(\alpha, Z, J, g_l)$ of type $l$ will be Killing if and only if $l$ is a $Z$-basic function. From the harmonic maps point of view we have the following theorem which is also a characterisation of K-contact structures of type 1.

THEOREM 4.2. *Let $(M, \alpha, Z, J, g_l)$ be a contact metric manifold of type $l$ and $T^1M$ be endowed with any contact metric structure $(\eta, S, J_{T^1M}, G_s)$ of type $s$ in the deformation class of the Sasaki contact metric structure determined by $g_l$. Then, the vector field*

$$Z: (M, \alpha, Z, J, g_l) \to (T^1M, \eta, S, J_{T^1M}, G_s)$$

*is a $J$-holomorphic map if and only if $l = 1$ and $h = 0$, that is, if and only if $(M, \alpha, Z, J, g_l)$ is a K-contact manifold of type 1.*

PROOF. Clearly, $Z_*Z = (Z, 0) = S(p, Z)$, so $Z$ maps the corresponding characteristic vector fields into one another. We need to see when $Z$ intertwines the partial almost complex operators $J$ and $J_{T^1M}$. Let $v \in T_pM$ be a tangent vector such that $\alpha(v) = 0$. On one hand

$$\begin{aligned} J_{T^1M}Z_*v &= J_{T^1M}(\pi_*Z_*v, \kappa Z_*v) \\ &= J_{T^1M}(v, {}^l\nabla_v Z) = (\frac{1}{l}Jv + Jhv + \frac{dl(Z)}{2l}J^2v, v). \end{aligned}$$

On the other hand

$$Z_*Jv = (Jv, {}^l\nabla_{Jv}Z) = (Jv, -\frac{1}{l}J^2v - JhJv - \frac{dl(Z)}{2l}J^3v)$$
$$= (Jv, \frac{1}{l}v - hv + \frac{dl(Z)}{2l}Jv).$$

So we see that $Z_*Jv = J_{T^1M}Z_*v$ if and only if $l = 1$ and $h = 0$, that is, if and only if $(M, \alpha, Z, J, g_l)$ is a K-contact manifold of type 1. □

As a consequence of Theorem 4.1, Theorem 4.2 and Corollary 4.1, one has the following result from [**RUK**].

PROPOSITION 4.1. *The characteristic vector field of a K-contact structure of type* 1 *is a minimal embedding and a harmonic map into the unit tangent sphere bundle with its Sasaki metric.*

PROOF. On a K-contact metric structure, one has $h = 0$ by definition. If the type is $l = 1$, then the conditions of Corollary 4.1 and Theorem 4.2 are satisfied. By Theorem 3.1, $Z$ is also a harmonic embedding. □

REMARK. Actually, Proposition 4.1 is still true if the Sasaki metric is replaced by any of its constant type deformations.

## 5. Application to 3-dimensional contact geometry

In Theorem 4.1, the condition for the characteristic vector field $Z$ to be a contact invariant embedding into $T^1M$ requires the tensor field $h^2$ (restricted to the contact distribution) to be multiplication by a constant or, in dimension 3, multiplication by a positive function which is constant along integral curves of $Z$. The K-contact structures ($h = 0$) meet the above requirement in any odd dimension and the flat contact metric structures do so in dimension 3 where $h^2X = X$ for any $X$ in the contact distribution (see [**RU1**]). One could ask if there are any nontrivial examples of contact metric structures such that $h^2$ is multiplication by a basic, nonconstant function. We will provide a class of such examples in dimension 3.

Recall [**RU1**] that on a closed, flat, 3-dimensional contact metric manifold $(M, \alpha, Z, J, g)$ of type 1, there is an orthonormal parallelisation $\{Z, E, JE\}$ where

(27) $$hE = -E, \; hJE = JE, \; [Z, E] = 0, \; \nabla(JE) = 0$$

(28) $$[Z, JE] = -2E, \; [E, JE] = 2Z.$$

$Z$ and $E$ determine a totally geodesic codimension 1 foliation of which $JE$ is a foliate vector field. Since $JE$ is a nontrivial parallel vector field, the 1-form dual to $JE$ is a harmonic nonsingular 1-form. It follows from Tischler's theorem [**TIS**] that $M$ fibers over $\mathbf{S}^1$. Moreover, in the situation at hand, a fibration can be found so that $Z$ is tangent to the fibers. Therefore, any smooth function on the base $\mathbf{S}^1$ lifts into a smooth function $f$ on $M$ satisfying $df(Z) = 0$. Given such a $Z$-basic function $f$, consider the deformation $g^f$ of the metric $g$, which is given by the following identities.

$$
\begin{aligned}
g^f(Z,Z) = 1,\ g^f(E,E) = 1 + f + \frac{1}{2}f^2, \\
g^f(E,JE) = \frac{1}{2}f^2,\ g^f(JE,JE) = 1 - f + \frac{1}{2}f^2 \\
g^f(Z,E) = 0 = g^f(Z,JE).
\end{aligned} \tag{29}
$$

A skew adjoint endomorphism $J_f$ of $TM$ is determined by setting

$$d\alpha(X,Y) = 2g^f(X, J_f Y).$$

Using the notation $h_f$ for $\frac{1}{2}L_Z J_f$, we prove the following.

PROPOSITION 5.1. *$(M, \alpha, Z, J_f, g^f)$ is a contact metric manifold of type 1 whose tensor field $h_f^2$ is multiplication by the basic function $(1 + f + \frac{1}{2}f^2)^2$ when restricted to kern $\alpha$.*

PROOF. In the basis $\{Z, E, JE\}$, the tensor field $J_f$ is given by the following identities.

$$J_f Z = 0 \tag{30}$$

$$J_f E = \frac{f^2}{2}E - (1 + f + \frac{1}{2}f^2)JE \tag{31}$$

$$J_f JE = (1 - f + \frac{1}{2}f^2)E - \frac{f^2}{2}JE. \tag{32}$$

The square of $J_f$ satisfies $J_f^2 Z = 0$, $J_f^2 E = -E$ and $J_f^2 JE = -JE$. This shows that $J_f$ is a partial almost complex operator associated with $\alpha$ and $(M, \alpha, Z, J_f, g^f)$ is a contact metric manifold of type 1. Next we compute $h_f E$ using identities (30), (31), (32), (27), and (28).

$$
\begin{aligned}
h_f E = \frac{1}{2}(L_Z J_f)E &= \frac{1}{2}([Z, J_f E] - J_f[Z,E]) \\
&= \frac{1}{2}[Z, \frac{f^2}{2}E - (1 + f + \frac{1}{2}f^2)JE] - 0 \\
&= \frac{1}{2}(1 + f + \frac{1}{2}f^2)[Z, JE] = -(1 + f + \frac{1}{2}f^2)E.
\end{aligned}
$$

Thus, $h_f^2 E = (1 + f + \frac{1}{2}f^2)^2 E$. Since $M$ is 3-dimensional, this means that $h_f^2$ (restricted to the contact distribution) is multiplication by the basic function $(1 + f + \frac{1}{2}f^2)^2$. □

Here is an application of Theorem 4.1 and Proposition 5.1 to the 3-dimensional contact geometry. In [**RU2**] (see also [**RUK**]), it was shown that on a flat contact metric manifold $(M, \alpha, Z, J, g)$ of type 1, the characteristic vector field $Z$ was a minimal embedding into the corresponding unit tangent bundle, but, as it was pointed out, $Z$ is not a contact invariant embedding. As a consequence of Theorem 4.1 and Proposition 5.1, there is a class of metrics parameterized by smooth basic functions on $M$ for which $Z$ is a minimal and contact invariant embedding.

PROPOSITION 5.2. *Let $(M, \alpha, Z, J, g)$ be a 3-dimensional contact metric manifold of type 1 whose tensor field $h^2$ (restricted to the contact distribution) is multiplication by a basic function $\lambda^2$. Let $(\alpha, Z, J, g_l)$ be the deformed contact metric structure of type $l = \frac{1}{\sqrt{\lambda^2+1}}$ and $(\eta, S, J_{T^1M}, G_s)$ a contact metric structure of type $s$ on $T^1M$ with $ds(S) = 0$. Further, suppose that $(\eta, S, J_{T^1M}, G_s)$ is in the deformation class of the Sasaki contact metric structure derived from $g_l$. Then the vector field*

$$Z: M \to (T^1M, \eta, S, J_{T^1M})$$

*is a contact invariant embedding. In case $s$ is constant, then*

$$Z: M \to (T^1M, G_s)$$

*is also a minimal embedding.*

PROOF. Under the hypothesis, $(\frac{1}{l^2} - 1)X = (\lambda^2 + 1 - 1)X = \lambda^2 X = h^2 X$ for any section $X$ of the contact distribution. Therefore, the condition of Theorem 4.1 is satisfied and $Z$ is a contact invariant embedding. If $s$ is constant, then the condition of Corollary 4.1 is met, so $Z$ is a minimal embedding in this case. □

## References


[BLA] D. Blair, *Riemannian geometry of contact and symplectic manifolds*, vol. 203, Progress in Mathematics, Birkhäuser, Boston, Mass., 2002.

[BVA] E. Boeckx, L. Vanhecke, *Harmonic and minimal vector fields on tangent and unit tangent bundles*, Diff. Geom. Appl. **13** (2000), 77–93.

[DOM] P. Dombrowski, *On the geometry of the tangent bundles*, J. Reine Angew. Math. **210** (1962), 73–88.

[EEL] J. Eells, L. Lemaire, *Selected topics in harmonic maps, CBMS*, vol. 50, AMS, Providence, RI, 1983.

[GIM] O. Gil-Medrano, *On the volume functional in the manifold of unit vector fields*, Publ. Depto. Geometria y Topologia, Univ. Santiago de Compostela **89** (1998), 155–163.

[GIO] O. Gil-Medrano, *Relationship between volume and energy of vector fields*, Diff. Geom. Applic. **15** (2001), 137–152.

[IAN] S. Ianus, A. M. Pastore, *Harmonic maps on contact metric manifolds*, Ann. Math. Blaise Pascal **2 (2)** (1995), 45–53.

[OLS] Z. Olszak, *On contact metric manifolds*, Tôhoku Math. J. **31** (1979), 247–253.

[POO] W. Poor, *Differential geometric structures*, McGraw-Hill Book Company, New York, 1981.

[RU2] P. Rukimbira, *Criticality of unit contact vector fields*, Proc. Intern. Conf. Infinite dimensional Lie groups and representation theory, Howard University, Washington, D.C., August 2000, World Sci. Publ., River Edge, NJ, 2002, pp. 105–115.

[RUK] P. Rukimbira, *Criticality of K-contact vector fields*, Jour. Geom. Physics **40** (2002), 209–214.

[RU1] P. Rukimbira, *A characterization of flat contact metric geometry*, Hous. J. Math. **24** (1998), 409–414.

[TAN] S. Tanno, *The topology of contact riemannian manifolds*, Illinois J. Math. **12** (1968), 700–717.

[TIS] D. Tischler, *On fibering certain foliated manifolds over* $\mathbf{S}^1$, Topology **9** (1970), 153–154.

[TON] P. Tondeur, *Structure presque kahlerienne naturelle sur le fibré des vecteurs covariants d'une variété riemannienne*, C.R. Acad. Sci. Paris **254** (1962), 407–408.



DEPARTMENT OF MATHEMATICS, FLORIDA INTERNATIONAL UNIVERSITY, MIAMI, FLORIDA 33199

*E-mail address*: `rukim@ fiu.edu`

Contemporary Mathematics
Volume **337**, 2003

# Holomorphically Planar Conformal Vector Fields On Almost Hermitian Manifolds

Ramesh Sharma

ABSTRACT. We introduce the notion of a holomorhically planar conformal vector (HPCV) field on an almost Hermitian manifold. First we generalize a result of Goldberg, showing that an HPCV on a Kaehler manifold is homothetic and analytic. Then we have proved some results on Kaehler and almost Kaehler manifolds whose curvature/Weyl conformal tensor is harmonic. In particular, we have characterized an HPCV field on the Euclidean $C, S^2, C^n$, and the nearly Kaehler $S^6$.

## 1. Introduction

Let $M$ denote a real $2n$-dimensional smooth manifold with a $(1,1)$- tensor field $J$ and a Riemannian metric $g$ such that $J^2 = -I$ and $g(JX, JY) = g(X, Y)$ for arbitrary vector fields $X, Y$ on $M$. Such a manifold is said to be an almost Hermitian manifold with almost complex structure $J$ and Hermitian metric $g$. An almost Hermitian manifold is called an almost Kaehler manifold if the fundamental 2-form $\Omega$ defined by $\Omega(X, Y) = g(X, JY)$, is closed, i.e. $d\Omega = 0$. An almost Kaehler manifold whose almost complex structure is integrable (i.e. the Nijenhuis tensor of $J$ vanishes on $M$), is known as a Kaehler manifold. It is well known that an almost Kaehler manifold is Kaehler iff $\nabla J = 0$, where $\nabla$ is the Riemannian connection of $g$. One may note that an almost Kaehler manifold is a symplectic manifold with $\Omega$ as the symplectic form. In the literature we find that almost Kaehler manifolds have been studied assuming certain curvature conditions and/or compactness (for details we refer to the survey article [**AAD**]). On the other hand we have the following result based on the assumption of a conformal symmetry: "A conformal vector field on a compact almost Kaehler manifold $M$ is Killing, and is also analytic if $dim.M > 2$" (stated in [**YA1**]). For a closed conformal vector field on a Kaehler (not necessarily compact) manifold we have the following result (stated in [**GOL**]).

2000 *Mathematics Subject Classification.* Primary 53B35, Secondary 53A30.

*Key words and phrases.* Holomorphically planar conformal vector fields, Kaehler/almost Kaehler manifolds, Analytic vector field, Nullity vector fields.

The author was supported in part by the Arts & Sciences Initiative Grant of the University of New Haven. The author is thankful to Professor David E. Blair for helpful communications.

THEOREM G. *A closed conformal vector field $V$ on a Kaehler manifold $M$ is homothetic and holomorphic (analytic). Hence if $M$ is locally non-flat and complete, then $V$ generates a 1-parameter group of automorphisms.*

One of the goals of this paper is to generalize the aforementioned result by considering a special conformal vector field on an almost Hermitian manifold. Bearing this in mind we consider the simplest complex manifold $C$ (i.e. $E^2$) with the Euclidean metric $g$ and the complex structure $J$ defined by $J(\partial/\partial x) = \partial/\partial y, J(\partial/\partial y) = -\partial/\partial x$. For a vector field $V$ on $C$ the following three conditions are equivalent to one another: $(i)$ $V$ is conformal (i.e. $L_V g = 2ag$ for a smooth function $a$ on $C$, where $L_V$ denotes Lie-derivative operator along $V$), $(ii)$ $V$ is analytic (i.e. $L_V J = 0$, meaning that the components of $V$ satisfy the Cauchy-Riemann equations), $(iii) \nabla_X V = aX + bJX$, where $L_V$ denotes the Lie-derivative operator along $V$. Each of the three conditions can be expressed in terms of the coordinates $x, y$ as the following set of first order partial differential equations: $\partial_x V^1 = \partial_y V^2 (= a), \partial_x V^2 = -\partial_y V^1 (= b)$. However, the conditions $(i), (ii), (iii)$ are not equivalent to one another, on a general almost Hermitian manifold. Motivated by this observation, we would like to study vector fields satisfying condition $(iii)$ on certain almost Hermitian manifolds. We introduce such vector fields through the following definition.

DEFINITION 1.1. *A vector field $V$ on an almost Hermitian manifold $(M, g, J)$ is said to be a holomorphically planar conformal vector (abbreviated HPCV) field if*

$$\nabla_X V = aX + bJX \tag{1.1}$$

*where $X$ is an arbitrary vector field and $a, b$ are smooth functions on $M$.*

Obviously, an *HPCV* field is conformal, and its integral curves are holomorphically planar (see[**YA1**], p. 258).

## 2. Some Elementary Results

We first state the following lemma (proof is straightforward).

LEMMA 2.1. *If an almost Kaehler manifold admits an HPCV field $V$, then $L_V J = \nabla_V J$. Hence $V$ is almost analytic (i.e. $L_V J = 0$) iff $J$ is parallel along $V$.*

An immediate consequence of the above lemma is

PROPOSITION 2.1. *An HPCV field on a Kaehler manifold is analytic.*

Next, we have

LEMMA 2.2. *For an HPCV field $V$ on an almost Kaehler manifold $M$ of dimension $> 2$, the function $b$ is constant, and $dv = -2b\Omega$, where $v$ is the 1-form metrically equivalent to $V$.*

PROOF. Equation (1.1) immediately gives $dv = -2b\Omega$. Operating by $d$ and using the Poincare lemma: $d^2 = 0$, we get $(db) \wedge \Omega = 0$. Hence

$$(Xb)\Omega(Y, Z) + (Yb)\Omega(Z, X) + (Zb)\Omega(X, Y) = 0$$

for arbitrary vector fields $X, Y, Z$ on $M$. Factoring out $Z$ and operating by $J$ gives

$$(Xb)Y - (Yb)X + < X, JY > J(Db) = 0$$

where $<,>$ denotes the inner product and $D$ the gradient operator with respect to the Hermitian metric $g$. Contracting this equation with respect to $Y$ one gets $2(n-1)Xb = 0$. As $2n > 2$, by hypothesis, we conclude that $b$ is constant, proving the lemma. □

LEMMA 2.3. *If an almost Kaehler manifold admits an HPCV field $V$, then*

$$\nabla_V \Omega = -2a\Omega - d\omega \tag{2.1}$$

*where $\omega$ is a 1-form metrically equivalent to $JV$.*

PROOF. Lie-differentiating $g(X, JY) = \omega(X, Y)$ along $V$, and noting that $V$ is conformal, we have

$$2ag(X, JY) + g(X, (L_V J)Y) = (L_V \Omega)(X, Y) \tag{2.2}$$

From the formula: $L_V \Omega = d i_V \Omega + i_V d\Omega$( $i_V$ denoting interior product with $V$), and the closedness of $\Omega$, one has $L_V \Omega = d i_V \Omega$. But $(i_V \Omega)X = \Omega(V, X) = -\omega(X)$, where $\omega(X) = g(JV, X)$. Hence we obtain $i_V \Omega = -\omega$, and $L_V \Omega = -d\omega$. Use of this in (2.2) and Lemma 2.1 yields (2.1), completing the proof. □

## 3. *HPCV* Fields On Kaehler Manifolds

Noting that a closed conformal vector field is a special case ($b = 0$) of an HPCV field, we present the following generalization of Goldberg's theorem (stated as Theorem G in section 1).

THEOREM 3.1. *An HPCV field on a Kaehler manifold $M$ of dimension $> 2$, is homothetic and analytic.*

PROOF. By virtue of lemma 2.2 we already know that $b$ is constant. Next, we will show that $a$ is constant. For, if $M$ is Kaehler, then (2.1) implies $d\omega = -2a\Omega$. Taking its exterior derivative gives $(da) \wedge \Omega = 0$. Arguing as in the proof of lemma 2.2 and noting that dim.$M > 2$, we conclude that $a$ is constant, i.e. $V$ is homothetic. That $V$ is analytic, follows from proposition 2.1. This completes the proof. □

REMARK. If, in the hypothesis of Theorem 3.1, we further assume that $M$ is complete and locally non-flat, then $V$ reduces to a Killing vector field and hence generates a 1-parameter group of automorphisms of $M(g, J)$.

We are now going to state and prove the following result that provides an example of an HPCV field on the unit 2-sphere (which is a Kaehler manifold with its standard metric).

THEOREM 3.2. *An HPCV field on the unit sphere $S^2$ can be represented as $-(Da + JDb)$, where $a, b$ are eigenfunctions of the Laplacian corresponding to the first eigenvalue* 2.

PROOF. Using equation (1.1) we have

$$R(X, Y)V = (Xa)Y - (Ya)X + (Xb)JY - (Yb)JX$$

where $R$ denotes the curvature tensor. Contracting the above equation with respect to $X$ we get

$$S(Y, V) = -Ya - g(JDb, Y)$$

where $S$ denotes the Ricci tensor. But $S = g$ on $S^2$. Therefore we obtain

$$V = -(Da + JDb) \tag{3.1}$$

Use of this in (1.1) shows

$$\nabla_X Da - J\nabla_X Db = -(aX + bJX) \tag{3.2}$$

Let us take a local orthonormal $J$-adapted frame $(e_i) = (e, Je)$. Substituting $X = e_i$ in (3.2), taking inner product with $e_i$, and summing over $i$, we get $-\nabla^2 a = 2a$, i.e. $a$ is an eigenfunction of the Laplacian $-\nabla^2$ corresponding to the first eigenvalue 2. Consequently, (see Tanno [**TA1**]) we conclude that $a, b$ satisfy the Obata's equation (see [**OBA**])

$$\nabla_X Da = -aX, \nabla_X Db = -bX$$

One may note that $a$ and $b$ are any two of the 3 linearly independent eigenfunctions corresponding to eigenvalue 2. □

Next we consider the flat $C^n$ with coordinates $(x^i, y^i)$ where $i = 1, 2, 3..., n$ and Kaehler structure $J$ defined by $J(\partial/\partial x^i) = \partial/\partial y^i, J(\partial/\partial y^i) = -\partial/\partial xi$. Solving the HPCV equation (1.1) and noting that $a, b$ are constants ( in view of Lemma 2.2 and Theorem 3.1, as $C^n$ is Kaehler) we can show that $V = (aI + bJ)\mathbf{x} + \mathbf{c}$ where $\mathbf{x}$ is the position vector of a point $(x^i, y^i)$ and $\mathbf{c}$ is a constant vector field.

Another motivation for considering $HPCV$ fields lies in the following problem "If a vector field $V$ on a Riemannian manifold $M$ is closed and homothetic (in particular, parallel), then $V$ is a nullity vector field, i.e. satisfies: $R(X, Y)V = 0$ (see [**CLM**] and [**TA1**] for details). Conversely, how do we characterize a nullity vector field on a Riemannian manifold in terms of its covariant derivative?" If a vector field $V$ on a Kaehler manifold is $HPCV$, then a straightforward computation (using the fact that $a, b$ are constants, as implied by Lemma 2.2 and Theorem 3.1) reveals that $V$ is a nullity vector field. Thus we have a partial answer to the aforementioned question, i.e. a solution to the nullity equation $R(X, Y)V = 0$ on a Kaehler manifold is equation (1.1), which is more general than $V$ being homothetic and closed.

In the passing we note that the Lie-bracket of two $HPCV$ fields on a Kaehler manifold is a parallel vector field (easy to verify). Hence the set of all HPCV fields on a Kaehler manifold $M$ forms a sub-algebra of the conformal algebra on $M$.

Our next result is

PROPOSITION 3.1. *A 4−dimensional Ricci-flat Kaehler manifold $M$ admitting a non-zero HPCV field is locally flat.*

PROOF. As remarked earlier, we have $R(X, Y)V = 0$. By hypothesis, $M$ is Ricci-flat. Hence $C(X, Y)V = 0$, where $C$ denotes the Weyl conformal tensor. By hypothesis, $V$ is non-vanishing on an open neighborhood $N$ in $M$. Since dim.$M = 4$, we can choose a local orthonormal frame $(e_i), i = 1, 2, 3, 4$ on $N$ such that $e_4$ is collinear with $V$. In view of the well known symmetry and anti-symmetry properties of the Weyl conformal tensor in various slots, and its complete tracelessness, it follows by a straightforward combinatorial computation that $C = 0$ on $N$. But $N$ is Ricci-flat. Hence $N$ is flat, i.e. $M$ is locally flat, completing the proof. □

The following result can be proven by a straightforward calculation.

PROPOSITION 3.2. *If a locally symmetric Kaehler manifold $M$ admits an HPCV field $V$, then either $M$ is locally flat, or $V$ is parallel.*

REMARK. If we take the complex space-form for $M$ in Proposition 3.2, we observe (easy to verify) that the second alternative conclusion reduces to $V = 0$.

At this point we consider the curvature tensor $R$ as a cross-section $\hat{R}$ of $TM \otimes \wedge M \otimes \wedge^2 M$ and the exterior derivation $d$ and the codifferentiation $\delta$ to act on the last two components of $R$. Then $d\hat{R} = 0$ is the second Bianchi identity, and $\delta\hat{R} = 0$ is equivalent to

$$(\nabla_X S)(Y, Z) = (\nabla_Y S)(X, Z)$$

A Riemannian manifold is said to have harmonic curvature tensor if $\delta\hat{R} = 0$. (see Tanno [**TA2**]). This condition means that the Levi-Civita connection of such metrics is a Yang-Mills connection while keeping the metric on $M$ fixed (see Bourguignon [**BOU**]). For examples of such metrics we refer to Derdzinski [**DER**]. The analogous situation for harmonicity of Weyl conformal tensor is different. For harmonicity of the Weyl tensor it might seem necessary to have $d\hat{C} = 0$ and $\delta\hat{C} = 0$. Howevever, these two conditions are equivalent (see [**SHA**]). Thus we say that the Weyl tensor is harmonic if $\delta\hat{C} = 0$, i.e. $div.C = 0$.

Our next result is

PROPOSITION 3.3. *Let $V$ be a non-closed HPCV field on a Kaehler manifold $M$. If the Weyl conformal tensor of $M$ is harmonic, then $M$ is Ricci-flat.*

PROOF. As stated earlier, $R(X, Y)V = 0$. This implies $S(Y, V) = 0$. Differentiating it we have

$$(\nabla_X S)(Y, V) + aS(X, Y) + bS(Y, JX) = 0$$

The use of the above equation in the hypothesis, $div.C = 0$, i.e.

$$(\nabla_X S)(Y, Z) - (\nabla_Y S)(X, Z) = 1/2(2n-1)[(Xr) < Y, Z > X - (Yr) < X, Z >]$$

provides

$$b[S(X, JY) - S(Y, JX)] = 1/2(2n-1)[(Xr) < Y, V > -(Yr) < X, V >]$$

Replacing $X$ by $JX$, and noting the $J$-invariance of $S$ (for a Kaehler manifold) we obtain

$$2bQX = 1/2(2n-1)[((JX)r)V - < JX, V > Dr] \tag{3.3}$$

Since $b$ is non-zero, by hypothesis, the above equation shows that $Q$ vanishes at a point where $V$ vanishes. So assume $V$ vanishes nowhere on some open neighborhood $N$ in $M$. Sustituting $X = JV$, and bearing in mind that $QJV = JQV = 0$, yields $Dr = fV$ for some function $f$ on $N$. Substituting this in (3.3) shows $Q = 0$ on $N$. By continuity, $Q = 0$ on $M$, completing the proof. □

COROLLARY. *Proposition 3.3 holds in particular, when $M$ has harmonic curvature.*

REMARK. Since the 2-form $\rho$, defined by $\rho(X,Y) = S(X,JY)$, on a Kaehler manifold is closed, it therefore represents the first Chern class $C_1$. Complex manifolds with a Ricci-flat Kaehler metric are called Calabi-Yau manifolds. The Calabi-Yau manifolds have their application in superstring theory which is based on a 10-dimensional manifold $V_4 \times M$, where $V_4$ is the 4-dimensional spacetime and $M$ is a 6-dimensional Calabi-Yau manifold (see [**CAH**]). Also, Lebrun ([**LEB**]) has shown the existence of a complete non-flat Ricci-flat Kaehler metric on $C^n$.

## 4. *HPCV* Fields On Almost Kaehler Manifolds

We begin with the following result (mentioned on p. 164 of Yano [**YA1**]) "A conformal vector field on a compact almost Kaehler manifold $M$ is Killing for $dim.M > 2$, and is almost analytic for $dim.M \geq 2$". We prove

PROPOSITION 4.1. *If a compact almost Kaehler manifold $M$ admits an HPCV field $V$, then $V$ is parallel.*

PROOF. In view of the result stated before the proposition, we have that $V$ is Killing, i.e. $a = 0$. Thus $\nabla_X V = bJX$. Also, as $b$ is constant, we derive

$$R(X,Y)V = b[(\nabla_X J)Y - (\nabla_Y J)X]$$

Contracting it and noting that $div.J = 0$ on an almost Kaehler manifold (actually, it follows on contracting the identity $(\nabla_{JX} J)Y = -J(\nabla_X J)Y$, see [**AAD**]), we have $S(Y,V) = 0$. Now we recall the following integral formula for a compact Riemannian manifold (see p. 41 of [**YA2**]):

$$\int_M [S(V,V) + (\nabla^j V^i)(\nabla_i V_j) - (div.V)^2] dv = 0$$

where $V^i$ are the components of $V$ with respect to a local cordinate system $(x^i)$. Here we have the first and last terms of the integrand vanishing on $M$. The middle term turns out to be $-2nb^2$. Consequently, we end up with having $b = 0$, completing the proof. □

We now turn our attention to $HPCV$ fields on an (*not necessarily compact*) almost Kaehler manifold and prove

THEOREM 4.1. *Let an almost Kaehler manifold $M$ ($dim.M > 2$) admit a non-closed HPCV field $V$. Then $V$ is homothetic and almost analytic. Also, $M$ is Kaehler iff $V$ is a nullity vector field.*

PROOF. As $b$ is constant (by Lemma 2.2), and by hypothesis $V$ is non-closed, we have $b \neq 0$. If $Da = 0$ on $M$, then $a$ is constant. So, let $Da \neq 0$ on some open neighborhood $N$ in $M$. Equation (1.1) implies

$$R(X,Y)V = (Xa)Y - (Ya)X + b[(\nabla_X J)Y - (\nabla_Y J)X]$$

Using the closedness of $\Omega$ in the above equation we get

$$R(V,X)Y = (Ya)X - (Xa)Y - b(\nabla_X J)Y \tag{4.1}$$

Substituting $V$ for $X$ readily yields

$$b(\nabla_V J)Y = (Ya)V - < Y,V > Da \tag{4.2}$$

Applying $J$ on (4.2) and separately substituting $JY$ for $Y$ in the above equation we get two equations, adding them and using $(\nabla_V J)J + J(\nabla_V J) = 0$ (which follows upon differentiating $J^2 = -I$ along $V$) we find

$$(4.3) \qquad (Ya)JV + ((JY)a)V = < Y, V > JDa - < JV, Y > Da$$

Contraction of the above equation with respect to $Y$ implies $(JV)a = 0$. Substituting $V$ for $Y$ in (4.3), using the foregoing equation and operating by $J$ gives $|V|^2 Da = (Va)V$. Let $O$ be a neighborhood in $N$ on which $V \neq 0$. Then we have

$$(4.4) \qquad Da = fV$$

on $O$, where $f = (Va)/|V|^2$. Now we go back to (4.1) and contract it with respect to $X$ to get

$$(4.5) \qquad S(Y, V) = (1 - 2n)f < Y, V >$$

As $V$ is conformal, taking Lie-derivative of (4.5) along $V$ and using the following integrability condition (see p. 28 of [**YA2**])

$$(4.6) \qquad (L_V S)(X, Y) = -(2n - 2) < \nabla_X Da, Y > -(\nabla^2 a) < X, Y >$$

for a conformal vector field, we obtain

$$(4.7) \qquad 2(1 - n)\nabla_V Da = [\nabla^2 a + (1 - 2n)(Vf + 2af)]V$$

Next, we Lie-differentiate (4.4) along $V$ and get

$$(4.8) \qquad L_V Da = (Vf)V$$

Also, taking the divergence of (4.4) we have

$$\nabla^2 a = Vf + 2anf$$

Using this along with (1.1), (4.7) and (4.8) we end up having that $bfJV = 0$. As $b$ is non-zero and constant, and $V$ is non-zero on $O$, it follows that $f = 0$ and hence from (4.4) that $Da = 0$ on $O$. This contradicts our assumption that $Da \neq 0$ on $O$ which is a part of $N$. Thus we conclude that $a$ is constant on $M$, i.e. $V$ is homothetic. Hence (4.2) implies $\nabla_V J = 0$, which, in turn, implies through Lemma 2.1, that $V$ is almost analytic. Further, (4.1) reduces to

$$R(V, X)Y = -b(\nabla_X J)Y$$

Hence, $M$ is Kaehler iff $R(V, X)Y = 0$, i.e. iff $V$ is a nullity vector field. This completes the proof. □

REMARK. We observed (Theorem 3.1) in the case of a Kaehler manifold (of *dim.* > 2 that an *HPCV* field is homothetic and analytic, and furthermore a nullity vector field. Theorem 4.1 shows that a non-closed *HPCV* field on an almost Kaehler manifold $M$ is homothetic and almost analytic, and remarkably that it is a nullity vector field iff $M$ is Kaehler (thus tying up the integrability of the almost complex structure of $M$ with the non-closed *HPCV* field being a nullity vector field).

Our next result is

THEOREM 4.2. *Let $V$ be a non-zero non-closed non-Killing HPCV field on an almost Kaehler manifold $M$ with harmonic Weyl conformal tensor. Then $M$ is Ricci-flat.*

PROOF. As $b \neq 0$, we know from the previous theorem that $a$ is constant, and hence equation (4.1) reduces to

$$R(V,X)Y = -b(\nabla_X J)Y \tag{4.9}$$

which implies

$$S(Y,V) = 0 \tag{4.10}$$

The hypothesis, $div.C = 0$, is equivalent to

$$\begin{aligned}(\nabla_X S)(Y,Z) - (\nabla_Y S)(X,Z) &= 1/2(2n-1)[(Xr) < Y,Z > \\ &\quad - (Yr) < X,Z >]\end{aligned} \tag{4.11}$$

Computing $(\nabla_X S)(Y,V) - (\nabla_Y S)(X,V)$ using (4.10) and (4.11) one has

$$b(QJ + JQ)X = 1/2(2n-1)[< X,V > Dr - (Xr)V] \tag{4.12}$$

From (4.11) one also gets

$$(\nabla_X Q)V - (\nabla_V)X = 1/2(2n-1)[(Xr)V - (Vr)X] \tag{4.13}$$

As $V$ is homothetic, we have $L_V Q = -2aQ$ and $L_V r = -2ar$. The first equation yields

$$\nabla_V Q = b(JQ - QJ) - 2aQ$$

Substituting this in (4.13) we get an expression for $(\nabla_X Q)V$, and differentiating $QV = 0$ gives another expression. Equating those two expressions shows

$$(bJQ - aQ)X = 1/2(2n-1)[(Vr)X - (Xr)V] \tag{4.14}$$

If $Q = 0$ on $M$, then we are done. Let $Q \neq 0$ on some open neighborhood $N$ in $M$. Considering a unit eigenvector $E$ of $Q$ with eigenvalue $\lambda$, and taking inner product of (4.14) with $QE$ (also, noting $Vr = -2ar$) we obtain $a\lambda(\lambda - \frac{r}{2n-1}) = 0$. By hypothesis, $V$ is non-Killing, hence $a \neq 0$. Thus, either $\lambda = 0$ or $\frac{r}{2n-1}$. If $r$ is constant, then (4.14) shows that $Q = 0$, contradicting our assumption. So, we consider $r$ as non-constant. From (4.14) it follows that the multiplicity of the eigenvalue 0 of $Q$ is 1, and the corresponding eigenvector is $V$ (upto a scalar multiple). Consequently, we have

$$QX = r/(2n-1)[X - (v(X)/|V|^2)V]$$

where $v$ is the 1-form metrically equivalent to $V$. Making use of the above equation in (4.14) and operating by $J$ we have that

$$2br|V|^2 X = (|V|^2 Xr + 2arv(X))JV + 2brv(X)V$$

Since $dim.M > 2$, we take $X$ orthogonal to $V$ and $JV$, and take its inner product with the above equation, and hence get $br = 0$. This implies $r = 0$, which contradicts our assumption. Thus we conclude that $M$ is Ricci-flat, completing the proof. □

## 5. *HPCV* Field On Nearly Kaehler $S^6$

An almost Hermitian manifold $(M, g, J)$ is said to be a nearly Kaehler manifold if

$$(\nabla_X J)Y + (\nabla_Y J)X = 0 \tag{5.1}$$

It is known that the unit 6-sphere $S^6$ does carry a nearly Kaehler structure, for example, the almost complex structure on $S^6$ as an orientable hypersurface of $R^7$ induced by the Cayley numbers (see [**GRA**] for details). We state and prove the following result.

THEOREM 5.1. *An HPCV field on the nearly Kaehler $S^6$ is closed.*

PROOF. Using (1.1), (5.1) and that $S^6$ has constant curvature 1, we obtain

$$< X, V + Da > Y - < Y, V + Da > X = (Yb)JX - (Xb)JY - 2b(\nabla_X J)Y \tag{5.2}$$

Contracting this with respect to $Y$ yields

$$V = -(1/5)(5Da + JDb) \tag{5.3}$$

Differentiating it and using (1.1) gives

$$-(1/5)[5\nabla_X Da + (\nabla_X J)Db + J\nabla_X Db] = aX + bJX$$

Taking its inner product with $Y$, symmetrizing with respect to $X, Y$, and factoring out $Y$ yields

$$\nabla_X Da = -aX + (1/10)(\nabla_{JX} Db - J\nabla_X Db)$$

We now take a local orthonormal $J$-adapted frame $(e_i, Je_i)$, $i = 1, 2, 3$, and compute

$$\begin{aligned} \nabla^2 a &= < \nabla_{e_i} Da, e_i > + < \nabla_{Je_i} Da, Je_i > \\ &= -6a + (1/10)(< \nabla_{Je_i} Db - J\nabla_{e_i} Db, e_i > - < \nabla_{e_i} Db + J\nabla_{Je_i} Db, Je_i >) \end{aligned}$$

which reduces to $-\nabla^2 a = 6a$, i.e. $a$ is the eigenfunction of the Laplacian $-\nabla^2$ on $S^6$ with first eigenvalue 6. This implies (see [**TA1**])

$$\nabla_X Da = -aX \tag{5.4}$$

The use of (5.3) in (5.2) and taking inner product with $Y$ gives

$$< Y, JDb >< X, Y > - < X, JDb > |Y|^2 = 5(Yb) < JX, Y > \tag{5.5}$$

Let us refer to the local $J$-adapted frame and set $Y = e_i, X = Je_i$ in (5.5) (leaving $i$ unsummed). Then we get $e_i b = 0$ for $i = 1, 2, 3$. Next, set $Y = Je_i, X = e_i$ in (5.5) so that we get $(Je_i)b = 0$. Hence $b$ is constant, which reduces (5.3) to

$$V = -Da \tag{5.6}$$

Differentiating (5.6) and using (5.4) we obtain

$$\nabla_X V = aX \tag{5.7}$$

This shows that $b = 0$, i.e. $V$ is closed, completing the proof. □

## References


[AAD] V. Apostolov, J. Armstrong, T. Draghici, *Local models and integrability of certain almost Kaehler 4-manifolds*, Math. Ann. **323** (2002), 633-666.

[BOU] J.P. Bourguignon, *Harmonic curvature for gravitational and yang-Mills fields*, Lecture Notes in Math., Vol. 949, Springer-Verlag, Berlin, 1982, 35-47.

[CAH] P. Candelas, G. Horowitz, A. Strominger, E. Witten, *Vacuum configurations for superstrings*, Nucl. Phys. B **258** (1985), 46-74.

[CLM] Y.H. Clifton, R. Maltz, *The k-nullity space of curvature operator*, Mich. Math. J. **17** (1970), 85-89.

[DER] A. Derdzinski, *Compact Riemannian manifolds with harmonic curvature and non-parallel Ricci tensor*, Lecture Notes in Math., Vol. 838, Springer-Verlag, Berlin, 1981.

[GOL] S.I. Goldberg, *Curvature and Homology*, Academic Press, N.Y., 1962.

[GRA] A. Gray, *Some examples of almost Hermitian manifolds*, Illinois J. Math. **10** (1966), 353-366.

[LEB] C. Lebrun, *Complete Ricci-flat Kaehler metrics on $C^n$ need not be flat*, Proc. Symp. Pure Math. 52 (Part 2) (1991), 297-304.

[OBA] M. Obata, *Riemannian manifolds admitting a solution of a certain system of differential equations*, Proc. U.S.-Japan Sem. in Diff. Geom., Kyoto, Japan, 101-114.

[SHA] R. Sharma, *Proper conformal symmetries of spacetimes with divergence-free Weyl conformal tensor*, J. Math. Phys. **34** (1993), 3582-3587.

[TA1] S. Tanno, *Some differential equations on Riemannian manifolds*, J. Math. Soc. Japan **30** (1978), 509-531.

[TA2] S. Tanno, *Promenades on spheres*, Department of Mathematics, Tokyo Institute of Technology, 1996.

[YA1] K. Yano, *Differential geometry of complex and almost complex spaces*, Pergamon Press, N.Y., 1965 .

[YA2] K. Yano, *Integral formulas in Riemannian geometry*, Marcel-Dekker, N.Y., 1970.



Department of Mathematics, University of New Haven, West Haven, CT 06516
*E-mail address*: rsharma@newhaven.edu

Contemporary Mathematics
Volume **337**, 2003

# Fundamental Inequalities and Strongly Minimal Submanifolds

Bogdan D. Suceavă

ABSTRACT. Over the last decade, B.-Y.Chen's fundamental inequalities have been investigated by many authors from various viewpoints. The motivation of the present study is Chen's fundamental inequality for Kähler submanifolds in complex space forms (see **[12]**). We provide an extension of this inequality and study a few characterizations of strongly minimal complex surfaces in the three dimensional complex space. The last part of this work is dedicated to the study of strong minimality through examples and a discussion of the classification of strongly minimal complex surfaces in the three dimensional complex space.

## 1. Introduction to Chen's Fundamental Inequalities: Riemannian context

In the geometry of submanifolds, a fundamental problem is to establish simple relationships between the main intrinsic invariants and the main extrinsic invariants of the submanifolds. The first result in this respect was Gauss' Theorema Egregium, which in 1827 asserted that the Gaussian curvature is an intrinsic invariant. In respect to this fundamental problem, Chen's fundamental inequalities obtained in **[5]**, **[10]** are the starting point for many recent papers done by various geometers during the last ten years or so, as for example one may see in **[8]**, **[9]**, **[13]**, **[17]**, **[18]**, **[19]**, **[20]**, **[26]**, **[33]**, **[34]**, **[35]**, **[38]**.

Let $h$ denote the second fundamental form of an isometric immersion of a Riemannian $n$-manifold $M^n$ into an ambient Riemannian space $\bar{M}^{n+m}$. Then the mean curvature vector field is $H = (1/n)\,\mathrm{trace}\,h$. The immersion is called minimal if its mean curvature vector field $H$ vanishes identically.

A classical problem in Riemannian geometry is to study *when does a given Riemannian manifold $M$ admit (or does not admit) a minimal immersion into a Euclidean space of arbitrary dimension ?* It is well known that for a minimal submanifold $M$ in a Euclidean space the Gauss equation implies that the Ricci

2000 *Mathematics Subject Classification.* Primary 53C42, Secondary 53C40, 53C55.

*Key words and phrases.* Curvature invariants, Chen's fundamental inequalities, Kähler submanifolds, Strongly minimal submanifolds.

The author thanks Professors David E. Blair and Bang-Yen Chen for many valuable discussions about the present work, and the editors and the referee for their useful suggestions.

tensor satisfies

$$(1.1)\qquad Ric(X,X) = -\sum_{i=1}^{n} |h(X,e_i)|^2 \leq 0,$$

where $e_1, ..., e_n$ is an orthonormal local frame field on $M$. This gives rise to the first solution to the problem mentioned above; namely, the Ricci tensor of a minimal submanifold $M$ of a Euclidean space is negative semi-definite, and a Ricci-flat minimal submanifold of a Euclidean space is totally geodesic. S.-S. Chern pointed out in [**15**] that up to that moment this was the only known Riemannian obstruction to minimal immersion. The second solution to this problem was obtained by B.-Y. Chen as an immediate application of his fundamental inequality and his invariants (see [**5**], [**10**]).

Let $M^n$ be a Riemannian $n$-manifold. For any orthonormal basis $e_1, ..., e_n$ of the tangent space $T_pM$, the scalar curvature is defined to be $scal(p) = \sum_{i<j} sec(e_i \wedge e_j)$. For any $r$-dimensional subspace of $T_pM$ denoted $L$ with orthonormal basis $e_1, ..., e_r$ one may define

$$(1.2)\qquad scal(L) = \sum_{1\leq i<j\leq r} sec(e_i \wedge e_j).$$

In [**10**], Chen considered the finite set $S(n)$ of $k$-tuples $(n_1, ..., n_k)$ with $k \geq 0$ which satisfy the conditions: $n_1 < n$, $n_i \geq 2$, and $n_1 + ... + n_k \leq n$. For each $(n_1, ..., n_k) \in S(n)$ he introduced the following Riemannian invariants:

$$(1.3)\qquad \delta(n_1, ..., n_k)(p) = scal(p) - \inf\{scal(L_1) + ... + scal(L_k)\}(p),$$

where the infimum is taken for all possible choices of orthogonal subspaces $L_1, ..., L_k$, satisfying $n_j = \dim L_j$, $(j = 1, ..., k)$. Note that the Chen invariant with $k = 0$ is nothing but the scalar curvature.

As in [**10**], we put

$$c(n_1, ..., n_k) = \frac{n^2(n+k-1-\sum n_j)}{2(n+k-\sum n_j)},$$

$$b(n_1, ..., n_k) = \frac{1}{2}\Big\{(n(n-1) - \sum_{j=1}^{k} n_j(n_j-1)\Big\}.$$

Chen's fundamental inequalities obtained in [**10**] can be stated as follows:

THEOREM 1.1. *For any $n$-dimensional submanifold $M$ of a Riemannian space form $R^{n+m}(\varepsilon)$ of constant sectional curvature $\varepsilon$ and for any $k$-tuple $(n_1, ..., n_k) \in S(n)$, we have*

$$(1.4)\qquad \delta(n_1, ..., n_k) \leq c(n_1, ..., n_k)|H|^2 + b(n_1, ..., n_k)\varepsilon.$$

*The equality case of the inequality above holds at a point $p \in M$ if and only if there exists an orthonormal basis $e_1, ..., e_{n+m}$ at $p$ such that the shape operators of $M$ in $R^{n+m}(\varepsilon)$ at $p$ take the following forms: $S_r = diag\,(A_1^r, ..., A_k^r, \mu_r, ..., \mu_r)$ for $r = n+1, ..., m$, where each $A_j^r$ is a symmetric $n_j \times n_j$ submatrix such that*

$$trace(A_1^r) = ... = trace(A_k^r) = \mu_r.$$

The invariants $\delta(n_1, \ldots, n_k)$ became known as the Chen invariants in literature and inequality (1.4) as Chen's fundamental inequality. Chen's fundamental inequality has many nice applications; for example, in [**10**] it is pointed out the following important result as an immediate consequence.

THEOREM 1.2. *Let $M$ be a Riemannian n-manifold. If there exists a k-tuple $(n_1, ..., n_k)$ in $S(n)$ and a point $p \in M$ such that*

$$\delta(n_1, ..., n_k)(p) > \frac{1}{2}\{n(n-1) - \sum n_j(n_j - 1)\}\varepsilon, \tag{1.5}$$

*then $M$ admits no minimal isometric immersion into any Riemannian space form $R^m(\varepsilon)$ with arbitrary codimension. In particular, if $\delta(n_1, ..., n_k)(p) > 0$ at a point for some k-tuple $(n_1, ..., n_k) \in S(n)$, then $M$ admits no minimal isometric immersion into any Euclidean space for any codimension.*

In the context of Theorem 1.2, we were interested (see [**38**]) to study if there exist manifolds with $Ric < 0$, for which one of the Chen invariants is positive, both conditions being satisfied at every point. Such an example would point out that there exists a class of manifolds obstructed to minimal immersion by Theorem 1.2, but not obstructed by the classical $Ric$ condition. The question of existence of manifolds with some $\delta(n_1, ..., n_k)(p) > 0$ and $Ric < 0$ is similar to problem 4 mentioned by Peter Petersen's list of problems in [**1**]: **Scalar versus Ricci curvature problem.** *Are there examples of simply connected manifolds which admit Riemannian metrics of positive scalar curvature, but do not admit Riemannian metrics of positive Ricci curvature ?*

The Ricci vs. Chen invariant problem is solved in [**38**] by considering multiwarped products of hyperbolic planes with suitable warping functions. As far as we know, the scalar vs. Ricci curvature problem is still open.

The purpose of the above introductory presentation was to lay down the background for the discussion of Chen's fundamental inequality in Kählerian context, bearing in mind that the origin of this study is in Riemannian context.

## 2. Chen's Fundamental Inequality for Kähler Submanifolds

**2.1. New Kählerian Invariants.** Let $M^n$ be a Kähler manifold of complex dimension $n$ and denote by $J$ its complex structure. We denote by $sec(X \wedge Y)$ the sectional curvature of the plane determined by the vectors $X$ and $Y$, and by $scal(p)$ the scalar curvature at the point $p$. Consider $U$ a coordinate chart on $M$, and $e_1$,..., $e_n$,$e_1^* = Je_1$,...,$e_n^* = Je_n$ a local orthonormal frame on $U$. Then, at $p \in U$, we have

$$scal(p) = \sum_{i<j} sec(e_i \wedge e_j), \quad i, j = 1, ..., n, 1^*, ..., n^*. \tag{2.1}$$

Let $\pi \subset T_pM$ be a plane section. Then $\pi$ is called totally real if $J\pi$ is perpendicular to $\pi$. For each real number $k$, B.-Y.Chen's Kählerian invariant of order 2 and coefficient $k$ at $p \in M$ is defined by

$$\delta_k^r(p) = scal(p) - k \inf sec(\pi^r), \tag{2.2}$$

where $\inf sec(\pi^r)$ is taken over all totally real plane sections in $T_pM$. In [**12**] the following theorem is proved:

THEOREM 2.1. *For any Kähler submanifold $M^n$ of complex dimension $n \geq 2$ in a complex space form $\bar{M}^{n+p}(4c)$, the following statements hold:*

(1) *For each $k \in (-\infty, 4]$ we have*

$$\delta_k^r \leq (2n^2 + 2n - k)c. \tag{2.3}$$

(2) *Inequality (2.3) fails for every $k > 4$.*

(3) $\delta_k^r = (2n^2 + 2n - k)c$ *holds identically for some* $k \in (-\infty, 4)$ *if and only if* $M^n$ *is a totally geodesic Kähler submanifold of* $\bar{M}^{n+p}(4c)$.

The theorem describes completely, in a forth claim, the pointwise equality situation in the case $k = 4$.

**2.2. Strongly Minimal Submanifolds.** It is known ( see for example [**28**]) that the shape operator of a Kähler submanifold $M^n$ in $\bar{M}^{n+p}$ satisfies:

$$A_{J\xi_r} = JA_r, \quad JA_r = -A_r J, \tag{2.4}$$

for $r = 1, ..., p, 1^*, ..., p^*$, where we use the well-known convention $A_r = A_{\xi_r}$. Therefore the shape operator of $M^n$ takes the form

$$A_\alpha = \begin{pmatrix} A'_\alpha & A''_\alpha \\ A''_\alpha & -A'_\alpha \end{pmatrix}, \quad A_{\alpha^*} = \begin{pmatrix} -A''_\alpha & A'_\alpha \\ A'_\alpha & A''_\alpha \end{pmatrix}, \quad \alpha = 1, ..., p \tag{2.5}$$

where $A'_\alpha$ and $A''_\alpha$ are $n \times n$ matrices. The condition (2.5) implies that every Kähler submanifold $M^n$ is minimal, i.e. *trace* $A_\alpha =$ *trace* $A_{\alpha^*} = 0$, $\alpha = 1, ..., p$.

DEFINITION 2.2. A Kähler submanifold $M^n$ of a Kähler manifold $\bar{M}^{n+p}$ is called *strongly minimal* if at each point there exists an orthonormal frame $e_1$,..., $e_n$,$e_1^* = Je_1$,...,$e_n^* = Je_n$ such that the shape operator satisfies the conditions

$$trace\ A'_\alpha = trace\ A''_\alpha = 0, \quad \alpha = 1, ..., p.$$

This class of submanifolds was introduced and studied by B.-Y.Chen in [**12**], motivated by the following results.

THEOREM 2.3. [**12**] *A complete Kähler submanifold* $M^n$ $(n \geq 2)$ *in* $\mathbb{CP}^{n+p}(4c)$ *satisfies the equality*

$$\delta_4^r = 2(n^2 + n - 2)c \tag{2.6}$$

*identically if and only if*

(1) $M^n$ *is a totally geodesic Kähler submanifold, or*

(2) $n = 2$ *and* $M^2$ *is a strongly minimal Kähler surface in* $\mathbb{CP}^{2+p}(4c)$.

THEOREM 2.4. [**12**] *A complete Kähler submanifold* $M^n$ $(n \geq 2)$ *of* $\mathbb{C}^{n+p}$ *satisfies* $\delta_4^r = 0$ *identically if and only if*

(1) $M^n$ *is a complex n-plane of* $\mathbb{C}^{n+p}$, *or*

(2) $M^n$ *is a complex cylinder over a strongly minimal Kähler surface* $M^2$ *in* $\mathbb{C}^{n+p}$ *(i.e.* $M^n$ *is the product submanifold of a strongly minimal Kähler surface* $M^2$ *in* $\mathbb{C}^{p+2}$ *and the identity map of the complex Euclidean* $(n-2)$*-space* $\mathbb{C}^{n-2}$).

In [**12**] it is proved that the complex surface $N^2$ in $\mathbb{C}^3$ given by the equation $z_1^2 + z_2^2 + z_3^2 = 1$ is a strongly minimal Kähler surface in $\mathbb{C}^3$. The argument presented in [**12**] implies that $z_1^2 + z_2^2 + z_3^2 = \kappa$ is strongly minimal in $\mathbb{C}^3$ for any complex constant $\kappa$, since the construction of the basis where the strong minimality condition holds does not depend on $\kappa$. We prove in the present work that $z_1 + z_2 + z_3^2 = \kappa$ is also strongly minimal in $\mathbb{C}^3$. The above mentioned results and examples motivate our present study of strongly minimal submanifolds. A classification of strongly minimal surfaces in $\mathbb{C}^3$, $\mathbb{C}^{n+p}$ or $\mathbb{CP}^n(4c)$ would mean a better understanding of the "best fit" in the respective complex space forms, in the sense of fundamental inequalities described above (see [**12**]). We focus here on the class of strongly minimal surfaces in $\mathbb{C}^3$.

**2.3. An Extension of Chen's Fundamental Inequality with Kählerian Invariants.** In the present section we extend the inequality (2.3) to orders higher than 2. First let us present the reasons for developing this extension. As we have mentioned before, the first form of B.-Y.Chen's fundamental inequality in Riemannian context was given in [**5**], and the string of B.-Y.Chen's fundamental inequalities was presented in [**10**]. It is natural to ask what could be the most general statement one could get from B.-Y.Chen's approach for Kähler submanifolds in space forms, as it is presented in [**12**].

An $l$-dimensional linear subspace $L_l \subset T_pM$ is called *totally real* if $JL_l$ is orthogonal to $L$. For each real number $k$ we can extend the above presented curvature invariants to *the Kählerian invariant of order $l$ and coefficient $k$* defined by

$$\delta^r_{l,k}(p) = scal(p) - \frac{k}{l-1} \inf_{L^r_l \subset T_pM} [scal(L_l(p))], \tag{2.7}$$

where $L^r_l$ runs over all $l$-dimensional totally real linear subspaces in $T_pM$, and the scalar curvature of a linear subspace is

$$scal[L_l(p)] = \sum_{1 \le i < j \le l} sec(\eta_i \wedge \eta_j),$$

for the orthonormal basis $\eta_1, ..., \eta_l$ in $L_l$. Assume that $M^n$ is a Kähler submanifold of complex dimension $n$ in a complex space form $\bar{M}^{n+p}(4c)$. We follow the notations from [**12**], unless stated otherwise. Consider the orthonormal basis $e_1,..., e_n, e^*_1 = Je_1,...,e^*_n = Je_n$ such that $L_l = \text{span}\{ e_1,..., e_l\}$, where $L_l$ achieves the infimum for $scal[L]$, with $\dim L = l$. If $R$ is the curvature tensor of $M$, then the Gauss equation is

$$\begin{aligned} \langle R(X,Y)Z,W\rangle &= \langle h(X,W), h(Y,Z)\rangle - \langle h(X,Z), h(Y,W)\rangle + \\ &+ c\{\langle X,W\rangle \langle Y,Z\rangle - \langle X,Z\rangle \langle Y,W\rangle + \langle JY,Z\rangle \langle JX,W\rangle - \\ &- \langle JX,Z\rangle \langle JY,W\rangle + 2\langle X,JY\rangle \langle JZ,W\rangle\}, \end{aligned} \tag{2.8}$$

where $h$ is the second fundamental form. The result we prove is the following.

THEOREM 2.5. *Let $M^n$ be a Kähler submanifold of complex dimension $n \geq 2$ in a complex space form $\bar{M}^{n+p}(4c)$. Let $\delta^r_{k,l}$ be the Chen's Kählerian invariant of order $l$ and coefficient $k$. Then we have:*

(1) *For any $2 \le l \le n$, the following inequality holds*

$$\delta^r_{4,l} \le (2n^2 + 2n - \tbinom{l}{2})c. \tag{2.9}$$

*The equality case for $l = 2$ has been described in* [**12**]. *Equality holds at every point for a fixed $l \geq 3$ if and only if $M^n$ is a totally geodesic submanifold.*

(2) *For any $k \in [0, 4]$ the following inequality holds*

$$\delta^r_{k,l} \le \left[2n^2 + 2n - \frac{k}{4}\tbinom{l}{2}\right] c. \tag{2.10}$$

*Equality holds at every point for a fixed $l \geq 3$ if and only if $M^n$ is a totally geodesic submanifold.*

PROOF. Let us discuss first claim (1) of the theorem. Taking $X = W = e_k$ and $Y = Z = e_s$ for $1 \le k, s \le t$ in the Gauss' equation one gets

$$sec(e_k \wedge e_s) = \sum_{\alpha=1}^{p} \left[h^\alpha_{kk} h^\alpha_{ss} - (h^\alpha_{ks})^2 + h^{\alpha*}_{kk} h^{\alpha*}_{ss} - (h^{\alpha*}_{ks})^2\right] + c. \tag{2.11}$$

Consider $l \in \{2, 3, ..., n\}$, the dimension of the totally real space $L_l^r$. Then

$$scal(L_l^r) = \sum_{1\le k<s\le l} \langle R(e_k, e_s)e_s, e_k\rangle = \tag{2.12}$$

$$= \sum_{\alpha=1}^{p} \sum_{1\le k<s\le l} \left\{ \left[h_{ii}^{\alpha} h_{jj}^{\alpha} - (h_{ij}^{\alpha})^2\right] + \left[h_{ii}^{\alpha*} h_{jj}^{\alpha*} - (h_{ij}^{\alpha*})^2\right]\right\} + \binom{l}{2} c.$$

For the following computations, we use $\left(h_{kk}^{\alpha}\right)^2 + \left(h_{ss}^{\alpha}\right)^2 \ge -2h_{kk}^{\alpha} h_{ss}^{\alpha}$, the similar relations for $\alpha^*$ and $l \ge 2$ imply

$$2\left(h_{ks}^{\alpha}\right)^2 \ge \frac{2}{l-1}\left(h_{ks}^{\alpha}\right)^2. \tag{2.13}$$

As in relation (3.5) from [**12**] one may compute the quantity

$$4n(n+1)c - 2scal = 4\sum_{\alpha=1}^{p} \{||A'_\alpha||^2 + ||A''_\alpha||^2\} \ge \tag{2.14}$$

$$\ge 4\sum_{\alpha=1}^{p}\left\{\sum_{i=1}^{l}(h_{ii}^{\alpha})^2 + 2\sum_{1\le i<j\le l}(h_{ij}^{\alpha})^2 + \sum_{i=1}^{l}(h_{ii}^{\alpha*})^2 + 2\sum_{1\le i<j\le l}(h_{ij}^{\alpha*})^2\right\} =$$

$$= 4\sum_{\alpha=1}^{p}\left\{\sum_{1\le i<j\le l}\left[\frac{(h_{ii}^{\alpha})^2}{l-1} + \frac{(h_{jj}^{\alpha})^2}{l-1}\right] + 2\sum_{i<j}(h_{ij}^{\alpha})^2\right\} +$$

$$+4\sum_{\alpha=1}^{p}\left\{\sum_{1\le i<j\le l}\left[\frac{(h_{ii}^{\alpha*})^2}{l-1} + \frac{(h_{jj}^{\alpha*})^2}{l-1}\right] + 2\sum_{i<j}(h_{ij}^{\alpha*})^2\right\} \ge$$

$$\ge 4\sum_{\alpha=1}^{p}\left\{\sum_{1\le i<j\le l}\left[\frac{-2h_{ii}^{\alpha}h_{jj}^{\alpha}}{l-1} + 2(h_{ij}^{\alpha})^2\right]\right\} +$$

$$+4\sum_{\alpha=1}^{p}\left\{\sum_{1\le i<j\le l}\left[\frac{-2h_{ii}^{\alpha*}h_{jj}^{\alpha*}}{l-1} + 2(h_{ij}^{\alpha*})^2\right]\right\} \ge$$

$$4\sum_{\alpha=1}^{p}\sum_{1\le i<j\le l}\left\{\frac{-2}{l-1}[h_{ii}^{\alpha}h_{jj}^{\alpha} - (h_{ij}^{\alpha})^2]\right\} +$$

$$+4\sum_{\alpha=1}^{p}\sum_{1\le i<j\le l}\left\{\frac{-2}{l-1}[h_{ii}^{\alpha*}h_{jj}^{\alpha*} - (h_{ij}^{\alpha*})^2]\right\} =$$

$$= -\frac{8}{l-1}\sum_{\alpha=1}^{p}\sum_{1\le i<j\le l}\left[h_{ii}^{\alpha}h_{jj}^{\alpha} - (h_{ij}^{\alpha})^2 + h_{ii}^{\alpha*}h_{jj}^{\alpha*} - (h_{ij}^{\alpha*})^2\right] =$$

$$= -8\left[scal(L_l^r) + \tbinom{l}{2}c\right].$$

Therefore, we proved

$$2n(n+1)c - \tbinom{l}{2}c \ge scal - \frac{4}{l-1}scal(L_l^r) \tag{2.15}$$

for every totally real $l$-dimensional space $L_l^r$. Therefore

$$\delta_{4,l}^r \le [2n(n+1) - \tbinom{l}{2}]c. \tag{2.16}$$

From the sequence of inequalities above, it is clear that equality holds everywhere if and only if $(h^\alpha_{ii})^2 + (h^\alpha_{jj})^2 = -2h^\alpha_{ii}h^\alpha_{jj}$, for any disjoint pair $i, j$ from 1 to $l$ and $2 = \frac{2}{l-1}$. The equality case for $l = 2$ was completely discussed by B.-Y.Chen in [**12**]. For $l \geq 3$ the fact that $h_{ii} = 0$ is immediate.

Now let's prove the claim (2) of the theorem 2.5. It is proved in [**12**] that

$$scal \leq (2n^2 + 2n)c. \tag{2.17}$$

We multiply this inequality by $p > 0$ and we add it to (2.9) term by term. We get

$$(p+1)scal - \frac{4}{l-1} inf\ scal(L^r_l) \leq [(p+1)(2n^2 + 2n) - \tbinom{l}{2}]c \tag{2.18}$$

Dividing both sides by $p + 1 > 0$, we get

$$scal - \frac{4}{(l-1)(p+1)} inf\ scal(L^2_l) \leq [(2n^2 + 2n) - \frac{1}{p+1}\tbinom{l}{2}]c. \tag{2.19}$$

By denoting $\frac{4}{p+1} = k$, we get $p = \frac{4}{k} - 1 = \frac{4-k}{k}$. Using this in (2.19) we can write the result as

$$scal - \frac{k}{l-1} inf\ scal(L^2_l) \leq [(2n^2 + 2n) - \frac{k}{4}\tbinom{l}{2}]c, \tag{2.20}$$

which is the claimed inequality. From the equality case in relation (3.3) in [**12**], claim 3 of Theorem 2 from [**12**] and the claim (1) of present theorem, if the equality holds at every point for some $k \in [0, 4]$ and some $l$ between 2 and $n$, then $M^2$ is a totally geodesic submanifold. In detail, for $k \in (0, 4)$, the argument is:

$$\left[2n^2 + 2n - \frac{k}{4}\tbinom{l}{2}\right] c = \left(1 - \frac{k}{4}\right) \delta^r_{0,l} + \frac{k}{4}\delta^r_{4,l} \leq \tag{2.21}$$

$$\leq \left(1 - \frac{k}{4}\right)(2n^2 + 2n)c + \frac{k}{4}\left[2n^2 + 2n - \tbinom{l}{2}\right] c = \left[2n^2 + 2n - \frac{k}{4}\tbinom{l}{2}\right] c.$$

From this equality, in particular, one gets $\delta^r_{0,l} = (2n^2+2n)c$. It is shown in [**12**] that this equality at every point implies that the submanifold $M^2$ is totally geodesic. □

**2.4. Characterizations of Strongly Minimal Surfaces in the Complex Three Dimensional Space.** In this section we consider a complex surface $M^2 \subset \mathbb{C}^3$. The coordinates of the ambient space are $z_1$, $z_2$, $z_3$; for $j = 1, 2, 3$. Put $z_j = x_j + iy_j$. We suppose $M$ is embedded so that there exists $\phi \in hol(\mathbb{C}^3)$ such that $M = \{z \in \mathbb{C}^3 | \, \phi(z) = 0\} = V(\phi)$ and

$$\frac{\partial \phi}{\partial z} = \left(\frac{\partial \phi}{\partial z_1}, \frac{\partial \phi}{\partial z_2}, \frac{\partial \phi}{\partial z_3}\right)$$

never vanishes on $M$.

Assume that $p = (z^0_1, z^0_2, z^0_3)$ is a nonsingular point on $M$. Two unit normal vectors at $p$ are $\xi$ and $J\xi$, where

$$\xi = \frac{1}{||\partial\phi/\partial z||}\frac{\bar{\partial\phi}}{\partial z}.$$

By definition, $M^2$ is called strongly minimal in $\mathbb{C}^3$ if the second fundamental form can be written pointwise as we describe below. There exist an open neighborhood $U \subset M$ of $p$ such that there exists two orthogonal unit length vector fields $X$ and

$Y$ on $U$, such that the second fundamental form with respect to the orthonormal basis $\{X, Y, JX, JY\}$ can be written in the form:

$$A_\xi = \begin{pmatrix} a(z) & b(z) & c(z) & d(z) \\ b(z) & -a(z) & d(z) & -c(z) \\ c(z) & d(z) & -a(z) & -b(z) \\ d(z) & -c(z) & -b(z) & a(z) \end{pmatrix}$$

and, respectively,

$$A_{J\xi} = \begin{pmatrix} -c(z) & -d(z) & a(z) & b(z) \\ -d(z) & c(z) & b(z) & -a(z) \\ a(z) & b(z) & c(z) & d(z) \\ b(z) & -a(z) & d(z) & -c(z) \end{pmatrix}$$

where $a, b, c, d$ are real analytic functions on $U$. Suppose that the strongly minimal submanifold is realized on $U$ as a graph manifold, i.e., $z_3 = f(z_1, z_2)$. Consider an open set $V \subset \mathbb{C}^2$ and $\omega : V \to \mathbb{C}^3$, such that $\omega(z_1, z_2) = (z_1, z_2, f(z_1, z_2))$. We also have $\phi(z_1, z_2, z_3) = f(z_1, z_2) - z_3$. Then

$$e_1 = \frac{\partial \omega}{\partial z_1} = \left(1, 0, \frac{\partial f}{\partial z_1}\right) \quad e_2 = \frac{\partial \omega}{\partial z_2} = \left(0, 1, \frac{\partial f}{\partial z_2}\right),$$

and $Je_j = ie_j$, where $j = 1, 2$. To express $f$ as a function of $a(z), b(z), c(z), d(z)$, we use:

$$\begin{aligned} a(z) &= \langle A_\xi X, X\rangle = \langle h(X,X), \xi\rangle = \left\langle \bar{\nabla}_X X - \nabla_X X, \xi\right\rangle, \\ b(z) &= \langle A_\xi Y, X\rangle = \langle h(Y,X), \xi\rangle = \left\langle \bar{\nabla}_Y X - \nabla_Y X, \xi\right\rangle, \\ c(z) &= \langle A_\xi JX, X\rangle = \langle h(JX,X), \xi\rangle = \left\langle \bar{\nabla}_X JX - \nabla_X JX, \xi\right\rangle, \\ d(z) &= \langle A_\xi JY, X\rangle = \langle h(JY,X), \xi\rangle = \left\langle \bar{\nabla}_X JY - \nabla_X JY, \xi\right\rangle. \end{aligned}$$

We use these equations below, in the proof of the parametric equations of a strongly minimal surface.

Consider the real and complex parts of the function $f$ :

$$z_3 = f(z_1, z_2) = u(x_1, x_2, y_1, y_2) + iv(x_1, x_2, y_1, y_2).$$

We use the notation:

$$\frac{\partial u}{\partial x_1} = u_{x_1}, \quad \frac{\partial u}{\partial x_2} = u_{x_2},$$

and the other similar notations. Since $f$ is holomorphic with respect to both variables

$$u_{x_j} = v_{y_j} \qquad v_{x_j} = -u_{y_j}. \tag{2.22}$$

We use the notation suggested by the example discussed in Proposition 5 from [**12**]. Consider two vector fields $X = \alpha + i\beta$ and $Y = \gamma + i\delta$ on the open neighborhood $U$. We put

$$\alpha = (\alpha_1, \alpha_2, \alpha_3), \;\; \beta = (\beta_1, \beta_2, \beta_3), \;\; \gamma = (\gamma_1, \gamma_2, \gamma_3), \;\; \delta = (\delta_1, \delta_2, \delta_3).$$

The fact that $X$ and $Y$ are tangent vector fields can be expressed as

$$X = (\alpha_1 + i\beta_1)\left(1, 0, \frac{\partial f}{\partial z_1}\right) + (\alpha_2 + i\beta_2)\left(0, 1, \frac{\partial f}{\partial z_2}\right), \tag{2.23}$$

$$Y = (\gamma_1 + i\delta_1)\left(1, 0, \frac{\partial f}{\partial z_1}\right) + (\gamma_2 + i\delta_2)\left(0, 1, \frac{\partial f}{\partial z_2}\right). \tag{2.24}$$

For the third component, we get by direct computations

$$\alpha_3 = \alpha_1 u_{x_1} + \beta_1 u_{y_1} + \alpha_2 u_{x_2} + \beta_2 u_{y_2}, \tag{2.25}$$

$$\beta_3 = \beta_1 u_{x_1} - \alpha_1 u_{y_1} + \beta_2 u_{x_2} - \alpha_2 u_{y_2}. \tag{2.26}$$

The conditions $g(X,Y) = g(JX,Y) = 0$ can be written

$$\alpha_1\gamma_1 + \alpha_2\gamma_2 + \alpha_3\gamma_3 + \beta_1\delta_1 + \beta_2\delta_2 + \beta_3\delta_3 = 0, \tag{2.27}$$

$$-\gamma_1\beta_1 - \gamma_2\beta_2 - \gamma_3\beta_3 + \alpha_1\delta_1 + \alpha_2\delta_2 + \alpha_3\delta_3 = 0, \tag{2.28}$$

where $\alpha_j, \beta_j, \gamma_j, \delta_j$, are real analytic functions on the open set $U \subset M$, for $j = 1, 2, 3$. (The conditions $g(X, JY) = g(JX, JY) = 0$ are insured by the relations above and the conditions $g(X, JX) = g(Y, JY) = 0$ are a consequence.)

The fact that $X$ and $Y$ are unit vector fields can be written as

$$\alpha_1^2 + \alpha_2^2 + \alpha_3^2 + \beta_1^2 + \beta_2^2 + \beta_3^2 = 1, \tag{2.29}$$

$$\gamma_1^2 + \gamma_2^2 + \gamma_3^2 + \delta_1^2 + \delta_2^2 + \delta_3^2 = 1. \tag{2.30}$$

We may state and prove now *the parametric equations of strongly minimal complex surfaces in* $\mathbb{C}^3$.

THEOREM 2.6. *Let $u$ be the real part of a holomorphic function $f(z_1, z_2, z_3)$. Then the complex surface $V(f) = M$ is strongly minimal if, for every point $p \in M$, there exist an open coordinate neighborhood $U \subset M$ of $p$ such that on $U$ there exist four real analytic functions $a$, $b$, $c$, $d$ and two orthonormal tangent vector fields $X$ and $Y$ such that $g(X,X) = g(Y,Y) = 1$, $g(X,Y) = g(X,JY) = 0$, $X = \alpha + i\beta$, $Y = \gamma + i\delta$,*

$$\alpha = (\alpha_1, \alpha_2, \alpha_3), \beta = (\beta_1, \beta_2, \beta_3), \gamma = (\gamma_1, \gamma_2, \gamma_3), \delta = (\delta_1, \delta_2, \delta_3),$$

*such that on $U$ we have*

$$\begin{aligned}(2.31)\quad & -a(z)(1 + u_{x_1}^2 + u_{x_2}^2 + u_{y_1}^2 + u_{y_2}^2)^{1/2} = (\alpha_1^2 - \beta_1^2)u_{x_1x_1} + 2\alpha_1\beta_1 u_{x_1y_1} + \\ & (\alpha_1\alpha_2 - \beta_1\beta_2)u_{x_1x_2} + (\alpha_1\beta_2 + \alpha_2\beta_1)u_{y_1x_2} + (\alpha_2^2 - \beta_2^2)u_{x_2x_2} + 2\alpha_2\beta_2 u_{x_2y_2} = \\ & -(\gamma_1^2 - \delta_1^2)u_{x_1x_1} - 2\gamma_1\delta_1 u_{x_1y_1} - (\gamma_1\gamma_2 - \delta_1\delta_2)u_{x_1x_2} - \\ & (\gamma_1\delta_2 + \gamma_2\delta_1)u_{y_1x_2} - (\gamma_2^2 - \delta_2^2)u_{x_2x_2} - 2\gamma_2\delta_2 u_{x_2y_2},\end{aligned}$$

$$\begin{aligned}(2.32)\quad & -b(z)(1 + u_{x_1}^2 + u_{x_2}^2 + u_{y_1}^2 + u_{y_2}^2)^{1/2} = (\alpha_1\gamma_1 - \beta_1\delta_1)u_{x_1x_1} + \\ & (\beta_1\gamma_1 + \alpha_1\delta_1)u_{x_1y_1} + \frac{1}{2}(\alpha_2\gamma_1 + \alpha_1\gamma_2 - \beta_2\delta_1 - \beta_1\delta_2)u_{x_1x_2} + \\ & \frac{1}{2}(\beta_2\gamma_1 + \beta_1\gamma_2 + \alpha_2\delta_1 + \alpha_1\delta_2)u_{y_1x_2} + (\alpha_2\gamma_2 - \beta_2\delta_2)u_{x_2x_2} + (\beta_2\gamma_2 + \alpha_2\delta_2)u_{x_2y_2},\end{aligned}$$

$$\begin{aligned}(2.33)\quad & -c(z)(1 + u_{x_1}^2 + u_{x_2}^2 + u_{y_1}^2 + u_{y_2}^2)^{1/2} = -2\alpha_1\beta_1 u_{x_1x_1} + (\alpha_1^2 - \beta_1^2)u_{x_1y_1} - \\ & (\alpha_1\beta_2 + \alpha_2\beta_1)u_{x_1x_2} + (\alpha_1\alpha_2 - \beta_1\beta_2)u_{y_1x_2} - 2\alpha_2\beta_2 u_{x_2x_2} + (\alpha_2^2 - \beta_2^2)u_{x_2y_2} = \\ & 2\gamma_1\delta_1 u_{x_1x_1} + (\delta_1^2 - \gamma_1^2)u_{x_1y_1} + (\gamma_1\delta_2 + \gamma_2\delta_1)u_{x_1x_2} + \\ & (\delta_1\delta_2 - \gamma_1\gamma_2)u_{y_1x_2} + 2\gamma_2\delta_2 u_{x_2x_2} + (\delta_2^2 - \gamma_2^2)u_{x_2y_2},\end{aligned}$$

$$(2.34)\quad d(z)(1 + u_{x_1}^2 + u_{x_2}^2 + u_{y_1}^2 + u_{y_2}^2)^{1/2} = (\alpha_1\delta_1 + \beta_1\gamma_1)u_{x_1x_1} -$$

$$(\alpha_1\gamma_1-\beta_1\delta_1)u_{x_1y_1}+\frac{1}{2}(\alpha_1\delta_2+\alpha_2\delta_1+\beta_1\gamma_2+\beta_2\gamma_1)u_{x_1x_2}+$$

$$\frac{1}{2}(\beta_2\delta_1+\beta_1\delta_2-\alpha_2\gamma_1-\alpha_1\gamma_2)u_{y_1x_2}+(\alpha_2\delta_2+\beta_2\gamma_2)u_{x_2x_2}+(\beta_2\delta_2-\alpha_2\gamma_2)u_{x_2y_2}.$$

PROOF. The normal unit vector field $\xi$ has the form (see for example [**41**])

$$\xi=(1+u_{x_1}^2+u_{x_2}^2+u_{y_1}^2+u_{y_2}^2)^{-1/2}(u_{x_1},u_{x_2},-1,u_{y_1},u_{y_2},0) \tag{2.35}$$

Consider the expression which yields the first entry in the second fundamental form operator: $g(A_\xi X,X)=a(z)$. Now, we use a computational idea presented as relation (3.2) in [**41**], respectively relation (5.12) in [**12**], to get

$$g\left(-||\frac{\partial f}{\partial z}||^{-1}\{\bar{X}\bar{H_{jk}}\}^{tan},X\right)=a, \tag{2.36}$$

where

$$H_{jk}=\frac{\partial^2\phi}{\partial z_j\partial z_k}.$$

This can be written in detail as

$$-||\frac{\partial\phi}{\partial z}||^{-1}g\left(\left[\begin{pmatrix}\alpha_1-i\beta_1\\ \alpha_2-i\beta_2\\ \alpha_3-i\beta_3\end{pmatrix}\begin{pmatrix}\frac{\partial^2\bar\phi}{\partial z_1^2} & \frac{\partial^2\bar\phi}{\partial z_1\partial z_2} & 0\\ \frac{\partial^2\bar\phi}{\partial z_1\partial z_2} & \frac{\partial^2\bar\phi}{\partial z_2^2} & 0\\ 0 & 0 & 0\end{pmatrix}\right]^{tan},\begin{pmatrix}\alpha_1+i\beta_1\\ \alpha_2+i\beta_2\\ \alpha_3+i\beta_3\end{pmatrix}\right)=a(z).$$

We can use as basis of the tangent fibre bundle on $U$ the orthonormal frame $\{X,Y,JX,JY\}$. This means

$$v^{tan}=\langle v,X\rangle X+\langle v,Y\rangle Y+\langle v,JX\rangle JX+\langle v,JY\rangle JY.$$

In fact we need only $g(v^{tan},X)=\langle v,X\rangle$. (We denote consistently by $\langle\ ,\ \rangle$ the scalar product in $\mathbb{R}^6$.)

In this context, we have

$$\bar{X}\bar{H_{jk}}=(\alpha_1u_{x_1x_1}+\beta_1u_{x_1y_1}+\frac{\alpha_2}{2}u_{x_1x_2}+\frac{\beta_2}{2}u_{y_1x_2}, \tag{2.37}$$

$$\frac{\alpha_1}{2}+\frac{\beta_1}{2}u_{y_1x_2}+\alpha_2u_{x_2x_2}+\beta_2u_{x_2y_2},0,$$

$$\alpha_1u_{x_1y_1}-\beta_1u_{x_1x_1}-\frac{\beta_2}{2}u_{x_1x_2}+\frac{\alpha_2}{2}u_{y_1x_2},\frac{\alpha_1}{2}u_{y_1x_2}-\frac{\beta_1}{2}u_{x_1x_2}+\alpha_2u_{x_2y_2}-\beta_2u_{x_2x_2},0).$$

Computing the 6-dimensional scalar product we get the claimed equation. Similar computations prove the other analogous equalities. □

Now, we study the Gauss and Codazzi equations of a strongly minimal complex surface into $\mathbb{C}^3$ (see [**36**] and [**27**]). The symmetric covariant tensors $h$ and $k$ and the tensor field $s$ of type (0,1) are defined and studied in [**36**]; Gauss and Weingarten formulae are

$$\bar\nabla_XY=\nabla_XY+h(X,Y)\xi+k(X,Y)J\xi, \tag{2.38}$$

$$\bar\nabla_X\xi=-AX+s(X)J\xi. \tag{2.39}$$

With these notations, Gauss and Codazzi equations (see [**36**]) are

$$R(X,Y,Z,W)=g(AX,Z)g(AY,W)-g(AX,W)g(AY,Z)+ \tag{2.40}$$

$$+g(JAX,W)g(JAY,Z) - g(JAX,Z)g(JAY,W),$$

$$(\nabla_X A)Y - (\nabla_Y A)X = s(X)JAY - s(Y)JAX \tag{2.41}$$

Gauss equation is used to prove Proposition 6 in [**12**]. It can also be used in the following.

REMARK 2.7. Let $\{X, Y\}$ a pair of orthonormal tangent vectors $X, Y \in T_pM$, such that $g(X,Y) = g(X, JY) = 0$, with respect to which the shape operator of the manifold $M = V(\phi)$ on the open set $U$ satisfies the strong minimality condition. Then we have

$$Ric(X,X) = Ric(Y,Y) = -2(a^2 + b^2 + c^2 + d^2).$$

PROOF. We compute, from Gauss equation

$$sec(X \wedge Y) = g(A_\xi X, X)g(A_\xi Y, Y) - g(A_\xi X, Y)g(A_\xi Y, X) +$$
$$+g(JA_\xi X, Y)g(JA_\xi Y, X) - g(JA_\xi X, X)g(JA_\xi Y, Y).$$

Using either this relation or (5) from [**36**], we get the claimed fact. □

The condition $[A'_\xi, A''_\xi] = 0$, proved also in Proposition 6 from [**12**], is satisfied identically once we prescribe the shape operator in the form $A_\xi$, as we did.

REMARK 2.8. Let U be an open neighborhood of a regular point M such that on U there exists a pair of orthonormal tangent vector fields $X$ and $Y$, with the property that at every point $g(X,Y) = g(X,JY) = 0$, satisfying the strong minimality condition. If $s$ is the tensor field of type $(0,1)$ defined by the Weingarten formula

$$\bar{\nabla}_X \xi = -A_\xi X + s(X)J\xi,$$

then the following relations hold:

$$X(c(z)) - JX(a(z)) = s(X)a(z) + s(JX)c(z); \tag{2.42}$$

$$X(a(z)) + JX(c(z)) = -s(X)c(z) + s(JX)a(z); \tag{2.43}$$

$$Y(c(z)) - JY(a(z)) = s(Y)a(z) + s(JY)c(z); \tag{2.44}$$

$$Y(a(z)) + JY(c(z)) = -s(Y)c(z) + s(JY)a(z). \tag{2.45}$$

$$X(d(z)) - JX(b(z)) = s(X)b(z) + s(JX)d(z); \tag{2.46}$$

$$X(b(z)) + JX(d(z)) = -s(X)d(z) + s(JX)b(z); \tag{2.47}$$

$$Y(d(z)) - JY(b(z)) = s(Y)b(z) + s(JY)d(z); \tag{2.48}$$

$$Y(b(z)) + JY(d(z)) = -s(Y)d(z) + s(JY)b(z). \tag{2.49}$$

PROOF. It is convenient to work with the following form of the Codazzi equation:

$$\nabla_X(A_\xi Y) - \nabla_Y(A_\xi X) + A_\xi([Y,X]) = s(x)JA_\xi Y - s(Y)JA_\xi X. \tag{2.50}$$

As a matter of example, we prove the third equation from the ones stated above. We write the Codazzi equation in $Y$ and $JY$ and dot product to the right with $Y$. We get

$$\langle \nabla_Y(A_\xi JY), Y\rangle - \langle \nabla_{JY}(A_\xi Y), Y\rangle = s(Y)\langle JA_\xi JY, Y\rangle - s(JY)\langle JA_\xi Y, Y\rangle. \tag{2.51}$$

Using $JY = iY$ and the metric property of the Riemannian connection on the submanifold $U$, we have

$$\begin{aligned} Y \langle A_\xi Y, JY \rangle - \langle A_\xi Y, \nabla_Y JY \rangle - JY \langle A_\xi Y, Y \rangle + \langle A_\xi Y, \nabla_{JY} Y \rangle \\ = -s(Y)a(z) - s(JY)c(z). \end{aligned}$$

Now, one can use the fact that: $J\nabla_Y Y = \nabla_Y JY = \nabla_Y iY = i\nabla_Y Y$ to simplify the expression. Furthermore, the shape operator $A_\xi$ has a prescribed form on the considered basis. Therefore, we find

$$\begin{aligned} Y(-c(z)) - \langle A_\xi Y, J\nabla_Y Y \rangle - JY(-a(z)) + \langle A_\xi Y, i\nabla_Y Y \rangle \\ = -s(Y)a(z) - s(JY)c(z). \end{aligned}$$

This proves the relation (2.44). Similarly one can prove the other equations. □

REMARK 2.9. Let U be an open neighborhood of a regular point M such that on U there exists a pair of orthonormal tangent vector fields $X$ and $Y$, with the property that at every point $g(X, Y) = g(X, JY) = 0$, satisfying the strong minimality condition. If $s$ is the tensor field of type $(0, 1)$ defined by the Weingarten formula

$$\bar{\nabla}_X \xi = -A_\xi X + s(X) J\xi,$$

then we have the following relations:

$$\begin{aligned} (X+Y)(d(z)+b(z)) + (JX+JY)(d(z)-b(z)) = \\ (s(X)+s(Y))[(b(z)-d(z)) + i(b(z)+d(z))], \end{aligned} \tag{2.52}$$

$$\begin{aligned} (X+Y)(c(z)+a(z)) + (JX+JY)(c(z)-a(z)) = \\ (s(X)+s(Y))[(a(z)-c(z)) + i(c(z)+a(z))], \end{aligned} \tag{2.53}$$

$$\begin{aligned} (X+Y)(a(z)+b(z)+c(z)+d(z)) + (JX+JY)(c(z)+d(z)-a(z)-b(z)) = \\ (s(X)+s(Y))[(a(z)+b(z)-c(z)-d(z)+i(a(z)+b(z)+c(z)+d(z))]. \end{aligned} \tag{2.54}$$

One could obtain these relations by straightforward computations from the previous remark.

REMARK 2.10. Let U be an open neighborhood of a regular point M such that on U there exists a pair of orthonormal tangent vector fields $X$ and $Y$, with the property that at every point $g(X, Y) = g(X, JY) = 0$, satisfying the strong minimality condition. Suppose that at least one of the analytic functions $a, b, c, d$ is nonvanishing everywhere on $U$, say $a \neq 0$ on $U$. Then we have

$$X(d(z) - ib(z)) = \frac{b(z) + id(z)}{a(z) + ic(z)} X(c(z) - ia(z)), \tag{2.55}$$

$$Y(d(z) - ib(z)) = \frac{b(z) + id(z)}{a(z) + ic(z)} Y(c(z) - ia(z)). \tag{2.56}$$

PROOF. We have the relations:

$$\begin{aligned} X(d(z)) - JX(b(z)) = s(X)(b(z) + id(z)), \\ X(c(z)) - JX(a(z)) = s(X)(a(z) + ic(z)). \end{aligned}$$

Solving the second equation for $s(X)$ and replacing it in the first, then using $JX = iX$ and keeping in mind that $a \neq 0$ everywhere on $U$, we get the claimed result. Similar relations hold in $Y$. □

## 3. A Study of Strong Minimality Through Examples

In the previous section we have seen a few characterizations of strongly minimal complex surfaces in $\mathbb{C}^3$. We keep the same notations in the present section, which are consistent with the notations from [**12**].

In [**12**] it is proved that $z_1^2+z_2^2+z_3^2=1$ is a strongly minimal surface. (Actually the proof in [**12**] implies that $z_1^2+z_2^2+z_3^2=\kappa$, with $\kappa\in\mathbb{C}$, is strongly minimal in $\mathbb{C}^3$, since the basis obtained does not depend on $\kappa$. A direct proof of this claim would have to follow the same steps as for $\kappa=1$.) We compute the functions $a, b, c, d$, to see how these analytic functions look like for this strongly minimal surface. The computation is local, in the sense described below. First, we have

$$f(z_1,z_2)=(1-z_1^2-z_2^2)^{1/2},$$

on the principal branch of the complex radical. Since $z_j=x_j+iy_j$, $j=1,2,3$, we will denote by $\zeta$ the expression

$$\zeta=1-z_1^2-z_2^2=(1-x_1^2-x_2^2+y_1^2+y_2^2)+i(-2x_1y_1-2x_2y_2).$$

If we denote by $\theta$ the polar angle of the complex number $\zeta$, then, on the principal branch of the complex radical, we get

$$\zeta^{(1/2)}=\sqrt{|\zeta|}\Big(\cos\frac{\theta}{2}+i\sin\frac{\theta}{2}\Big),$$

or, by a direct computation

$$f(z_1,z_2)=\frac{1}{\sqrt{2}}(\sqrt{|\zeta|+Re\,\zeta}+i\sqrt{|\zeta|-Re\,\zeta}),$$

where $Re\,\zeta=1-x_1^2-x_2^2+y_1^2+y_2^2$. Therefore, we find

$$u(x_1,x_2,y_1,y_2)=\frac{1}{\sqrt{2}}\sqrt{|\zeta|+Re\,\zeta},\qquad v(x_1,x_2,y_1,y_2)=\frac{1}{\sqrt{2}}\sqrt{|\zeta|-Re\,\zeta}. \tag{3.1}$$

It was established by Chen in [**12**] that, for a position vector $x=(a_1+ib_1,a_2+ib_2,a_3+ib_3)\in V(f)$, the basis realizing the strong minimality condition is given by

$$e_1=\frac{1}{\sqrt{2}}\left(\frac{b_2}{\sqrt{a_1^2+b_2^2}},\frac{ia_1}{\sqrt{a_1^2+b_2^2}},1\right), \tag{3.2}$$

$$e_2=\frac{1}{\sqrt{2}}\left(\frac{ib_2}{\sqrt{a_1^2+b_2^2}},\frac{a_1}{\sqrt{a_1^2+b_2^2}},i\right). \tag{3.3}$$

Therefore we get

$$a(z)=-||\frac{\partial f}{\partial z}||^{-1}\left(\frac{b_2^2-a_1^2}{a_1^2+b_2^2}+1\right), \tag{3.4}$$

$$b(z)=-||\frac{\partial f}{\partial z}||^{-1}, \tag{3.5}$$

$$c(z)=0, \tag{3.6}$$

$$d(z)=-||\frac{\partial f}{\partial z}||^{-1}\left(\frac{b_2^2-a_1^2}{a_1^2+b_2^2}-1\right), \tag{3.7}$$

at every point $x=(a_1+ib_1,a_2+ib_2,a_3+ib_3)\in V(\phi)$.

EXAMPLE 3.1. The Kähler submanifold given by

$$M^2 = \{z \in \mathbb{C}^3 / z_1 + z_2 + z_3^2 = \kappa, \kappa \in \mathbb{C}\}$$

is strongly minimal in $\mathbb{C}^3$.

PROOF. We use consistently the notations from [**12**]. Let $f(z_1, z_2, z_3) = z_1 + z_2 + z_3^2$, then $\frac{\partial f}{\partial z} = (1, 1, 2z_3)$. For $z = (z_1, z_2, z_3) = (a_1 + ib_1, a_2 + ib2, a_3 + ib_3)$, we get $||\frac{\partial f}{\partial z}|| = (2 + 4(a_3^2 + b_3^2))^{1/2} > 0$. We compute also $\frac{\partial^2 f}{\partial z_j \partial z_k} = diag(0, 0, 2)$. The submanifold is strongly minimal if and only if at every point $z$ there exists two vectors $X$ and $Y$ such that

$$g(X, X) = g(Y, Y) = 1, \tag{3.8}$$

$$g(X, Y) = g(X, JY) = 0, \tag{3.9}$$

$$g(X, \xi) = g(X, J\xi) = g(Y, \xi) = g(Y, J\xi) = 0, \tag{3.10}$$

$$g(A_\xi X, X) + g(A_\xi Y, Y) = 0, \tag{3.11}$$

$$g(A_J \xi X, X) + g(A_J \xi Y, Y) = 0. \tag{3.12}$$

From the first equation (3.8) we can keep only $g(X, X) = g(Y, Y)$, and normalize the obtained basis at the end. With this adjustment and using the same notation convention as before, i.e. $X = (\alpha_1, \alpha_2, \alpha_3, \beta_1, \beta_2, \beta_3)$, $Y = (\gamma_1, \gamma_2, \gamma_3, \delta_1, \delta_2, \delta_3)$, the system becomes

$$\alpha_1^2 + \alpha_2^2 + \alpha_3^2 + \beta_1^2 + \beta_2^2 + \beta_3^2 = \gamma_1^2 + \gamma_2^2 + \gamma_3^2 + \delta_1^2 + \delta_2^2 + \delta_3^2, \tag{3.13}$$

$$\langle \alpha, \gamma \rangle + \langle \beta, \delta \rangle = 0, \tag{3.14}$$

$$\langle \alpha, \delta \rangle = \langle \beta, \gamma \rangle, \tag{3.15}$$

$$\alpha_1 + \alpha_2 + 2a_3\alpha_3 - 2b_3\beta_3 = 0, \tag{3.16}$$

$$\beta_1 + \beta_2 + 2a_3\beta_3 + 2b_3\alpha_3 = 0, \tag{3.17}$$

$$\gamma_1 + \gamma_2 + 2a_3\gamma_3 - 2b_3\delta_3 = 0, \tag{3.18}$$

$$\delta_1 + \delta_2 + 2a_3\delta_3 + 2b_3\gamma_3 = 0, \tag{3.19}$$

$$\alpha_3^2 - \beta_3^2 + \gamma_3^2 - \delta_3^2 = 0, \tag{3.20}$$

$$\alpha_3\beta_3 + \gamma_3\delta_3 = 0. \tag{3.21}$$

A solution for this system is obtained if one takes $\beta_3 = \gamma_3 = 0$. With this choice, we eliminate the unknowns

$$\alpha_1 = -\alpha_2 - 2a_3\alpha_3 \qquad \beta_1 = -\beta_2 - 2b_3\alpha_3, \tag{3.22}$$

$$\gamma_1 = -\gamma_2 + 2b_3\delta_3 \qquad \delta_1 = -\delta_2 - 2a_3\delta_3. \tag{3.23}$$

Setting $\alpha_2 = 0$, $\alpha_3 = 1$ and $\delta_3 = -1$ we get by direct computation:

$$\alpha = (-2a_3, 0, 1), \tag{3.24}$$

$$\beta = (-b_3 - \frac{\sqrt{2 + 4b_3^2}}{2}, -b_3 + \frac{\sqrt{2 + 4b_3}}{2}, 0), \tag{3.25}$$

$$\gamma = (-b_3 + \frac{\sqrt{2+4b_3^2}}{2}, -b_3 - \frac{\sqrt{2+4b_3^2}}{2}, 0), \tag{3.26}$$

$$\delta = (0, 2a_3, -1). \tag{3.27}$$

One could verify directly that the system (3.13)-(3.21) is satisfied by the above solution. Therefore, after normalization, the basis satisfying the strong minimality condition is

(3.28)

$$X = (2 + 4a_3^2 + 4b_3^2)^{-1/2} \left( -2a_3 - i(b_3 + \frac{\sqrt{2+4b_3^2}}{2}), i(-b_3 + \frac{\sqrt{2+4b_3^2}}{2}), 1 \right),$$

$$Y = (2 + 4a_3^2 + 4b_3^2)^{-1/2} \left( -b_3 + \frac{\sqrt{2+4b_3^2}}{2}, -b_3 - \frac{\sqrt{2+4b_3^2}}{2} + 2a_3 i, -i \right). \tag{3.29}$$

□

REMARK 3.2. Therefore, the problem of classification of strongly minimal complex surfaces in $\mathbb{C}^3$ reduces to the classification of polynomial systems of type (3.8)-(3.12) that admit a solution. Such systems have ten algebraic equations and twelve unknowns. A complete classification of these polynomial systems that admit a solution would complete the study of Chen's fundamental inequality in complex space forms, in the sense of (2) from Theorem 2.4 from [**12**], and, in our view, represents an interesting problem for future research.

## References


[1] G. Besson, J. Lohkamp, P. Pansu, P. Petersen: *Riemannian Geometry*, American Mathematical Society, 1996.

[2] D. E. Blair: *Riemannian Geometry of Contact and Symplectic Manifolds,* Birkhäuser, 2002.

[3] B.-Y. Chen: *Geometry of Submanifolds,* Marcel Dekker, New York, 1973.

[4] B.-Y. Chen: *Geometry of Submanifolds and Its Applications*, Science University of Tokyo, 1981.

[5] B.-Y. Chen: *Some pinching and classification theorems for minimal submanifolds,* Arch. Math., **60** (1993), 568-578.

[6] B.-Y.Chen: *A Riemannian invariant and its applications to submanifold theory,* Results Math., **27** (1995), 17-26.

[7] B.-Y. Chen: *Mean curvature and shape operator of isometric immersions in real-space-forms,* Glasgow Math. J., **38** (1996), 87-97.

[8] B.-Y.Chen: *Strings of Riemannian invariants, inequalities, ideal immersions and their applications,* Third Pacific Rim Geom. Conf., pp. 7-60, (Intern. Press, Cambridge, MA), 1998.

[9] B.-Y. Chen: *Relations between Ricci curvature and shape operator for submanifolds with arbitrary codimension,* Glasgow Math. J., **41** (1999), 33-41.

[10] B.-Y. Chen: *Some new obstructions to minimal and Lagrangian isometric immersions,* Japanese J. Math., **26** (2000), 105-127.

[11] B.-Y. Chen: *Riemannian Submanifolds,*, in *Handbook of Differential Geometry,* Vol. I, pp. 187-418, North Holland, 2000.

[12] B.-Y. Chen: *A series of Kählerian invariants and their applications to Kählerian geometry,* Beiträge Algebra Geom. **42** (2001), 165-178.

[13] B.-Y. Chen: *On isometric minimal immersions from warped products into real space forms*, Proc. Edinburgh Math. Soc., **45** (2002), 579-587.

[14] B.-Y. Chen: *Non-immersion theorems for warped products in complex hyperbolic spaces,* Proc. Japan Acad. Ser. A Math. **78** (2002), 96-100.

[15] S.-S. Chern: *Minimal submanifolds in a Riemannian manifold,* University of Kansas, 1968.

[16] S.-S. Chern, M. P. do Carmo, S. Kobayashi: *Minimal submanifolds of a sphere with second fundamental form of constant length,* Functional Analysis and Related Fields, Springer, pp 59-75, 1970.

[17] M. Dajczer, L. A. Florit: *On Chen's basic equality,* Illinois J.Math., **42** (1998), 97-106.

[18] F. Defever, I. Mihai, L. Verstraelen: *B.Y.Chen's inequality for C-totally real submanifolds in Sasakian space forms,* Boll. Un. Mat. Ital., Ser. B, **11** (1997), 365-374.

[19] F. Dillen, M. Petrovic, L, Verstraelen: *Einstein, conformally flat and semi-symmetric submanifolds satisfying Chen's equality,* Israel J. Math., **100** (1997), 163-169.

[20] F. Dillen, Franki, L. Vrancken: *Totally real submanifolds in $S^6(1)$ satisfying Chen's equality,* Trans. Amer. Math. Soc., **348** (1996), 1633-1646.

[21] Ph. Griffiths and J. Harris: *Principles of Algebraic Geometry*, John Wiley & Sons, Inc., 1978.

[22] K. Kodaira: *Complex Manifolds and Deformation of Complex Structures*, Springer-Verlag, 1986.

[23] S. Kobayashi, K. Nomizu: *Foundations in Differential Geometry*, vol. I (1963), vol. II (1969), Wiley-Interscience, New York.

[24] P.-F. Leung: *On a relation between the topology and the intrinsic and extrinsic geometries of a compact submanifold*, Proc. Edinburgh Math. Soc., **28** (1985), 305-311.

[25] H. B. Lawson: *Some intrinsic characterizations of minimal surfaces,* J. d'Analyse Math., **24** (1971), 151-161.

[26] G. Li: *Semi-parallel, semi-symmetric immersions and Chen's equality,* Results Math. **40** (2001), 257-264.

[27] K. Nomizu and B. Smyth: *Differential geometry of complex hypersurfaces II*, J. Math. Soc. Japan, **20** (1968), 498-521.

[28] K. Ogiue: *Differential geometry of Kähler submanifolds*, Adv. in Math., **13** (1974), 73-114.

[29] R. Osserman: *Minimal surfaces in the large,* Comment. Math. Helv., **35** (1961), 65-76.

[30] R. Osserman: *Minimal varieties,* Bull. Amer. Math. Soc., **75** (1969), 1092-1120.

[31] T. Otsuki: *Minimal hypersurfaces in a Riemannian manifold of constant curvature,* Amer. J. Math., **92** (1970), 145-173.

[32] P. Petersen: *Riemannian Geometry*, Springer Verlag, 1998.

[33] T. Sasahara: *CR-submanifolds in a complex hyperbolic space satisfying an equality of Chen*, Tsukuba J. Math., **23** (1999), 565-583.

[34] T. Sasahara: *Chen invariant of CR-submanifolds*, in Geometry of submanifolds, 114-120, Kyoto, 2001.

[35] T. Sasahara: *On Chen invariant of CR-submanifolds in a complex hyperbolic space*, Tsukuba J. Math. **26** (2002), 119-132.

[36] B. Smyth: *Differential geometry of complex hypersurfaces I*, Ann. of Math., **85** (1967), 246-266.

[37] B. Suceavă: *Some remarks on B.-Y.Chen's inequality involving classical invariants*, Anal. Sti. Univ. "Al.I.Cuza" Iasi, s.I.a, Math., **64** (1999), 405-412.

[38] B. Suceavă: *The Chen invariants of warped products of hyperbolic planes and their applications to immersibility problems*, Tsukuba J.Math., **25** (2001), 311- 320.

[39] T. Takahashi: *Minimal immersions of Riemannian manifolds,* J. Math. Soc. Japan, **18** (1966), 380-385.

[40] Y. Tsukamoto: *On Kählerian manifolds with positive holomorphic sectional curvature,* Proc. Japan Acad., **33** (1957), 333-335.

[41] A. Vitter: *On the Curvature of Complex Hypersurfaces,* Indiana Univ. Math. J., **23** (1974), 813-826.

[42] R. O. Wells: *Differential Analysis on Complex Manifolds,* Second Edition, Spinger-Verlag, 1980.

[43] K. Yano and M. Kon: *Structures on Manifolds*, World Scientific, 1984.



Department of Mathematics, California State University at Fullerton, P.O.Box 6850, Fullerton, CA, 92834-6850, U.S.A.

*E-mail address*: `bsuceava@fullerton.edu`

Contemporary Mathematics
Volume **337**, 2003

# Linear perturbations of spatially locally homogeneous spacetimes

Masayuki Tanimoto

ABSTRACT. Methods and properties regarding the linear perturbations are discussed for some spatially closed (vacuum) solutions of Einstein's equation. The main focus is on two kinds of spatially locally homogeneous solution; one is the Bianchi III (Thurston's $H^2 \times \mathbf{R}$) type, while the other is the Bianchi II (Thurston's Nil) type. With a brief summary of a previous results on the Bianchi III perturbations, asymptotic solutions for the gauge-invariant variables for the Bianchi III are shown, with which (in)stability of the background solution is also examined. The issue of linear stability for a Bianchi II solution is still an open problem. To approach it, appropriate eigenfunctions are presented for an explicitly compactified Bianchi II manifold and based on that, some field equations on the Bianchi II background spacetime are studied. Differences between perturbation analyses for Bianchi class B (to which Bianchi III belongs) and class A (to which Bianchi II belongs) are stressed for an intention to be helpful for applications to other models.

## 1. Introduction

A classification of three dimensional homogeneous (Riemannian) manifolds is known in relativity as the Bianchi classification, which is a classification of the three dimensional simply transitive Lie groups by isomorphism. A four dimensional spacetime $(X \times \mathbf{R}, \mathrm{g}_{ab})$ is said to be spatially homogeneous if a Bianchi group $G$ acts spacelike and makes the metric $\mathrm{g}_{ab}$ invariant. Cosmological models based on such a spacetime structure, especially with isotropy, have long been a basis for cosmology, and (in case of being anisotropic) have recently been subjects for more mathematical study in relativity (e.g.[**WE**]). Departure from a homogeneity however belongs to a very recent development in the field (except for linear perturbations of rather special homogeneous solutions such as those with isotropy or ones that are not isotropic but can become so in a limit). In particular, whether the solution is stable or not is unknown for almost all such solutions.

---

2000 *Mathematics Subject Classification.* Primary 83C25; Secondary 53C50, 58J37.

*Key words and phrases.* gauge-invariant perturbations, cosmological models, asymptotic stability, 3-dimensional geometry and topology, hyperbolic, nilgeometry.

The author was supported by the Japan Society for the Promotion of Science (JSPS). The author is grateful to Vincent Moncrief and Katsuhito Yasuno for collaborations related to this work and helpful discussions.

This article discusses linear perturbations of two Bianchi type solutions; one is Bianchi III, the other is Bianchi II. The spatial manifold of each solution is assumed to be *closed* (meaning 'compact without boundary'), introducing an appropriate compactification. (As a result, the solution becomes *locally homogeneous.*) One of the common features of these two models is that both of the spatial manifolds belong to so-called Yamabe type $-1$, which is of special interest in relativity [**FM**]. One of the biggest differences is that while Bianchi III belongs to so-called class B, Bianchi II belongs to class A. (The class A or B is defined with respect to whether the trace of the structure constant of a Bianchi group vanishes or not. [**EM**]) In fact, we will see some of the largest differences between Bianchi III and II amount to those between class A and B. Therefore, in particular, the basic methods for Bianchi II will be applicable to other class A Bianchi models, too. (Difficulties in Bianchi II will also be common for all Bianchi A models.)

We deal with the Bianchi III perturbations first. In this case, we have almost complete results. The technical details of fundamental analysis of Bianchi III vacuum perturbations, including the derivation of the wave equations for the gauge-invariant variables, have already been given in [**TMY**] and is also briefly summarized in subsections 2.1 to 2.3 of the present article. The asymptotic stability of the perturbations has been discussed in [**YT**] in a framework of dynamical system approach, where one does not have to appeal to an exact background solution. In the present article we discuss the asymptotic stability from a different point of view in subsections 2.4 and 2.5, using the exact background solution. The issue of stability of the Bianchi II solution is on the other hand still an open question. In this article we discuss properties of some simpler field equations on that background in the subsequent section, stressing differences from the Bianchi III background.

## 2. Bianchi III vacuum perturbations

**2.1. The background solution.** First, let us describe the background geometry briefly.

The Bianchi III algebra is generated by three vectors $\xi_I$ ($I = 1 \sim 3$) such that the commutation relation is given by

$$[\xi_1, \xi_2] = \xi_1, \quad [\xi_2, \xi_3] = [\xi_3, \xi_1] = 0. \tag{2.1}$$

Using coordinates, they can be represented as

$$\xi_1 = \frac{\partial}{\partial x} + y\frac{\partial}{\partial z}, \quad \xi_2 = \frac{\partial}{\partial y}, \quad \xi_3 = \frac{\partial}{\partial z}. \tag{2.2}$$

We will be able to think of these vectors as Killing vectors on our background spacetime. One can also consider invariant vectors $\chi_I$ and 1-forms $\sigma^I$ defined by

$$\mathcal{L}_{\xi_I}\chi_J = 0, \text{ and } \mathcal{L}_{\xi_I}\sigma^J = 0, \tag{2.3}$$

where $\mathcal{L}_{\xi_I}$ denotes the Lie derivative with respect to the vector $\xi_I$. Suppose $\chi_I$ and $\sigma^I$ are dual to each other; $\langle \chi_I, \sigma^J \rangle = \delta_I^J$. Then, using the coordinates they are represented as

$$\begin{gathered} \chi_1 = y\frac{\partial}{\partial x}, \quad \chi_2 = y\frac{\partial}{\partial y}, \quad \chi_3 = \frac{\partial}{\partial z}, \\ \sigma^1 = \frac{dx}{y}, \quad \sigma^2 = \frac{dy}{y}, \quad \sigma^3 = dz. \end{gathered} \tag{2.4}$$

For convenience we define a standard 2-dimensional hyperbolic metric

$$\tilde{h} \equiv \sigma^1 \otimes \sigma^1 + \sigma^2 \otimes \sigma^2, \tag{2.5}$$

and a standard 1-dimensional Euclid metric

$$\tilde{l} \equiv \sigma^3 \otimes \sigma^3. \tag{2.6}$$

Then, the (universal covering) vacuum solution we consider is given by

$$\tilde{g} = -N(t)^2 dt^2 + q_1(t)\tilde{h} + q_2(t)\tilde{l}, \tag{2.7}$$

where the two scale functions $q_1(t)$ and $q_2(t)$ and the lapse function $N(t)^2$ are explicitly given by

$$q_1(t) = (t+k)^2, \quad q_2(t) = (t-k)/(t+k), \quad N(t)^2 = 1/q_2(t), \tag{2.8}$$

where $k$ is a real parameter.

This type of solution is called *locally rotationally symmetric (LRS)*, because, due to the hyperbolic metric $\tilde{h}$ the spatial metric $\tilde{q} \equiv q_1\tilde{h} + q_2\tilde{l}$ has a fourth Killing vector (with degenerate points). We consider only this type of metric, because otherwise one cannot compactify the spatial manifold.

We can express the compactified solution as follows:

$$(M \times \mathbf{R}, \mathrm{g}) = (\mathbf{R}^4, \tilde{\mathrm{g}})/\Gamma, \tag{2.9}$$

where the spatial manifold $M$ is supposed to be the direct product $\Sigma_g \times S^1$, where $\Sigma_g$ $(g > 1)$ is a higher genus surface with genus $g$ while is $S^1$ the circle. For convenience, we call $\Sigma_g$ *the base* and $S^1$ *the fiber.* $\Gamma \subset \mathrm{Isom}(\mathbf{R}^4, \tilde{\mathrm{g}})$ is a discrete subgroup of the isometry group of $(\mathbf{R}^4, \tilde{\mathrm{g}})$, which acts on $(\mathbf{R}^4, \tilde{\mathrm{g}})$ from the left. We assume that $\Gamma$ is of *orthogonal* type[**TMY**], which means that (the action of) $\Gamma$ preserves the foliation by $H^2$ of the universal covering manifold. In this case one can write the spacetime metric g for the spatially compactified manifold in almost the same form as $\tilde{\mathrm{g}}$:

$$\mathrm{g} = -N(t)^2 dt^2 + q_1(t)h + q_2(t)l, \tag{2.10}$$

where $h$ is the *hyperbolic* metric on $\Sigma_g$ induced from the universal cover metric $\tilde{q}$, while $l$ is the Euclid metric on $S^1$ induced from the same metric.

**2.2. Eigenfunctions, vectors, and symmetric tensors.** We need to find complete sets of eigenfunctions, (eigen)vectors, and (eigen)symmetric tensors for the Laplacian so that we can mode-expand the field variables on our background. Thanks to the fact that our spatial manifold has the orthogonal direct product property, we can reduce this task to those for the base and fiber. Namely, once one finds eigenfunctions, vectors, and symmetric tensors on both the fiber and the base, the remaining task is just to make appropriate products. For details, see [**TMY**]. Let us in the following recall some basic definitions.

The most fundamental are the eigenfunctions on the fiber and the base. Let $c_m$ be a real eigenfunction on the fiber, corresponding to eigenvalue $m$;

$$(\chi_3)^2 c_m = -m^2 c_m. \tag{2.11}$$

We call $m$ *the fiber eigenvalue.* ($\chi_3 = \partial_z$ is the derivative operator along the fiber.) We also define another real eigenfunction $\bar{c}_m$ by

$$\chi_3 c_m = -m\bar{c}_m, \quad \chi_3 \bar{c}_m = m c_m. \tag{2.12}$$

Similarly, let $\hat{S}_\lambda$ be a real eigenfunction on the base, corresponding to eigenvalue $\lambda$;

$$\triangle_h \hat{S}_\lambda = -\lambda^2 \hat{S}_\lambda. \tag{2.13}$$

We call $\lambda$ *the base eigenvalue.* ($\triangle_h$ is the Laplacian with respect to the metric $h$.) The mode function on the spatial manifold $M = \Sigma_g \times S^1$ is given by the product $S_{m,\lambda} \equiv c_m \hat{S}_\lambda$;

$$\triangle_q S_{m,\lambda} = -(q_1^{-1}\lambda^2 + q_2^{-1}m^2) S_{m,\lambda}, \tag{2.14}$$

where $q = q_1 h + q_2 l$ is the spatial part of the metric g.

The mode vectors and symmetric tensors on $(M, q)$ are categorized, according to those for the base $(\Sigma_g, h)$, into four kinds; the *even* type, the *odd* type, the *harmonic* type, and the *transverse-traceless* (TT) type. The meaning of this is as follows.

The even vectors (or 1-forms) $\hat{S}_a$ on the base are equivalent to the exact (gradient) 1-forms on the base, $\hat{S}_a = \hat{D}_a \hat{S}$, where $\hat{D}_a$ is the covariant derivative operator on the base. $\hat{S}$ is generally a function on the base, but it is supposed to be the mode function $\hat{S} = \hat{S}_\lambda$ when thinking of $\hat{S}_a$ as a mode vector on the base. Similarly, the odd 1-forms $\hat{V}_a$ are equivalent to the (Hodge-)dual exact 1-forms, $\hat{V}_a = \varepsilon_a{}^b \hat{D}_b \hat{S}$, where $\varepsilon_{ab} = 2\sigma^1{}_{[a}\sigma^2{}_{b]}$ is the area 2-form on the base. (To raise indices, use $h^{ab}$, the inverse of $h_{ab}$.) The harmonic vectors $\hat{U}_a$ are the Hodge harmonic vectors defined by $d\hat{U} = \delta\hat{U} = 0$, or equivalently $\hat{D}_{[a}\hat{U}_{b]} = \hat{D}^a \hat{U}_a = 0$. The number of the independent harmonic one-forms is equivalent to the 1st-Betti number of the surface, which is given by $2g$ in case of the higher genus surface. The Hodge-decomposition tells us that an arbitrary 1-form on the base can be uniquely decomposed using these even, odd, and harmonic (mode) 1-forms.

The even symmetric tensors are symmetric tensors made from the even vectors or the metric $h_{ab}$ and the mode functions. (The mode functions themselves are supposed to be of even type.) We define the *even trace part* by $\hat{S}h_{ab}$, and the *even traceless part* $\hat{S}_{ab}$ as the traceless part of the gradient of $\hat{S}_a$. Similarly, we define the *odd symmetric tensors* $\hat{V}_{ab}$ and the *harmonic symmetric tensors* $\hat{U}_{ab}$ by the (symmetric) gradient of $\hat{V}_a$ and $\hat{U}_a$, respectively. (They are automatically traceless.) From York's decomposition [**YO**], we know that it is necessary for completeness to consider the *transverse-traceless (*TT*) tensors* $\hat{W}_{ab}$, which are defined by $\hat{D}^b \hat{W}_{ab} = h^{ab}\hat{W}_{ab} = 0$. The number of independent TT tensors on the higher genus hyperbolic surface is the same as the dimension of the so-called Teichmüller space, which is equal to $6g - 6$.

The symmetric mode tensors on the spatial manifold $M$ have, as a result, nine kinds, which are denoted as $(E_i)_{ab}$ $(i = 1, \cdots, 9)$. Four of them are of the even kind, which are defined as $(E_1)_{ab} = c_m \hat{S} h_{ab}$, $(E_2)_{ab} = c_m \hat{S}_{ab}$, $(E_3)_{ab} = c_m \hat{S} l_{ab}$, and $(E_4)_{ab} = 2\bar{c}_m \hat{S}_{(a}\sigma^3{}_{b)}$. (We omit the subscript $\lambda$ like $\hat{S} = \hat{S}_\lambda$.) Similarly, the odd symmetric mode tensors are defined as $(E_5)_{ab} = c_m \hat{V}_{ab}$, and $(E_6)_{ab} = 2\bar{c}_m \hat{V}_{(a}\sigma^3{}_{b)}$. The harmonic symmetric mode tensors are defined as $(E_7)_{ab} = c_m \hat{U}_{ab}$, and $(E_8)_{ab} = 2\bar{c}_m \hat{U}_{(a}\sigma^3{}_{b)}$. Finally, the TT symmetric mode tensors are defined as $(E_9)_{ab} = c_m \hat{W}_{ab}$. These are orthogonal to each other with respect to the standard $L^2$-norm. (The mode vectors on $M$ are also defined in a similar way.)

**2.3. Vacuum perturbations and wave equations.** Using the symmetric mode tensors defined above, we can expand the first variation $\delta q_{ab}$ of the spatial metric $q_{ab}$ as follows.

$$\delta q_{ab} = \sum \gamma^i (E_i)_{ab}, \tag{2.15}$$

where the sum is taken over $i = 1, \cdots, 9$, as well as all possible eigenvalues $\lambda$ and $m$. The coefficients $\gamma^i = \gamma^i(t)$ are functions of time that are supposed to be perturbation variables.

An important fundamental result that can be confirmed by direct computations is about how the perturbations decouple, which is stated as follows. *For given base eigenvalue $\lambda$ and the fiber eigenvalue $m$, the set of the even variables, the odd set, the harmonic set and* TT *set evolve independently from each other.* Note that this statement says two kinds of decoupling; first of all, different eigenmodes decouple from each other, and the four kinds of set of variables decouple from each other. Due to this decoupling property, we can deal with the perturbations one by one for each one of the four kinds for each mode.

The perturbation variables change their form by an infinitesimal diffeomorphism acting on the manifold. Certain combinations of the variables however remain invariant. Such a combination is known as the *gauge invariant variable.* In general, one of the central issues in perturbation analysis is to find these gauge invariant variables and determine the dynamics of them. Our choice of the gauge-invariant variables is as follows:

$$\begin{aligned} Q_{\mathrm{E}} &= -\Delta_1\gamma^1 - \Delta_2\gamma^2 + \gamma^3 - (2m/\lambda)\gamma^4, \\ Q_{\mathrm{O}} &= (\nu/\sqrt{2})\gamma^5 + \gamma^6, \\ Q_{\mathrm{H}} &= (m/2)\gamma^7 + \gamma^8, \\ Q_{\mathrm{T}} &= \gamma^9, \end{aligned} \tag{2.16}$$

where $\Delta_1 \equiv \dot{q}_2/\dot{q}_1$, $\Delta_2 \equiv (\lambda/\sqrt{2(\lambda^2+2)})(\Delta_1 + 2m^2/\lambda^2)$, and $\nu \equiv m/\sqrt{\lambda^2+2}$. The subscripts E, O, H, and T are attached to express it is for the even, odd, harmonic or TT perturbation, respectively.

Getting wave equations (equivalent to the linearized Einstein equation) is a substantial part of the work in the study of perturbations. While we show only the results of it in the following, it is worth mentioning that the actual computations were done in the Hamiltonian formalism, using so-called the method of generating function to find a desired canonical transformation that is necessary to split the system into gauge-dependent and independent parts. After lengthy computations [**TMY**], we get the wave equations for the gauge invariant variables, which are second order ODEs of the form

$$\ddot{Q}_{\mathrm{A}} + f_{\mathrm{A}}(t)\dot{Q}_{\mathrm{A}} + g_{\mathrm{A}}(t)Q_{\mathrm{A}} = 0. \quad (\mathrm{A} = \mathrm{E}, \mathrm{O}, \mathrm{H}, \text{ or } \mathrm{T}) \tag{2.17}$$

Each coefficient function $f_{\mathrm{A}}(t)$ or $g_{\mathrm{A}}(t)$ are given as follows:

**(i)** For the even perturbations:

$$f_{\mathrm{E}} = -2\left(\frac{(t-2k)}{t_+t_-} - \frac{X}{Z}\right), \quad g_{\mathrm{E}} = m^2\frac{t_+^2}{t_-^2} + \lambda^2\frac{1}{t_+t_-} + \frac{Y}{Z}, \tag{2.18}$$

where

$$\begin{aligned} X &\equiv 8\,m^2(t-2k){t_+}^2 + 2\,\lambda^2 m^2(2t-k)\,{t_+}^2 + 2\lambda^2(2\,t-3\,k) + \lambda^4(2\,t-k)t/t_+, \\ Y &\equiv 16\,m^4\,{t_+}^4 - 4\,k\,m^2\,t_+\left[\lambda^2\,(t-4k)/t_- - 8\right] + 2\,k^2\,\lambda^4\,(2\,t+k)/(t_-t_+^2), \\ Z &\equiv 4\,m^4\,{t_+}^6 + 8\,m^2\,t_-\,{t_+}^3 + 4\,\lambda^2\,m^2\,t\,{t_+}^3 + 2\,\lambda^2\,t_-\,t_+ + \lambda^4\,t^2. \end{aligned}$$

**(ii)** For the odd and harmonic perturbations ($\lambda \neq 0$ for the odd, $\lambda = 0$ for the harmonic)

$$f_{\rm O} = -\frac{2\nu^2(t-2k)t_+}{t_- u}, \quad g_{\rm O} = (\lambda^2+2)\frac{u}{t_-^2} + \frac{2(t-3k)}{t_+^2 t_-} - \frac{4(t-2k)}{t_+^3 u}, \tag{2.19}$$

where $u \equiv \nu^2 t_+^2 + t_- t_+^{-1}$.

**(iii)** For the TT perturbations

$$f_{\rm T} = -\frac{2(t-2k)}{t_+ t_-}, \quad g_{\rm T} = m^2 \frac{t_+^2}{t_-^2} + \frac{2(t-3k)}{t_+^2 t_-}. \tag{2.20}$$

**2.4. Asymptotic solutions.** Let us find future asymptotic solutions to discuss the future stability. A lot is known about how we can find asymptotic solutions for linear ODEs. The following simple result, taken from standard texts, is sufficient for our purpose.

THEOREM 2.1 (e.g., [**CH**]). *Consider the following second order ODE*

$$X'' + f(s)X = 0, \tag{2.21}$$

*where $f(s)$ is a function of $s$ that approaches a constant*

$$\lim_{s\to\infty} f(s) = C = \text{constant}. \tag{2.22}$$

*(Primes $'$ stand for $d/ds$.) If $C \neq 0$ and*

$$\int^\infty |f(s) - C|\, ds < \infty, \tag{2.23}$$

*then the equation (2.21) has a set of fundamental solutions $\left\{e^{\pm i\sqrt{C}s}(1+o(1))\right\}$ when $C > 0$, or $\left\{e^{\pm\sqrt{|C|}s}(1+o(1))\right\}$ when $C < 0$. If $C = 0$ and*

$$\int^\infty s\,|f(s)|\, ds < \infty, \tag{2.24}$$

*then the equation (2.21) has a set of fundamental solutions $\{1+o(1), s(1+o(1))\}$.*

With the aid of this theorem we can prove the following.

THEOREM 2.2 (Asymptotic solutions for the generic case). *Assume that the fiber eigenvalue $m$ does not vanish, $m \neq 0$. Then, the wave equations (2.17) for the gauge-invariant variables (2.16) possess the following fundamental solutions*

$$Q_{\rm A}(t) = \left\{t\, e^{\pm i(mt+2k\log t)}(1+o(1))\right\}. \quad (\text{A} = \text{E}, \text{O}, \text{H}, \textit{or } \text{T}) \tag{2.25}$$

PROOF. Define a new time coordinate $s$ by

$$\frac{ds}{dt} = \sigma(t) \equiv 1 + \frac{2k}{t} + O(\frac{1}{t^2}). \tag{2.26}$$

Rewrite the wave equation (2.17) in terms of $s$. Then, by taking

$$a_{\rm A} = \frac{1}{\sqrt{\sigma}} e^{-\frac{1}{2}\int f_{\rm A} dt}, \tag{2.27}$$

and putting $Q_{\rm A} = a_{\rm A} X_{\rm A}$, one has an alternative equation $X_{\rm A}'' + W_{\rm A}(s)X_{\rm A} = 0$. It is easy to check that $\int^\infty \left|W_{\rm A}(s) - m^2\right| ds < \infty$. Therefore, from the previous theorem, $X_A$ has the fundamental solutions $e^{\pm ims}(1+o(1))$, from which the claim follows. □

Similarly, we can prove the following for the $m=0$ case. This case is called the $U(1)$-symmetric case, because this kind of perturbation keeps the $U(1)$-symmetry of the background along the $S^1$ fibers.

THEOREM 2.3 (Asymptotic solutions for the $U(1)$-symmetric case). *Assume $m=0$. Define $\rho_\lambda = \lambda^2 - 1/4$, for convenience. Then, the wave equations (2.17) for the gauge-invariant variables (2.16) possess the following fundamental solutions:*
*For the even perturbations:*

$$Q_{\rm E}(t) = \begin{cases} \left\{ \frac{1}{\sqrt{t}} e^{\pm i\sqrt{\rho_\lambda}\log t}(1+o(1)) \right\} & (\lambda^2 > \frac{1}{4}) \\ \left\{ \frac{1}{\sqrt{t}}(1+o(1)), \frac{\log t}{\sqrt{t}}(1+o(1)) \right\} & (\lambda^2 = \frac{1}{4}) \\ \left\{ t^{-\frac{1}{2}\pm\sqrt{|\rho_\lambda|}}(1+o(1)) \right\}. & (\lambda^2 < \frac{1}{4}) \end{cases} \tag{2.28}$$

*For the odd perturbations:*

$$Q_{\rm O}(t) = \begin{cases} \left\{ \sqrt{t} e^{\pm i\sqrt{\rho_\lambda}\log t}(1+o(1)) \right\} & (\lambda^2 > \frac{1}{4}) \\ \left\{ \sqrt{t}(1+o(1)), \sqrt{t}\log t(1+o(1)) \right\} & (\lambda^2 = \frac{1}{4}) \\ \left\{ t^{\frac{1}{2}\pm\sqrt{|\rho_\lambda|}}(1+o(1)) \right\}. & (\lambda^2 < \frac{1}{4}) \end{cases} \tag{2.29}$$

PROOF. Define a new time coordinate $s$ by

$$\frac{ds}{dt} = \sigma^{(0)}(t) \equiv \frac{1}{t} + O(\frac{1}{t^2}). \tag{2.30}$$

Rewrite the wave equation (2.17) in terms of $s$. Then, by taking

$$a_A = \frac{1}{\sqrt{\sigma^{(0)}}} e^{-\frac{1}{2}\int f_{\rm A} dt}, \tag{2.31}$$

and putting $Q_{\rm A} = a_A X_{\rm A}$, one has an alternative equation $X_{\rm A}'' + W_{\rm A}(s)X_{\rm A} = 0$. (A=E or O.) It is straightforward to check that

$$\int^\infty |W_A(s) - \rho_\lambda|\, ds < \infty \quad (\lambda^2 \neq \frac{1}{4}) \tag{2.32}$$

and

$$\int^\infty s\,|W_A(s)|\, ds < \infty. \quad (\lambda^2 = \frac{1}{4}) \tag{2.33}$$

Theorem 2.1 therefore applies, and the claim follows after replacing the time coordinate back to $t$. □

REMARK 2.4. As seen from the last two proofs, the key point to find asymptotic solutions is the choice of a good new time variable (like the choice of $\sigma(t)$ or $\sigma^{(0)}(t)$ in the proofs.).

Interestingly enough, in the generic case the asymptotic solutions are common for all kinds of perturbations. Moreover, the asymptotic solution does not depend on the base eigenvalue $\lambda$. In this sense the asymptotic solutions are very universal.

On the other hand, for the $U(1)$-symmetric case, for each kind of perturbations the behavior of the asymptotic solutions is divided into three cases, depending upon the value of $\lambda$.

**2.5. Stability issue.** Although all the asymptotic solutions above show that our gauge-invariant perturbation variables are growing in time, one cannot say anything about stability of the solution from these results themselves. This is because they do not take into account the fact that the background solution is expanding. Because of the anisotropy of the background solution, it is a very subtle question how one can subtract this expansion effect. If the background was isotropic there exists a natural way to normalize the perturbation variables, because it has only one scale factor, but in an anisotropic case it is not clear how to find a right way of normalizing the variables, especially in a way such that the result is gauge invariant.

Nevertheless, one possible way may be to use the zero mode solution to normalize the gauge invariant variables. By a *zero mode solution* we mean a solution with the vanishing eigenvalues. Such a solution represents a perturbation from a locally homogeneous solution to another locally homogeneous solution, which is an effect we are not interested in. So, it may make sense to use the zero mode solution to do a normalization. Moreover, it is apparently gauge-invariant. (This scheme is essentially equivalent to the one discussed in [**YT**].)

All the zero mode (vacuum) solutions are already given in [**TMY**], from which one obtains the following growing rates for those solutions:

$$Q_{\mathrm{E}}^{(0)} = O(1), \quad Q_{\mathrm{H}}^{(0)} = O(t), \quad Q_{\mathrm{T}}^{(0)} = O(t^2). \tag{2.34}$$

(Although there does not exist a zero mode for the odd perturbations, we identify it with that of the harmonic ones, recalling the fact that the harmonic perturbation system is formally equivalent to the limit $\lambda \to 0$ of the odd system.)

We define *the stability measures by means of zero mode normalization* by

$$M_{\mathrm{A}} \equiv Q_{\mathrm{A}}/Q_{\mathrm{A}}^{(0)}, \quad (\text{A=E, O, H, or T}) \tag{2.35}$$

where we think of $Q_{\mathrm{O}}^{(0)} = Q_{\mathrm{H}}^{(0)}$ as noted above. From the asymptotic solutions for $Q_{\mathrm{A}}$ we immediately obtain the following behaviors for our stability measures.

THEOREM 2.5. *The stability measures by means of zero mode normalization for the vacuum orthogonal Bianchi III spacetime have the following asymptotic decaying or growing rates:*

*Generic ($m \neq 0$) case:*

$$|M_{\mathrm{A}}| = \begin{cases} O(t) & (A{=}E), \\ O(1) & (A{=}O \text{ or } H), \\ O(t^{-1}) & (A{=}T). \end{cases} \tag{2.36}$$

*(The rates are common for all eigenvalues (except $m = 0$), but depend on the kind of perturbations.)*

*$U(1)$-symmetric ($m = 0$) case:*

$$|M_{\mathrm{A}}| = \begin{cases} O(t^{-1/2}) & (\lambda^2 > \frac{1}{4}) \\ O(\frac{\log t}{\sqrt{t}}) & (\lambda^2 = \frac{1}{4}) \\ O(t^{-1/2+\sqrt{|\rho_\lambda|}}) & (\lambda^2 < \frac{1}{4}) \end{cases} \tag{2.37}$$

*for A=E or O. (The rates are common for the even and odd kinds, but there exists a critical eigenvalue $\lambda^2 = 1/4$ for each kind.)*

REMARK 2.6. One can replace (without any modifications) the time variable $t$ to the proper time $\tau$ in the rate formulas of the above theorem.

Note that we have three kinds of behaviors for the generic case; The TT perturbations are decaying, so we can say that the solution is stable against this kind of perturbations. The odd and harmonic perturbations approach a constant, so in this case the solution is quasi-stable. Finally, the even perturbations are growing, so the solution is *unstable* against the even perturbations.

On the other hand all the $U(1)$-symmetric perturbations are decaying, so we can say that the solution is stable against the $U(1)$-symmetric perturbations. Although this means that the $U(1)$-symmetric perturbations play no major role in the whole perturbations of our system, the existence of the critical base eigenvalue $\lambda^2 = 1/4$ is worth pointing out. A similar existence of critical value has been reported in a nonlinear (but $U(1)$-symmetric) analysis [**CBM**], too.

## 3. Field equations on a closed Bianchi II solution

**3.1. The background solution.** The Bianchi II (or Nil) group $G_{\rm II}$ is the 3-dimensional Lie group which consists of all real upper triangular matrices of the form $\begin{pmatrix} 1 & x & z \\ 0 & 1 & y \\ 0 & 0 & 1 \end{pmatrix}$ with the usual multiplication rule for matrices. To save space let us write an element $\mathbf{x} \in G_{\rm II}$ in (transposed) vector form $\mathbf{x} = (x, y, z)$. Then the multiplication rule is given by $(a, b, c) \circ (x, y, z) = (a + x, b + y, c + z + ay)$ for another $\mathbf{a} = (a, b, c) \in G_{\rm II}$. The group $G_{\rm II}$ acts on our universal covering spatial manifold $\tilde{M} = \mathbf{R}^3$ from the left, identifying the group manifold $G_{\rm II}$ with $\tilde{M}$. The Bianchi II algebra is generated by the following three vectors $\xi_I$ $(I = 1 \sim 3)$

$$(3.1) \qquad \xi_1 = \frac{\partial}{\partial x}, \quad \xi_2 = x\frac{\partial}{\partial x} + y\frac{\partial}{\partial y}, \quad \xi_3 = \frac{\partial}{\partial z},$$

which serve as Killing vectors on our background space or spacetime. The invariant vectors $\chi_I$ and their dual 1-forms $\sigma^I$ are given by

$$(3.2) \qquad \begin{gathered} \chi_1 = \frac{\partial}{\partial x}, \quad \chi_2 = \frac{\partial}{\partial y} + x\frac{\partial}{\partial z}, \quad \chi_3 = \frac{\partial}{\partial z}, \\ \sigma^1 = dx, \quad \sigma^2 = dy, \quad \sigma^3 = dz - xdy. \end{gathered}$$

Like the Bianchi type III solution the Bianchi type II vacuum solution can be expressed as

$$(3.3) \qquad (M \times \mathbf{R}, g_{ab}) = (\tilde{M} \times \mathbf{R}, \tilde{g}_{ab})/\Gamma,$$

using the universal cover solution $(\tilde{M} \times \mathbf{R}, \tilde{g}_{ab})$ and a covering map $\Gamma$ acting on it. The spatial manifold $M$ here is, for definiteness, specified to be the "circle bundle over the torus $(T^2)$ with Euler number $e = 1$." (See, e.g., [**HS**].) The fundamental group can be represented in the standard notation as

$$(3.4) \qquad \pi_1(M) = \langle g_1, g_2, g_3; [g_1, g_2] = g_3, [g_1, g_3] = 1, [g_2, g_3] = 1 \rangle,$$

where the brackets stand for commutators, $[a, b] \equiv aba^{-1}b^{-1}$. The vacuum metric $\tilde{g}_{ab}$ on the universal cover is given below. $\Gamma \subset G_{\rm II}$ is a discrete subgroup of the Bianchi II group $G_{\rm II}$, which acts on the universal covering spacetime $(\tilde{M} \times \mathbf{R}, \tilde{g}_{ab})$ from the left in the way that keeps the natural homogeneous slicings. Some details for the compactification are given in the next subsection. The metric $\mathrm{g}_{ab}$ on $M\times\mathbf{R}$ is the induced metric from $\tilde{\mathrm{g}}_{ab}$.

The exact vacuum solution $\tilde{\mathrm{g}}_{ab}$ is given [**TAU**] by

$$ds^2 = -N^2(t)dt^2 + q_1(t)(\sigma^1)^2 + q_2(t)(\sigma^2)^2 + q_3(t)(\sigma^3)^2, \tag{3.5}$$

where

$$N^2 = 1+\beta^2 t^{4p_3}, \quad q_1 = t^{2p_1}N^2, \quad q_2 = t^{2p_2}N^2, \quad q_3 = 16p_3^2\beta^2 t^{2p_3}/N^2. \tag{3.6}$$

$p_i (i=1,2,3)$ and $\beta$ are constant parameters such that $\beta>0$, $p_3\neq 0$, and

$$\Sigma p_i = \Sigma p_i^2 = 1. \tag{3.7}$$

When $p_1 = p_2$, or equivalently when $q_1(t) = q_2(t)$, the solution is said to be LRS as in the Bianchi III case. While there seem two possible such cases $(p_1,p_2,p_3) = (0,0,1)$ and $(2/3,2/3,-1/3)$, these two solutions represent the same one-parameter solution. When we consider a LRS solution, we may want to take $(p_1,p_2,p_3) = (2/3,2/3,-1/3)$, since the time coordinate $t$ in this solution approaches the proper time at future infinity, which is more favorable for comparisons with other models.

As shown in [**TKH**], $\Gamma$ is a four-parameter group. Our spatially closed solution (3.3) therefore comprises a six parameter solution (since the universal cover has, as we have seen, two independent parameters).

Notice the fact that $\Gamma$ is a subgroup of the Bianchi II group $G_{\mathrm{II}}$, not necessarily a subgroup of a larger group like in the Bianchi III case. This in particular means that *each* invariant vector $\chi_I$ is well defined not only on the universal cover $\tilde{M}$ but also on the compactified manifold $M$ (i.e., the induced vectors $\pi^*\chi_I$ for the covering map $\pi : \tilde{M}\to M$ are well defined on $M$. We omit $\pi^*$ for simplicity, though). This is because they are invariant under the action of $G_{\mathrm{II}}$, and so are under $\Gamma\subset G_{\mathrm{II}}$. We will see that because of this property the invariant vectors $\chi_I$ play a central role in developing calculus concerning mode expansions. This significance of the invariant vectors is common for all Bianchi class A models.

**3.2. Compactifications and eigenfunctions.** Consider an arbitrary locally homogeneous closed 3-manifold $(M, q_{ab})$ with $\pi_1(M)$ given by (3.4). Such a Riemannian manifold is isometric to the quotient $\mathcal{N}^3/A$, where $\mathcal{N}^3 = (\mathbf{R}^3, e^{2\alpha}q^{(0)}_{ab})$ is the universal cover with a metric that is conformal to the Bianchi II standard metric $q^{(0)} = (\sigma^1)^2 + (\sigma^2)^2 + (\sigma^3)^2$. $e^{2\alpha}$ is a constant scale factor. $A\subset \mathrm{Isom}\mathcal{N}^3$ is a three parameter infinite group generated by three generators $\mathbf{a}_i \in G_{\mathrm{II}}\subset \mathrm{Isom}\mathcal{N}^3$ $(i = 1\sim 3)$, which are of the form (cf. [**KTH**])

$$\mathbf{a}_1 = (u,\delta,0), \quad \mathbf{a}_2 = (0,2\pi v,0), \quad \mathbf{a}_3 = (0,0,2\pi uv). \tag{3.8}$$

We denote $A = \{\mathbf{a}_1,\mathbf{a}_2,\mathbf{a}_3\}$. The three parameters $u$, $v$, and $\delta$ are called the Teichmüller parameters.

Also, consider another Riemannian manifold $(\bar{M},\bar{q}_{ab})$, which can be represented as $\mathcal{N}^3/\bar{A}$, where $\bar{A} = \{\mathbf{a}_2,\mathbf{a}_3\}$. $(\bar{M},\bar{q}_{ab})$ is a covering of $(M,q_{ab})$. $\bar{M}$ is homeomorphic to the direct product $T^2\times\mathbf{R}$, where $T^2$ is the two-torus.

Note that the (scalar) Laplacian $\triangle$ with respect to the standard metric $q^{(0)}_{ab}$ can be expressed as $\triangle = (\chi_1)^2 + (\chi_2)^2 + (\chi_3)^2$. This operator is apparently well defined on both $\bar{M}$ and $M$, since so is each $\chi_I$. It is straightforward to confirm that $\triangle$ commutes with $\chi_3$. We can therefore diagonalize eigenfunctions with respect to both $\triangle$ and $\chi_3$. Consider another operator $\xi_2 = \partial/\partial y$, which we can find commutes with both $\triangle$ and $\chi_3$, so one can diagonalize the eigenfunctions with respect to $\xi_2$, too. This operator however is *not* well defined on $M$, but on $\bar{M}$. Because of this fact, it is convenient to first consider eigenfunctions on $\bar{M}$, and *then* construct those on $M$, taking linear combinations of the eigenfunctions on $\bar{M}$.

Let us define eigenvalues $\mu$ and $\nu$ for the operators $\chi_3$ and $\xi_2$ by the following relations

$$\chi_3\bar{\phi} = i\mu\bar{\phi}, \quad \xi_2\bar{\phi} = i\nu\bar{\phi}, \tag{3.9}$$

where $\bar{\phi}$ is an eigenfunction on $\bar{M}$. Also, we define $\lambda$ by $\triangle\bar{\phi} = -\lambda^2\bar{\phi}$. From the boundary condition $\bar{\phi}(\Gamma\mathbf{x}) = \bar{\phi}(\mathbf{x})$, it is found $\bar{\phi} = X(x)e^{i\mu z}e^{i\nu y}$, where

$$\begin{aligned} &\mu = m/(uv), \quad m = 0, \pm 1, \cdots, \pm\infty, \\ &\nu = n/v, \quad n = 0, \pm 1, \cdots, \pm\infty. \end{aligned} \tag{3.10}$$

We call $m$ (or $\mu$) the *fiber eigenvalue*, $n$ (or $\nu$) the *auxiliary eigenvalue*, $\lambda$ the *total eigenvalue*. The function $X(x)$ must satisfy the following equation (that is equivalent to the harmonic oscillator Schrödinger equation):

$$X'' + (\lambda^2 - \mu^2 - (\mu x + \nu)^2)X = 0. \tag{3.11}$$

This equation tells us in particular that like the Bianchi III case, one must deal with the generic ($m \neq 0$) case and the $U(1)$-symmetric ($m = 0$) case separately.

Let us focus on the generic ($m \neq 0$) case. One can define the eigenfunctions on $\bar{M}$ as

$$\bar{\phi}_{l,m,n}(\mathbf{x}) = D_l(\pm\sqrt{\frac{2uv}{|m|}}(\frac{m}{uv}x + \frac{n}{v}))e^{i\frac{m}{uv}z}e^{i\frac{n}{v}y}, \tag{3.12}$$

where $D_l(\zeta) = e^{-\frac{1}{4}\zeta^2}H_l(\zeta)$ is the parabolic cylinder function defined using the Hermite function $H_l(\zeta) = (-1)^l e^{\frac{1}{2}\zeta^2}\frac{d^l}{d\zeta^l}e^{-\frac{1}{2}\zeta^2}$. The $D_l$ factor in the above equation satisfies Eq.(3.11). The plus and minus signs, respectively, correspond to $m > 0$ and $m < 0$ cases. The index $l$ takes non-negative integers

$$l = 0, 1, \cdots, \infty. \tag{3.13}$$

This is related to the total eigenvalue through $\lambda^2 = |\mu|(2l + |\mu| + 1)$.

The eigenfunctions on $M$ are, as stated, expressed as infinite sums of these eigenfunctions on $\bar{M}$.

THEOREM 3.1. *Consider the generic ($m \neq 0$) modes. Using the eigenfunctions $\bar{\phi}_{l,m,n}(\mathbf{x})$ on the covering ($\bar{M} = T^2 \times \mathbf{R}, \bar{q}_{ab}$), the eigenfunctions on the closed manifold ($M, q_{ab}$) are represented as the infinite sum*

$$\phi_{l,m,n_0}(\mathbf{x}) = \sum_{k=-\infty}^{\infty} e^{i\delta(n_0 k + m\frac{k(k-1)}{2})}\bar{\phi}_{l,m,n_0+mk}(\mathbf{x}), \tag{3.14}$$

*where* $l = 0, 1, \cdots, \infty, \quad m = 1, 2, \cdots, \infty, \quad n_0 = 0, 1, \cdots, m.$

PROOF. Since the summand is well defined on $\bar{M}$, which means it is invariant under the action of $\mathbf{a}_2$ and $\mathbf{a}_3$, all one has to confirm is the invariance of the left hand side under the action of $\mathbf{a}_1$. It is easy to see the following transformation rule

$$(3.15)\quad \begin{aligned} \bar{\phi}_{l,m,n}(\mathbf{a}_1 \circ \mathbf{x}) &= D_l(\pm\sqrt{\frac{2uv}{|m|}}(\frac{m}{uv}(x+u)+\frac{n}{v}))e^{i\frac{m}{uv}(z+uy)}e^{i\frac{n}{v}(y+\delta)} \\ &= e^{i\frac{n}{v}\delta}D_l(\pm\sqrt{\frac{2uv}{|m|}}(\frac{m}{uv}x+\frac{m+n}{v}))e^{i\frac{m}{uv}z}e^{i\frac{m+n}{v}y} \\ &= e^{i\frac{\delta}{v}n}\bar{\phi}_{l,m,n+m}(\mathbf{x}). \end{aligned}$$

From this, one can confirm the invariance

$$(3.16)\quad \begin{aligned} \phi_{l,m,n_0}(\mathbf{a}_1 \circ \mathbf{x}) &= \sum_{k=-\infty}^{\infty} e^{i\frac{\delta}{v}(n_0k+m\frac{k(k-1)}{2})}e^{i\frac{\delta}{v}(n_0+mk)}\bar{\phi}_{l,m,n_0+m(k+1)}(\mathbf{x}) \\ &= \sum_{k=-\infty}^{\infty} e^{i\frac{\delta}{v}(n_0(k+1)+m\frac{k(k+1)}{2})}\bar{\phi}_{l,m,n_0+m(k+1)}(\mathbf{x}) \\ &= \phi_{l,m,n_0}(\mathbf{x}). \end{aligned}$$

(The last equality follows from the replacement $k \to k-1$.) □

Note that as a result of the complete compactification, the index $n_0$ for the eigenfunction on $M$ is bounded by $m$.

These eigenfunctions have the following properties.

THEOREM 3.2. *For the eigenfunctions for the generic ($m \neq 0$) modes, the following is fulfilled (indices $m$ and $n_0$ are for simplicity suppressed):*

$$(3.17)\quad \begin{aligned} \chi_1\phi_l &= -\sqrt{\frac{|\mu|}{2}}\left[\phi_{l+1} - l\phi_{l-1}\right], \\ \chi_2\phi_l &= \pm i\sqrt{\frac{|\mu|}{2}}\left[\phi_{l+1} + l\phi_{l-1}\right], \\ \chi_3\phi_l &= i\mu\phi_l, \end{aligned}$$

*where $\mu$ is defined in Eq.(3.10) and the plus and minus signs correspond, respectively, to $m > 0$ and $m < 0$.*

PROOF. Use $D_l'(\zeta) = -\frac{1}{2}(D_{l+1}(\zeta) - lD_{l-1}(\zeta))$. The remaining task is a straightforward computation. □

REMARK 3.3. These relations can also be derived in a purely algebraic manner.[1]

---

[1] In fact, defining $A_1 = \chi_1 + i\chi_2$ and $A_2 = \chi_1 - i\chi_2$, one can check commutation relations

$$(3.18)\quad [\triangle, A_1] = 2iA_1\chi_3, \quad [\triangle, A_2] = -2iA_2\chi_3,$$

which tell us, e.g., that

$$(3.19)\quad \triangle A_1\phi = ([\triangle, A_1] + A_1\triangle)\phi = (2iA_1\chi_3 + A_1\triangle)\phi = -(2\mu + \lambda^2)A_1\phi.$$

This means that $A_1\phi$ is an eigenfunction for eigenvalue $\lambda'^2 = \lambda^2 + 2\mu$, and the operator $A_1$ acts as an increment (decrement) operator for $m > 0$ ($m < 0$). Similarly, $A_2$ is a decrement (increment) operator with $\lambda'^2 = \lambda^2 - 2\mu$ for $m > 0$ ($m < 0$). Using these relations and following a similar argument for, e.g., the $SO(3)$ case to determine the ground state, one reaches at the following relations

$$(3.20)\quad \begin{aligned} A_1\phi_{l,m} &= -\sqrt{2\mu}\,\phi_{l+1,m}, \quad A_2\phi_{l,m} = \sqrt{2\mu}\,l\,\phi_{l-1,m}, \quad \text{for } m > 0 \\ A_1\phi_{l,m} &= \sqrt{2|\mu|}\,l\,\phi_{l-1,m}, \quad A_2\phi_{l,m} = -\sqrt{2|\mu|}\,\phi_{l+1,m}, \quad \text{for } m < 0 \end{aligned}$$

which are equivalent to the relations claimed.

**3.3. Scalar field.** As a direct application it is of great interest to see the *scalar field equation* on our Bianchi II background. For simplicity, let us consider a background solution where $\Gamma = A$, i.e., $\Gamma$ is specified by three parameters as in the form of $A$. We use the symbols $u$, $v$ and $\delta$ as in the $A$. These parameters are constants throughout the spacetime. (This is not to say the Teichmüller parameters for a Cauchy surface are constants of motion. [**TKH**]) In such a case, we can use the eigenfunctions obtained in the previous section without any modifications. We expand the scalar field $\Phi$ as follows:

$$\Phi(t, \mathbf{x}) = \sum_{l,m,n_0} a_{l,m,n_0}(t)\phi_{l,m,n_0}(\mathbf{x}). \tag{3.21}$$

Since the operators $\chi_I$ do not change $m$ and $n_0$ when acting on $\phi_{l,m,n_0}$, we suppress these subscripts and write simply $a_l$ and $\phi_l$, as far as no confusion occurs.

The scalar field equation $0 = \mathrm{g}^{ab}\nabla_a\nabla_b\Phi$ on our background reduces to

$$0 = \frac{-1}{\sqrt{-g}}(\sqrt{-g}N^{-2}\dot{\Phi})^{\cdot} + (q_1^{-1}(\chi_1)^2 + q_2^{-1}(\chi_2)^2 + q_3^{-1}(\chi_3)^2)\Phi, \tag{3.22}$$

where a dot stands for $d/dt$. Using the background spacetime solution for Bianchi II and the $\chi$-relations, we obtain the following wave equations:

$$\ddot{a}_l + \frac{1}{t}\dot{a}_l + \mu f(t) a_l = \frac{\mu}{2}(t^{-2p_1} - t^{-2p_2})[a_{l-2} + (l+2)(l+1)a_{l+2}], \tag{3.23}$$

where

$$f(t) \equiv \frac{2l+1}{2}(t^{-2p_1} + t^{-2p_2}) + \frac{\mu}{16(p_3)^2\beta^2}(1 + \beta^2 t^{4p_3})^2 t^{-2p_3}. \tag{3.24}$$

(In Eq.(3.23), $a_{l<0}$ should be considered as zero.)

A notable property of these equations is that they comprise a set of infinitely coupled equations, unless the solution is LRS. These couplings between different modes are an unavoidable feature for the field equations on a non-LRS spacetime solution, because for those backgrounds the mode functions defined with respect to a standard metric are no longer eigenfunctions with respect to the Laplacian for each spatial section. That is, those eigenfunctions defined on the standard Riemannian manifold do not remain eigenfunctions in the course of the anisotropic expansion. One could define another "time-dependent" eigenfunctions to get around this difficulty, but in that case time-derivatives of the eigenfunctions appear in the field equation, so in any case the complexities caused by anisotropic expansion are unavoidable.

The above wave equation gives a closed equation when the background is LRS. In this case we can again find asymptotic solutions:

THEOREM 3.4. *On the LRS Bianchi II vacuum solution with $p_1 = p_2 = 2/3$ and $p_3 = -1/3$, the scalar field equation (3.23) for the generic mode has the following fundamental solutions as $t \to \infty$:*

$$a_l(t) = \left\{ t^{-\frac{2}{3}} e^{\pm i \frac{3\mu}{4\beta}(\frac{3}{4}t^{4/3} + \beta^2 \log t)}(1 + o(1)) \right\}. \tag{3.25}$$

PROOF. As remarked in the previous section, it is a suitable choice of new time variable (denoted $s$) that is essential to have an asymptotic solution. An appropriate choice for the present equation is

$$\frac{ds}{dt} = \frac{3}{4}\left(\frac{t^{1/3}}{\beta} + \frac{\beta}{t}\right). \tag{3.26}$$

One can show the claim following the same procedure as in THEOREM 2.2 or 2.3. □

**3.4. Vectors and electromagnetic field.** Let us consider how we can perform separation of variable for *vector fields*. An advantage that our Bianchi II model has and is common for other class A spatially closed Bianchi models is that in principle, we do *not* have to define suitable eigenvectors for this purpose. This is because there exists a set of well-defined independent three invariant vectors $\chi_I$ on the compactified manifold $M$. Any vectors on $M$ can be expanded with respect to these invariant vectors (i.e., one can consider the components with respect to the frame $\{\chi_I\}$) in a group invariant way and so all one has to do is to mode-expand the components using the already defined eigenfunctions $\phi_{l,m,n_0}$. However, this straightforward expansion is found not to be very useful for a successful analysis.

A nicer way is to use an analogy with Regge-Wheeler's spherically symmetric case or one with Bianchi III hyperbolically symmetric case, where one can take advantage of the spherically or hyperbolically symmetric planes. In those cases, one is able to define so-called the 'even' vectors and 'odd' vectors on the symmetric plane (as seen in the previous section for the Bianchi III case), and the even and odd part equations for a linear vector field equation decouple from each other.

Although there do not exist similar symmetric planes for the Bianchi II case, we can, instead, consider the "plane field" spanned by $\chi_1$ and $\chi_2$. This plane field is not integrable, because the commutator $[\chi_1, \chi_2] = \chi_3$ is independent from $\chi_1$ and $\chi_2$. Nevertheless, the plane field can play a special role, because the metric can be LRS only when $q_1 = q_2$.

We define the area two-form $\varepsilon$ of the plane field by $\varepsilon_{ab} = 2\sigma^1{}_{[a}\sigma^2{}_{b]}$. We may define the even vectors by $S_a = (\chi_1\phi_l)\sigma^1{}_a + (\chi_2\phi_l)\sigma^2{}_a$, and odd vectors by $V_a = i\varepsilon_a{}^b\partial_b\phi_l = i((\chi_2\phi_l)\sigma^1{}_a - (\chi_1\phi_l)\sigma^2{}_a)$. (To raise an index for $\varepsilon_{ab}$ we use the standard Bianchi II metric.) We also need orthogonal vectors $S_a^\perp = (\chi_3\phi_l)\sigma^3{}_a$ and time-like vectors $S_a^{\mathrm{T}} = N^{-1}\phi_l(dt)_a$, which may be considered to belong to the even part. These four kinds of vectors complete our set of eigenvectors.

Because of the fact that the commutator of $\chi_1$ and $\chi_2$ does not belong to the tangent space spanned by $\chi_1$ and $\chi_2$, the even and odd parts defined this way do *not* decouple from each other, even when the background is LRS. Nonetheless, the use of even and odd vectors is found useful, especially when the background is LRS, since in that case each mode is found to decouple from the others. (Do not confuse couplings between the even and odd parts and ones between modes.) When the spacetime is not LRS, however, we will have couplings between different modes due to the anisotropic expansion.

Using these mode vectors, we can expand the *vector potential* for an electromagnetic field as follows:

$$A_a = \sum_{l,m,n_0} \alpha_0(t)S_a^{\mathrm{T}} + \alpha_1(t)S_a + \alpha_2(t)V_a + \alpha_3(t)S_a^\perp. \tag{3.27}$$

The four kinds of functions $\alpha_i(t)$ $(i = 0 \sim 3)$ serve as the field variables. Let us consider the *LRS case* below, on which background, as mentioned above, couplings between different modes vanish.

The quantities we are most interested in are the $(U(1)$-$)$gauge-invariant variables, which can be easily found by looking at components of the field strength $F_{ab} = \partial_a A_b - \partial_b A_a$. We obtain the following four independent gauge invariant variables:

$$Q_1 = \alpha_1 - \alpha_3,\ Q_2 = \alpha_2,\ P_1 = \dot{\alpha}_1 - \alpha_0,\ P_2 = \dot{\alpha}_2. \tag{3.28}$$

Although function $P_3 \equiv \dot{\alpha}_3 - \alpha_0$ is also gauge invariant, it is found that it can be (consistently) solved with the others, due to the constraint part of Maxwell's equation. The evolution equations for the LRS background are found as follows:

$$\begin{aligned} \dot{Q}_1 &= P_1 + \frac{1}{\mu}\frac{q_3}{q_1}((2l+1)P_1 + P_2), \quad \dot{Q}_2 = P_2, \\ \dot{P}_1 &= \left(\frac{\dot{N}}{N} - \frac{1}{2}\frac{\dot{q}_3}{q_3}\right) P_1 - \mu^2 \frac{N^2}{q_3} Q_1, \\ \dot{P}_2 &= \left(\frac{\dot{N}}{N} - \frac{1}{2}\frac{\dot{q}_3}{q_3}\right) P_2 - \mu^2 \frac{N^2}{q_3} Q_2 - \mu \frac{N^2}{q_1}(Q_1 + (2l+1)Q_2). \end{aligned} \tag{3.29}$$

To tackle the gravitational perturbation equation we need to generalize our framework so that it is applicable to systems with symmetric tensors. Again, thanks to the well-defined invariant vectors $\chi_I$, separations of variable for the tensors are straightforward in principle. However, because of the complexity of the linearized Einstein equation, to make our analysis successful we will need to find a well-arranged set of base mode tensors like the even and odd vectors used above.

## References


[CBM] Y. Choquet-Bruhat, and V. Moncrief, *Future global in time Einsteinian spacetimes with U(1) isometry group* Ann. Henri Poincaré **2** (2001) 1007–1064; *Non linear stability of einsteinian spacetimes with U(1) isometry group*, preprint (gr-qc/0302021).

[CH] L. Cesari, *Asymptotic behavior and stability problems in ordinary differential equations*, Springer, 3rd ed. (1971); P. Hartman, *Ordinary differential equations*, John Wiley (1964).

[EM] G.F.R. Ellis, and M.A.H. MacCallum, *A class of homogeneous cosmological models*, Commun. Math. Phys. **12** (1969) 108–141.

[FM] A.E. Fischer, and V. Moncrief, *The reduced Einstein equations and the conformal volume collapse of 3-manifolds*, Class. Quantum Grav. **18** (2001) 4493–4515; and reference therein.

[HS] J. Hempel, *3-manifolds*, Ann. of Math. Studies 86, Princeton University Press (1976); P. Scott, *The geometry of 3-manifolds*, Bull. London Math. Soc. **15** (1983) 401–487.

[KTH] T. Koike, M. Tanimoto, and A. Hosoya, *Compact Homogeneous Universes*, J. Math. Phys. **35** (1994) 4855–4888.

[TAU] A. H. Taub, *Empty space-times admitting a three-parameter group of motions*, Ann. Math. **53** (1952) 472–490.

[TKH] M. Tanimoto, T. Koike, and A. Hosoya, *Dynamics of compact homogeneous universes*, J. Math. Phys. **38** (1997) 350–368; *Hamiltonian structures for compact homogeneous universes*, J. Math. Phys. **38** (1997) 6560–6577.

[TMY] M. Tanimoto, V. Moncrief, and K. Yasuno, *Perturbations of spatially closed Bianchi III spacetimes*, Class. Quantum Grav. **20** (2003) 1879–1927.

[WE] J. Wainwright, and G.F.R. Ellis, *Dynamical Systems in Cosmology*, Cambridge University Press (1997).

[YT] K. Yasuno, and M. Tanimoto, *Asymptotic behavior of linearly perturbed spatially closed Bianchi III vacuum solutions*, in preparation.

[YO] J.W. York Jr., *Conformally invariant orthogonal decomposition of symmetric tensors on Riemannian manifolds and the initial-value problem of general relativity*, J. Math. Phys. **14** (1973) 456–464.



Department of Physics, Yale University, New Haven, CT 06511

*Current address*: Max-Planck-Institute für Gravitationsphysik, D-14476 Golm, Germany

Contemporary Mathematics
Volume **337**, 2003

# Certain basic inequalities for submanifolds in $(\kappa, \mu)$-spaces

Mukut Mani Tripathi

ABSTRACT. We obtain certain inequalities involving intrinsic invariants namely, scalar curvature, sectional curvature and Ricci curvature, and the main extrinsic invariant namely squared mean curvature for submanifolds in a $(\kappa, \mu)$-space. Applications of these inequalities give rise to several inequalities for slant, invariant, anti-invariant and *CR*-submanifolds. The equality cases are also discussed.

## 1. Introduction

Due to Nash's immersion theorem, each $n$-dimensional Riemannian manifold admits an isometric immersion into the Euclidean space $\mathbb{E}^{n(n+1)(3n+11)/2}$, and this enables us to consider any Riemannian manifold as a submanifold of a suitable Euclidean space. This phenomenon provides a natural motivation for the study of submanifolds of Riemannian manifolds.

In [**11**], B.-Y. Chen recalls one of the basic interests in the submanifold theory as "To establish simple relationships between the main intrinsic invariants and the main extrinsic invariants of a submanifold". The main intrinsic invariants include scalar curvature, sectional curvatures and Ricci curvature. The main extrinsic invariants are the squared mean curvature and the shape operator. Gauss-Bonnet Theorem, isoperimetric inequality and Chern-Lashof Theorem provide relations between intrinsic and extrinsic invariants for a submanifold in a Euclidean space. For more details we refer to [**12**].

On the other hand, the roots of contact geometry lie in differential equations as in 1872 Sophus Lie introduced the notion of contact transformation (Berührungstransformation) as a geometric tool to study systems of differential equations (for more details see [**1**], [**3**] and [**15**]). This subject has manifold connections with the other fields of pure mathematics, and substantial applications in applied areas such as mechanics, optics, phase space of a dynamical system, thermodynamics and control theory. The two large classes of examples of contact manifolds are the principal circle bundles of the Boothby-Wang fibration and the tangent sphere bundles (for

---

2000 *Mathematics Subject Classification.* Primary 53C40, Secondary 53D10.

*Key words and phrases.* Contact structure, $(\kappa, \mu)$-space, invariant submanifold, anti-invariant submanifold, slant submanifold, *CR*-submanifold.

The author is grateful to Elsevier Science BV for travel support to the 2003 Joint Mathematics Meetings in Baltimore, Maryland, USA.

details see [**3**]). Contact manifolds include the class of Sasakian manifolds. In [**4**], the authors studied a special class of contact manifolds as $(\kappa,\mu)$-spaces for which the characteristic vector field belongs to the $(\kappa,\mu)$-nullity distribution. Characteristic examples of non-Sasakian $(\kappa,\mu)$-spaces are the tangent sphere bundles of Riemannian manifolds of constant sectional curvature not equal to one.

Thus motivated sufficiently, in a previous paper [**25**], the authors established several basic inequalities involving intrinsic and extrinsic invariants of $C$-totally real submanifolds in a $(\kappa,\mu)$-contact space form. In the present paper, we establish several basic inequalities between intrinsic invariants and extrinsic invariants for submanifolds tangent to the structure vector field in non-Sasakian $(\kappa,\mu)$-spaces.

The paper is organized as follows. Section 2 contains a brief introduction to $(\kappa,\mu)$-spaces. In the section 3, we give some basic inequalities with the left hand side being scalar curvature and the right hand side containing squared mean curvature for a submanifold in a non-Sasakian $(\kappa,\mu)$-space. In the section 4, we establish an inequality involving intrinsic invariants, namely the sectional curvature and the scalar curvature of $M$; and the main extrinsic invariant, namely the squared mean curvature. In the section 5, we obtain an inequality involving Ricci curvature and squared mean curvature along with discussion of equality cases. In the section 6, a basic inequality between the $k$-Ricci curvature and the squared mean curvature is established. In the last section, some applications for invariant, anti-invariant, *CR* and slant submanifolds are discussed.

## 2. $(\kappa,\mu)$-spaces

A 1-form $\eta$ on a differentiable manifold $\tilde{M}$ of odd dimension $m$ $(m \geq 3)$ is called a *contact form* if $\eta \wedge (d\eta)^{(m-1)/2} \neq 0$ everywhere on $\tilde{M}$, and $\tilde{M}$ equipped with a contact form is a *contact manifold.* Since rank of $d\eta$ is $m-1$ on the Grassmann algebra $\bigwedge T_p^*\tilde{M}$ at each point $p \in \tilde{M}$, therefore there exists a unique global vector field $\xi$, called the *characteristic vector field*, such that

$$\eta(\xi) = 1, \qquad d\eta\,(\xi, X) = 0, \qquad X \in T\tilde{M}, \tag{2.1}$$

and consequently

$$\mathfrak{L}_\xi \eta = 0, \qquad \mathfrak{L}_\xi d\eta = 0, \tag{2.2}$$

where $\mathfrak{L}_\xi$ denotes the Lie differentiation by $\xi$. In 1953, S. S. Chern [**13**] proved that the structural group of an $m$-dimensional contact manifold can be reduced to $\mathcal{U}\left(\frac{m-1}{2}\right) \times 1$. An $m$-dimensional differentiable manifold $\tilde{M}$ is called an *almost contact manifold* [**16**] if its structural group can be reduced to $\mathcal{U}\left(\frac{m-1}{2}\right) \times 1$. Equivalently, there is an *almost contact structure* $(\varphi, \xi, \eta)$ [**21**] consisting of a $(1,1)$ tensor field $\varphi$, a vector field $\xi$, and a 1-form $\eta$ satisfying

$$\varphi^2 = -I + \eta \otimes \xi, \quad \eta(\xi) = 1, \quad \varphi\xi = 0, \quad \eta \circ \varphi = 0. \tag{2.3}$$

First and one of the remaining three relations of (2.3) imply the other two relations of (2.3).

An almost contact structure is said to be *normal* [**22**] if the induced almost complex structure $J$ on the product manifold $\tilde{M} \times \mathbb{R}$ defined by

$$J\left(X, \lambda \frac{d}{dt}\right) = \left(\varphi X - \lambda\xi, \eta\,(X)\,\frac{d}{dt}\right) \tag{2.4}$$

is integrable, where $X$ is tangent to $\tilde{M}$, $t$ the coordinate of $\mathbb{R}$ and $\lambda$ a smooth function on $\tilde{M}\times\mathbb{R}$. The condition for an almost contact structure being normal is equivalent to vanishing of the torsion tensor $[\varphi,\varphi]+2d\eta\otimes\xi$, where $[\varphi,\varphi]$ is the Nijenhuis tensor of $\varphi$. Let $\langle\,,\rangle$ be a compatible Riemannian metric with $(\varphi,\xi,\eta)$, i.e.,

$$\langle X,Y\rangle=\langle\varphi X,\varphi Y\rangle+\eta(X)\eta(Y) \tag{2.5}$$

or equivalently,

$$\Phi(X,Y)\equiv\langle X,\varphi Y\rangle=-\langle\varphi X,Y\rangle \quad\text{and}\quad \langle X,\xi\rangle=\eta(X) \tag{2.6}$$

for all $X,Y\in T\tilde{M}$. Then, $\tilde{M}$ becomes an *almost contact metric manifold* equipped with an *almost contact metric structure* $(\varphi,\xi,\eta,\langle\,,\rangle)$. An almost contact metric structure becomes a *contact metric structure* if $\Phi=d\eta$. A normal contact metric manifold is a *Sasakian manifold.* An almost contact metric manifold is Sasakian if and only if

$$(\tilde{\nabla}_X\varphi)Y=\langle X,Y\rangle\xi-\eta(Y)X,\qquad X,Y\in T\tilde{M}, \tag{2.7}$$

where $\tilde{\nabla}$ is the Levi-Civita connection. Also, a contact metric manifold $\tilde{M}$ is Sasakian if and only if the curvature tensor $\tilde{R}$ satisfies

$$\tilde{R}(X,Y)\xi=\eta(Y)X-\eta(X)Y,\qquad X,Y\in T\tilde{M}. \tag{2.8}$$

In a contact metric manifold $\tilde{M}$, the $(1,1)$-tensor field $h$ defined by $2h=\mathfrak{L}_\xi\varphi$ is symmetric and satisfies

$$h\xi=0,\quad h\varphi+\varphi h=0,\quad \tilde{\nabla}\xi=-\varphi-\varphi h,\quad \operatorname{trace}(h)=\operatorname{trace}(\varphi h)=0. \tag{2.9}$$

If the characteristic vector field $\xi$ satisfies

$$\tilde{R}(X,Y)\xi=\kappa(\eta(Y)X-\eta(X)Y)+\mu(\eta(Y)hX-\eta(X)hY), \tag{2.10}$$

where $(\kappa,\mu)\in\mathbb{R}^2$, then a contact metric manifold $\tilde{M}$ is called a *$(\kappa,\mu)$-space* [**4**]. For a $(\kappa,\mu)$-space one also has $h^2=(\kappa-1)\varphi^2$. Obviously, $\kappa\le 1$, equality holds if and only if the manifold is Sasakian. For $\kappa<1$, the curvature is completely determined for $(\kappa,\mu)$-spaces; in particular, they have constant scalar curvature. If the dimension of a non-Sasakian contact metric manifold $\tilde{M}$ is greater than three and in the equation (2.10) we assume that $\kappa$ and $\mu$ are some smooth functions on $\tilde{M}$ independent of the choice of vector fields $X$ and $Y$, then the functions $\kappa$ and $\mu$ must be constant [**19**]. Three dimensional $(\kappa,\mu)$-spaces are either Sasakian or locally isometric to one of the following Lie groups: $SO(3)$, $SL(2,R)$, $E(2)$, $E(1,1)$ with a left invariant metric. For more details see [**4**], [**18**], [**19**], and [**23**].

For a Sasakian manifold with constant $\varphi$-sectional curvature the formula of Riemann curvature tensor is well-known. Contrary to this for a non-Sasakian $(\kappa,\mu)$-space $\tilde{M}$, its Riemann curvature tensor $\tilde{R}$ is given explicitly in its $(0,4)$-form by

([5],[6])

$$\begin{aligned}(2.11)\qquad \tilde{R}(X,Y,Z,W) = &\left(1-\frac{\mu}{2}\right)\{\langle Y,Z\rangle\langle X,W\rangle - \langle X,Z\rangle\langle Y,W\rangle\}\\
&+ \langle Y,Z\rangle\langle hX,W\rangle - \langle X,Z\rangle\langle hY,W\rangle\\
&- \langle Y,W\rangle\langle hX,Z\rangle + \langle X,W\rangle\langle hY,Z\rangle\\
&+ \frac{1-(\mu/2)}{1-\kappa}\{\langle hY,Z\rangle\langle hX,W\rangle - \langle hX,Z\rangle\langle hY,W\rangle\}\\
&- \frac{\mu}{2}\{\langle \varphi Y,Z\rangle\langle \varphi X,W\rangle - \langle \varphi X,Z\rangle\langle \varphi Y,W\rangle\}\\
&+ \frac{\kappa-(\mu/2)}{1-\kappa}\{\langle \varphi hY,Z\rangle\langle \varphi hX,W\rangle - \langle \varphi hX,Z\rangle\langle \varphi hY,W\rangle\}\\
&+ \mu\langle \varphi X,Y\rangle\langle \varphi Z,W\rangle\\
&+ \eta(X)\eta(W)\{(\kappa-1+(\mu/2))\langle Y,Z\rangle + (\mu-1)\langle hY,Z\rangle\}\\
&- \eta(X)\eta(Z)\{(\kappa-1+(\mu/2))\langle Y,W\rangle + (\mu-1)\langle hY,W\rangle\}\\
&+ \eta(Y)\eta(Z)\{(\kappa-1+(\mu/2))\langle X,W\rangle + (\mu-1)\langle hX,W\rangle\}\\
&- \eta(Y)\eta(W)\{(\kappa-1+(\mu/2))\langle X,Z\rangle + (\mu-1)\langle hX,Z\rangle\}\end{aligned}$$

for all vector fields $X, Y, Z$ and $W$ on $\tilde{M}$. Throughout the paper a $(\kappa,\mu)$-space is assumed to be non-Sasakian unless specifically stated other wise.

## 3. Scalar curvature

Let $M$ be an $n$-dimensional submanifold in an $m$-dimensional manifold $\tilde{M}$ equipped with a Riemannian metric $\langle\, ,\rangle$. The Gauss and Weingarten formulae are given respectively by

$$\tilde{\nabla}_X Y = \nabla_X Y + \sigma(X,Y) \quad \text{and} \quad \tilde{\nabla}_X N = -A_N X + \nabla^{\perp}_X N$$

for all $X, Y \in TM$ and $N \in T^{\perp}M$, where $\tilde{\nabla}$, $\nabla$ and $\nabla^{\perp}$ are the Riemannian, induced Riemannian and induced normal connections in $\tilde{M}$, $M$ and the normal bundle $T^{\perp}M$ of $M$ respectively, and $\sigma$ is the second fundamental form related to the shape operator $A_N$ in the direction of $N$ by $\langle\sigma(X,Y),N\rangle = \langle A_N X, Y\rangle$. The Gauss equation is given by

$$(3.1)\quad \tilde{R}(X,Y,Z,W) = R(X,Y,Z,W) - \langle\sigma(X,W),\sigma(Y,Z)\rangle + \langle\sigma(X,Z),\sigma(Y,W)\rangle$$

for all $X, Y, Z, W \in TM$, where $\tilde{R}$ and $R$ are the curvature tensors of $\tilde{M}$ and $M$ respectively. The mean curvature vector $H$ at a point $p \in M$ is expressed by $nH = \operatorname{trace}(\sigma)$. A submanifold $M$ is *totally geodesic* in $\tilde{M}$ if $\sigma = 0$, and *minimal* if $H = 0$. If $\sigma(X,Y) = \langle X,Y\rangle H$ for all $X, Y \in TM$, then $M$ is *totally umbilical.* We put

$$\sigma^r_{ij} = \langle\sigma(e_i,e_j),e_r\rangle \quad \text{and} \quad \|\sigma\|^2 = \sum_{i,j=1}^{n}\langle\sigma(e_i,e_j),\sigma(e_i,e_j)\rangle,$$

where $\{e_1,...,e_n\}$ is an orthonormal basis of the tangent space $T_pM$ and $e_r$ belongs to an orthonormal basis $\{e_{n+1},...,e_m\}$ of the normal space $T_p^{\perp}M$.

The scalar curvature $\tau$ at $p \in M$ is given by $\tau = \sum_{i<j} K_{ij}$, where $K_{ij}$ is the sectional curvature of the plane section spanned by $e_i$ and $e_j$ at $p \in M$. If $\tilde{M}$ is an

almost contact metric manifold, we put

$$\|Q\|^2 = \sum_{i,j=1}^{n} \langle e_i, Qe_j\rangle^2, \quad Q \in \{P, h^T, (\varphi h)^T\},$$

where $PX$, $h^T X$ and $(\varphi h)^T X$ are the tangential parts of $\varphi X$, $hX$ and $\varphi hX$ respectively for all $X \in TM$. If $\tilde{M}$ is a non-Sasakian $(\kappa,\mu)$-space, we introduce

$$\begin{aligned}(3.2)\qquad \mathfrak{t}_M \equiv\ & \frac{1-(\mu/2)}{1-\kappa}\left\{\left(\operatorname{trace}\left(h^T\right)\right)^2 - \left\|h^T\right\|^2\right\} \\ &+ \frac{\kappa-(\mu/2)}{1-\kappa}\left\{\left(\operatorname{trace}\left((\varphi h)^T\right)\right)^2 - \left\|(\varphi h)^T\right\|^2\right\}.\end{aligned}$$

Moreover, if $\xi \in TM$, then in view of (2.11) and (3.1), we have

$$\begin{aligned}(3.3)\ K_{ij} =\ & \sum_{r=n+1}^{m} \left(\sigma_{ii}^r \sigma_{jj}^r - (\sigma_{ij}^r)^2\right) + 1 - \frac{\mu}{2} - \frac{3\mu}{2}\langle e_i, Pe_j\rangle^2 \\ &+ \{\kappa - 1 + (\mu/2)\}\left\{\eta(e_i)^2 + \eta(e_j)^2\right\} + \left\langle h^T e_i, e_i\right\rangle + \left\langle h^T e_j, e_j\right\rangle \\ &+ \frac{1-(\mu/2)}{1-\kappa}\left\{\left\langle h^T e_i, e_i\right\rangle\left\langle h^T e_j, e_j\right\rangle - \left\langle h^T e_i, e_j\right\rangle^2\right\} \\ &+ \frac{\kappa-(\mu/2)}{1-\kappa}\left\{\left\langle (\varphi h)^T e_i, e_i\right\rangle\left\langle (\varphi h)^T e_j, e_j\right\rangle - \left\langle (\varphi h)^T e_i, e_j\right\rangle^2\right\} \\ &+ (\mu-1)\{\eta(e_j)^2\left\langle h^T e_i, e_i\right\rangle + \eta(e_i)^2\left\langle h^T e_j, e_j\right\rangle - 2\eta(e_i)\eta(e_j)\left\langle h^T e_i, e_j\right\rangle\},\end{aligned}$$

from which it follows that the scalar curvature and the squared mean curvature of $M$ satisfy

$$\begin{aligned}(3.4)\qquad 2\tau =\ & \frac{1}{2}(n-1)\{(n-2)(2-\mu)+4\kappa\} - \frac{3\mu}{2}\|P\|^2 + \mathfrak{t}_M \\ &+ 2(\mu+n-2)\operatorname{trace}\left(h^T\right) + n^2\|H\|^2 - \|\sigma\|^2.\end{aligned}$$

In view of the above equation, we have the following basic inequality involving scalar curvature and squared mean curvature.

PROPOSITION 3.1. *For an $n$-dimensional submanifold $M$ in a non-Sasakian $(\kappa,\mu)$-space such that $\xi \in TM$, we have*

$$\begin{aligned}(3.5)\qquad \tau \leq\ & \frac{1}{2}n^2\|H\|^2 + \frac{1}{4}(n-1)\{(n-2)(2-\mu)+4\kappa\} - \frac{3\mu}{4}\|P\|^2 \\ &+ (\mu+n-2)\operatorname{trace}\left(h^T\right) + \frac{1}{2}\mathfrak{t}_M\end{aligned}$$

*with equality if and only if $M$ is totally geodesic.*

Now, we wish to improve the inequality (3.5). We need the following algebraic lemma.

LEMMA 3.2. *If $a_1, \ldots, a_n$ are $n$ $(n>1)$ real numbers then*

$$(3.6)\qquad \frac{1}{n}\left(\sum_{i=1}^{n} a_i\right)^2 \leq \sum_{i=1}^{n} a_i^2,$$

*with equality holding if and only if $a_1 = a_2 = \cdots = a_n$.*

PROOF. Since

$$0 \leq \sum_{i<j} (a_i - a_j)^2 = (n-1) \sum_i a_i^2 - 2 \sum_{i<j} a_i a_j, \tag{3.7}$$

therefore, we get

$$\left( \sum_{i=1}^{n} a_i \right)^2 = \sum_{i=1}^{n} a_i^2 + 2 \sum_{i<j} a_i a_j \leq n \sum_{i=1}^{n} a_i^2,$$

which gives (3.6). □

THEOREM 3.3. *For an $n$-dimensional submanifold $M$ in a non-Sasakian $(\kappa, \mu)$-space such that $\xi \in TM$, at each point $p \in M$, we have*

$$\begin{aligned} \tau \quad \leq \quad & \frac{n\,(n-1)}{2} \|H\|^2 + \frac{1}{4} (n-1) \{(n-2)(2-\mu) + 4\kappa\} - \frac{3\mu}{4} \|P\|^2 \\ & + \ (\mu + n - 2) \operatorname{trace}\left(h^T\right) + \frac{1}{2} \mathfrak{t}_M \end{aligned} \tag{3.8}$$

*with equality if and only if $p$ is a totally umbilical point.*

PROOF. We choose an orthonormal basis $\{e_1, ..., e_n, e_{n+1}, ..., e_m\}$ at $p$ such that $e_1, ..., e_n$ are tangential to $M$ at $p$ and $e_{n+1}$ is parallel to the mean curvature vector $H(p)$ and $e_1, ..., e_n$ diagonalize the shape operator $A_{n+1}$. Then the shape operators take the forms

$$A_{n+1} = \begin{pmatrix} a_1 & 0 & \cdots & 0 \\ 0 & a_2 & \cdots & 0 \\ \vdots & \vdots & \ddots & \vdots \\ 0 & 0 & \cdots & a_n \end{pmatrix}, \tag{3.9}$$

$$A_r = \left(\sigma_{ij}^r\right), \quad \operatorname{trace} A_r = \sum_{i=1}^{n} \sigma_{ii}^r = 0 \tag{3.10}$$

for all $i, j = 1, ..., n$ and $r = n+2, ..., m$; and from (3.4), we get

$$\begin{aligned} 2\tau = & \frac{1}{2} (n-1) \{(n-2)(2-\mu) + 4\kappa\} - \frac{3\mu}{2} \|P\|^2 + \mathfrak{t}_M \\ & + 2\,(\mu + n - 2) \operatorname{trace}\left(h^T\right) + n^2 \|H\|^2 - \sum_{i=1}^{n} a_i^2 - \sum_{r=n+2}^{m} \sum_{i,j=1}^{n} (\sigma_{ij}^r)^2. \end{aligned} \tag{3.11}$$

Using Lemma 3.2, we get

$$n \|H\|^2 \leq \sum_{i=1}^{n} a_i^2. \tag{3.12}$$

In view of (3.11) and (3.12), we get

$$\begin{aligned} \tau \leq & \frac{n\,(n-1)}{2} \|H\|^2 + \frac{1}{4} (n-1) \{(n-2)(2-\mu) + 4\kappa\} - \frac{3\mu}{4} \|P\|^2 \\ & + \ (\mu + n - 2) \operatorname{trace}\left(h^T\right) + \frac{1}{2} \mathfrak{t}_M - \frac{1}{2} \sum_{r=n+2}^{m} \sum_{i,j=1}^{n} (\sigma_{ij}^r)^2, \end{aligned} \tag{3.13}$$

which implies (3.8). If the equality in (3.8) holds, then from Lemma 3.2 and (3.13) it follows that

$$a_1 = \cdots = a_n \quad \text{and} \quad A_r = 0, \quad r = n+2, ..., m.$$

Therefore, $p$ is a totally umbilical point. The converse is straightforward. □

## 4. Sectional curvature

First, we recall the following algebraic Lemma due to B.-Y. Chen for later use.

LEMMA 4.1. [**9**] *If $a_1, \ldots, a_n, a_{n+1}$ are $n+1$ $(n > 1)$ real numbers such that*

$$\left(\sum_{i=1}^{n} a_i\right)^2 = (n-1)\left(\sum_{i=1}^{n} a_i^2 + a_{n+1}\right),$$

*then $2a_1a_2 \geq a_{n+1}$, with equality holding if and only if $a_1 + a_2 = a_3 = \cdots = a_n$.*

Let $\Pi \subset T_pM$ be a plane section spanned by an orthonormal basis $\{e_1, e_2\}$ for an $n$-dimensional submanifold in a contact metric manifold such that $\xi \in TM$. Then

$$\begin{aligned}
\alpha(\Pi) &\equiv \langle e_1, Pe_2\rangle^2, \\
\beta(\Pi) &\equiv (\eta(e_1))^2 + (\eta(e_2))^2, \\
\gamma(\Pi) &\equiv \eta(e_1)^2 \langle h^T e_2, e_2\rangle + \eta(e_2)^2 \langle h^T e_1, e_1\rangle - 2\eta(e_1)\eta(e_2)\langle h^T e_1, e_2\rangle
\end{aligned}$$

are real numbers, which do not depend on the choice of the orthonormal basis $\{e_1, e_2\}$. Of course, $\alpha(\Pi), \beta(\Pi) \in [0,1]$. We also introduce

$$\text{(4.1)} \qquad \mathfrak{t}_1(\Pi) \equiv \frac{1-(\mu/2)}{1-\kappa} \det(h|_\Pi) + \frac{\kappa - (\mu/2)}{1-\kappa} \det((\varphi h)|_\Pi).$$

In [**9**], B.-Y. Chen established a sharp inequality for submanifolds $M$ in a real space form involving intrinsic invariants, namely the sectional curvature and the scalar curvature of $M$; and the main extrinsic invariant, namely the squared mean curvature as follows:

THEOREM 4.2. [**9**] *Let $M$ be an $n$-dimensional $(n \geq 3)$ submanifold in a real space form $R^m(c)$. Then*

$$\text{(4.2)} \qquad \tau - \inf K(\Pi) \leq \frac{n^2(n-2)}{2(n-1)} \|H\|^2 + \frac{1}{2}(n+1)(n-2)c.$$

*Equality holds iff, with respect to suitable orthonormal frame fields $e_1, \ldots, e_n$, $e_{n+1}$, $\ldots, e_m$, the shape operators $A_r = A_{e_r}$, $r = n+1, \ldots, m$ take the following forms:*

$$\text{(4.3)} \qquad A_{n+1} = \begin{pmatrix} a & 0 & 0 \\ 0 & b & 0 \\ 0 & 0 & (a+b)I_{n-2} \end{pmatrix},$$

$$\text{(4.4)} \qquad A_r = \begin{pmatrix} c_r & d_r & 0 \\ d_r & -c_r & 0 \\ 0 & 0 & 0_{n-2} \end{pmatrix}, \qquad r \in \{n+2, \ldots, m\}.$$

He also established similar inequality in Theorem 2 of [**10**] for a submanifold in a complex space form. Similar inequalities for different kind of submanifolds of Sasakian space forms are established in [**8**], [**14**], [**17**] and [**26**]. Now, we prove the following contact version of Theorem 2 of [**10**].

THEOREM 4.3. *Let $M$ be an $n$-dimensional $(n \geq 3)$ submanifold in an $m$-dimensional non-Sasakian $(\kappa, \mu)$-space $\tilde{M}$ such that $\xi \in TM$. Then, for each point $p \in M$ and each plane section $\Pi \subset T_pM$, we have*

$$
\begin{aligned}
(4.5)\quad \tau - K(\Pi) \leq{} & \frac{n^2(n-2)}{2(n-1)} \|H\|^2 + \frac{1}{4}(n-1)\{(n-2)(2-\mu) + 4\kappa\} + \frac{\mu}{2} - 1 \\
& - \frac{3\mu}{4}\left\{\|P\|^2 - 2\alpha(\Pi)\right\} - \{\kappa - 1 + (\mu/2)\}\beta(\Pi) - (\mu-1)\gamma(\Pi) \\
& - \mathfrak{t}_1(\Pi) - \operatorname{trace}(h|_\Pi) + (\mu + n - 2)\operatorname{trace}\left(h^T\right) + \frac{1}{2}\mathfrak{t}_M.
\end{aligned}
$$

*The equality in* (4.5) *holds at $p \in M$ if and only if there exist an orthonormal basis $\{e_1, \dots, e_n\}$ of $T_pM$ and an orthonormal basis $\{e_{n+1}, \dots, e_m\}$ of $T_p^\perp M$ such that* **(a)** $\Pi = Span\{e_1, e_2\}$ *and* **(b)** *the forms of shape operators $A_r \equiv A_{e_r}$, $r = n+1, \dots, m$, become* (4.3) *and* (4.4).

PROOF. Let $\Pi \subset T_pM$ be a plane section at $p \in M$. We choose an orthonormal basis $\{e_1, e_2, \dots, e_n\}$ for $T_pM$ and $\{e_{n+1}, \dots, e_m\}$ for the normal space $T_p^\perp M$ at $p$ such that $\Pi = \operatorname{Span}\{e_1, e_2\}$ and the mean curvature vector $H$ is in the direction of the normal vector to $e_{n+1}$. We rewrite (3.4) as

$$
(4.6)\quad \left(\sum_{i=1}^{n} \sigma_{ii}^{n+1}\right)^2 = (n-1)\left(\sum_{i=1}^{n}\left(\sigma_{ii}^{n+1}\right)^2 + \sum_{i \neq j}\left(\sigma_{ij}^{n+1}\right)^2 + \sum_{r=n+2}^{m}\sum_{i,j=1}^{n}\left(\sigma_{ij}^{r}\right)^2 + \rho\right),
$$

where

$$
\begin{aligned}
(4.7)\quad \rho ={}& 2\tau - \frac{n^2(n-2)}{n-1}\|H\|^2 + \frac{3\mu}{2}\|P\|^2 - 2(\mu + n - 2)\operatorname{trace}\left(h^T\right) \\
& - \frac{1}{2}(n-1)\{(n-2)(2-\mu) + 4\kappa\} - \mathfrak{t}_M.
\end{aligned}
$$

Now, applying Lemma 4.1 to (4.6), we obtain

$$
(4.8)\quad 2\sigma_{11}^{n+1}\sigma_{22}^{n+1} \geq \rho + \sum_{i \neq j}\left(\sigma_{ij}^{n+1}\right)^2 + \sum_{r=n+2}^{m}\sum_{i,j=1}^{n}\left(\sigma_{ij}^{r}\right)^2.
$$

Using $K(\Pi) = K_{12}$ from equation (3.3) it also follows that

$$
\begin{aligned}
(4.9)\quad K(\Pi) ={}& 1 - \frac{\mu}{2} - \frac{3\mu}{2}\alpha(\Pi) + \{\kappa - 1 + (\mu/2)\}\beta(\Pi) + (\mu-1)\gamma(\Pi) + \mathfrak{t}_1(\Pi) \\
& + \operatorname{trace}(h|_\Pi) + \sigma_{11}^{n+1}\sigma_{22}^{n+1} - (\sigma_{12}^{n+1})^2 + \sum_{r=n+2}^{m}(\sigma_{11}^{r}\sigma_{22}^{r} - (\sigma_{12}^{r})^2).
\end{aligned}
$$

Thus, from (4.8) and (4.9) we have

$$
\begin{aligned}
(4.10)\quad K(\Pi) \geq{}& 1 - \frac{\mu}{2} - \frac{3\mu}{2}\alpha(\Pi) + \{\kappa - 1 + (\mu/2)\}\beta(\Pi) + (\mu-1)\gamma(\Pi) \\
& + \mathfrak{t}_1(\Pi) + \operatorname{trace}(h|_\Pi) + \frac{1}{2}\rho + \sum_{r=n+1}^{m}\sum_{j>2}\{(\sigma_{1j}^{r})^2 + (\sigma_{2j}^{r})^2\} \\
& + \frac{1}{2}\sum_{i \neq j > 2}(\sigma_{ij}^{n+1})^2 + \frac{1}{2}\sum_{r=n+2}^{m}\sum_{i,j>2}(\sigma_{ij}^{r})^2 + \frac{1}{2}\sum_{r=n+2}^{m}(\sigma_{11}^{r} + \sigma_{22}^{r})^2,
\end{aligned}
$$

or

$$K(\Pi) \geq 1-\frac{\mu}{2}-\frac{3\mu}{2}\alpha(\Pi)+\{\kappa-1+(\mu/2)\}\beta(\Pi) + (\mu-1)\gamma(\Pi)+\mathfrak{t}_1(\Pi)+\operatorname{trace}(h|_\Pi)+\frac{1}{2}\rho. \tag{4.11}$$

In view of (4.7) and (4.11), we get (4.5).

If the equality in (4.5) holds, then the inequalities given by (4.8) and (4.10) become equalities. In this case, we have

$$\begin{cases} \sigma_{1j}^{n+1}=0,\ \sigma_{2j}^{n+1}=0,\ \sigma_{ij}^{n+1}=0, & i\neq j>2; \\ \sigma_{1j}^{r}=\sigma_{2j}^{r}=\sigma_{ij}^{r}=0,\ r=n+2,\dots,m; & i,j=3,\dots,n; \\ \sigma_{11}^{n+2}+\sigma_{22}^{n+2}=\cdots=\sigma_{11}^{m}+\sigma_{22}^{m}=0. \end{cases} \tag{4.12}$$

Now, we choose $e_1$ and $e_2$ so that $\sigma_{12}^{n+1}=0$. Applying Lemma 4.1 we also have

$$\sigma_{11}^{n+1}+\sigma_{22}^{n+1}=\sigma_{33}^{n+1}=\cdots=\sigma_{nn}^{n+1}. \tag{4.13}$$

Thus, after choosing a suitable orthonormal basis $\{e_1,\dots,e_m\}$, the shape operator of $M$ becomes of the form given by (4.3) and (4.4). The converse is easy to follow. □

If the structure vector field $\xi$ is tangential to $M$, then we write the orthogonal direct decomposition $TM=\{\xi\}\oplus\{\xi\}^\perp$, where $\{\xi\}$ is the distribution spanned by $\xi$. Next, we prove the following theorem.

THEOREM 4.4. *Let $M$ be an $n$-dimensional $(n\geq 3)$ submanifold isometrically immersed in an $m$-dimensional non-Sasakian $(\kappa,\mu)$-space $\tilde{M}$ such that $\xi\in TM$. Then, for each point $p\in M$ and each plane section $\Pi\subset\{\xi\}_p^\perp$, we have*

$$\begin{aligned}\tau-K(\Pi) \leq{}& \frac{n^2(n-2)}{2(n-1)}\|H\|^2+\frac{1}{4}(n-1)\{(n-2)(2-\mu)-4\kappa\} \\ &-\frac{3\mu}{4}\{\|P\|^2-2\alpha(\Pi)\}+(\mu+n-2)\operatorname{trace}(h^T) \\ &+\frac{\mu}{2}-1-\operatorname{trace}(h|_\Pi)-\mathfrak{t}_1(\Pi)+\frac{1}{2}\mathfrak{t}_M.\end{aligned} \tag{4.14}$$

*The equality in (4.14) holds at $p\in M$ if and only if there exist an orthonormal basis $\{e_1,\dots,e_n\}$ of $T_pM$ and an orthonormal basis $\{e_{n+1},\dots,e_m\}$ of $T_p^\perp M$ such that* **(a)** *$e_n=\xi$,* **(b)** *$\Pi=Span\{e_1,e_2\}$ and* **(c)** *the forms of shape operators $A_r\equiv A_{e_r}$, $r=n+1,\dots,m$, become (4.4) and*

$$A_{n+1}=\begin{pmatrix} a & 0 & 0 \\ 0 & -a & 0 \\ 0 & 0 & 0I_{n-2}\end{pmatrix}. \tag{4.15}$$

PROOF. Let $\Pi\subset\{\xi\}_p^\perp$ be a plane section at $p\in M$. We choose an orthonormal basis $\{e_1,e_2,\dots,e_n=\xi\}$ for $T_pM$ and $\{e_{n+1},\dots,e_m\}$ for the normal space $T_p^\perp M$ at $p$ such that $\Pi=\operatorname{Span}\{e_1,e_2\}$ and the mean curvature vector $H(p)$ is parallel to $e_{n+1}$. Using $\eta(e_1)=0=\eta(e_2)$, we get $\beta(\Pi)=0=\gamma(\Pi)$. Thus, the proof of (4.14) is similar to that of (4.5). In equality case, using (5.3), (4.13) becomes

$$\sigma_{11}^{n+1}+\sigma_{22}^{n+1}=\sigma_{33}^{n+1}=\cdots=\sigma_{nn}^{n+1}=0, \tag{4.16}$$

and thus (4.3) is modified to (4.15). □

## 5. Ricci curvature

For an $n$-dimensional submanifold in an $m$-dimensional Riemannian manifold, let $\{e_1, ..., e_n\}$ be a local orthonormal basis for $T_pM$ and $\{e_{n+1}, ..., e_m\}$ an orthonormal basis for the normal space $T_p^\perp M$. Then, the squared second fundamental form and the squared mean curvature satisfy

$$\begin{aligned}(5.1)\quad \|\sigma\|^2 &= \frac{1}{2}n^2\|H\|^2 + \frac{1}{2}\sum_{r=n+1}^{m}(\sigma_{11}^r - \sigma_{22}^r - \cdots - \sigma_{nn}^r)^2 \\ &\quad + 2\sum_{r=n+1}^{m}\sum_{j=2}^{n}(\sigma_{1j}^r)^2 - 2\sum_{r=n+1}^{m}\sum_{2\le i<j\le n}(\sigma_{ii}^r\sigma_{jj}^r - (\sigma_{ij}^r)^2).\end{aligned}$$

The Ricci curvature $\mathrm{Ric}(e_1)$ is defined by $\mathrm{Ric}(e_1) = \sum_{j=2}^{n} K_{1j}$. Next, the *relative null space* of $M$ at a point $p \in M$ is defined by [**11**]

$$N_p = \{X \in T_pM \mid \sigma(X, Y) = 0 \text{ for all } Y \in T_pM\}.$$

In [**11**], B.-Y. Chen established a sharp relationship between Ricci curvature and the squared mean curvature for a submanifold in a real space form as follows.

THEOREM 5.1. *Let $M$ be an $n$-dimensional submanifold in a real space form $R^m(c)$. Then,*
**1.** *For each unit vector $X \in T_pM$, we have*

$$(5.2)\qquad \|H\|^2 \ge \frac{4}{n^2}\{\mathrm{Ric}(X) - (n-1)c\}.$$

**2.** *If $H(p) = 0$, then a unit vector $X \in T_pM$ satisfies the equality case of* (5.2) *if and only if $X$ lies in the relative null space $N_p$ at $p$.*
**3.** *The equality case of* (5.2) *holds for all unit vectors $X \in T_pM$, if and only if either $p$ is a totally geodesic point or $n = 2$ and $p$ is a totally umbilical point.*

A similar inequality is proved in [**25**] for $C$-totally real submanifolds [**27**] in a $(\kappa, \mu)$-contact space form. In this section, we prove a contact version of Theorem 5.1 for submanifolds tangent to the structure vector field in a non-Sasakian $(\kappa, \mu)$-space. First, we prove a Lemma.

LEMMA 5.2. *Every totally umbilical submanifold $M$ of a contact metric manifold such that $\xi \in TM$ is minimal and consequently totally geodesic.*

PROOF. From (2.9), we have $0 = -\varphi\xi - \varphi h\xi = \tilde{\nabla}_\xi\xi = \nabla_\xi\xi + \sigma(\xi, \xi)$, which gives

$$(5.3)\qquad \nabla_\xi\xi = 0 \quad \text{and} \quad \sigma(\xi, \xi) = 0.$$

If $M$ is totally umbilical then $H = \langle\xi, \xi\rangle H = \sigma(\xi, \xi) = 0$. □

We define the following 1-form by

$$\begin{aligned}(5.4)\quad \mathfrak{t}_2(X) &\equiv \frac{1-(\mu/2)}{1-\kappa}\left\{\left\|h^TX\right\|^2 - \mathrm{trace}(h^T)\langle h^TX, X\rangle\right\} \\ &\quad + \frac{\kappa-(\mu/2)}{1-\kappa}\left\{\left\|(\varphi h)^TX\right\|^2 - \mathrm{trace}((\varphi h)^T)\langle(\varphi h)^TX, X\rangle\right\}.\end{aligned}$$

Now, we prove the main result of this section.

THEOREM 5.3. *Let $M$ be an $n$-dimensional submanifold in an $m$-dimensional non-Sasakian $(\kappa,\mu)$-space such that $\xi \in TM$. Then*
**(i)** *For each unit vector $X \in T_pM$, we have*

$$(5.5)\quad \begin{aligned}\mathrm{Ric}(X) \le\ & \frac{1}{4}n^2\|H\|^2 + \kappa + \frac{1}{2}(n-2)(2-\mu) + \mathrm{trace}\left(h^T\right)\\ & - \frac{3\mu}{2}\|PX\|^2 - \mathfrak{t}_2(X) + (\mu+n-3)\left\langle h^TX, X\right\rangle\\ & + \left\{(n-2)(\kappa - 1 + (\mu/2)) + (\mu-1)\,\mathrm{trace}\left(h^T\right)\right\}\eta(X)^2.\end{aligned}$$

**(ii)** *For $H(p) = 0$, a unit tangent vector $X \in T_pM$ satisfies the equality case of* (5.5) *if and only if $X$ belongs to the relative null space $N_p$.*
**(iii)** *The equality in* (5.5) *holds identically for all unit tangent vectors at $p$ if and only if $p$ is a totally geodesic point.*

PROOF. We choose an orthonormal basis $\{e_1, ..., e_n, e_{n+1}, ..., e_m\}$ such that $e_1, ..., e_n$ are tangential to $M$ at $p$. From (3.4) and (5.1), we get

$$(5.6)\quad \begin{aligned}\frac{1}{4}n^2\|H\|^2 =\ & \tau - \frac{1}{4}(n-1)\{(n-2)(2-\mu)+4\kappa\} + \frac{3\mu}{4}\|P\|^2 - \frac{1}{2}\mathfrak{t}_M\\ & - (\mu+n-2)\,\mathrm{trace}\left(h^T\right) + \frac{1}{4}\sum_{r=n+1}^{m}\left(\sigma_{11}^r - \sigma_{22}^r - \cdots - \sigma_{nn}^r\right)^2\\ & + \sum_{r=n+1}^{m}\sum_{j=2}^{n}(\sigma_{1j}^r)^2 - \sum_{r=n+1}^{m}\sum_{2\le i<j\le n}\left(\sigma_{ii}^r\sigma_{jj}^r - (\sigma_{ij}^r)^2\right).\end{aligned}$$

From (3.3), we also have

$$(5.7)\quad \begin{aligned}\sum_{2\le i<j\le n} K_{ij} =\ & \sum_{r=n+1}^{m}\sum_{2\le i<j\le n}\left(\sigma_{ii}^r\sigma_{jj}^r - (\sigma_{ij}^r)^2\right)\\ & - \frac{3\mu}{4}\left\{\|P\|^2 - 2\|Pe_1\|^2\right\}\\ & - \frac{1}{4}(n-2)((n-3)(\mu-2) - 4\kappa) + \frac{1}{2}\mathfrak{t}_M + \mathfrak{t}_2(e_1)\\ & + (\mu+n-3)\left\{\mathrm{trace}\left(h^T\right) - \left\langle h^Te_1, e_1\right\rangle\right\}\\ & - \{(n-2)(\kappa - (1-(\mu/2))) + (\mu-1)\mathrm{trace}(h^T)\}\eta(e_1)^2.\end{aligned}$$

From (5.6) and (5.7) we get

$$(5.8)\quad \begin{aligned}\mathrm{Ric}(e_1) =\ & \frac{1}{4}n^2\|H\|^2 + \kappa + \frac{1}{2}(n-2)(2-\mu) + \mathrm{trace}\left(h^T\right)\\ & - \frac{3\mu}{2}\|Pe_1\|^2 - \mathfrak{t}_2(e_1) + (\mu+n-3)\left\langle h^Te_1, e_1\right\rangle\\ & + \left\{(n-2)\left(\kappa - 1 + \frac{\mu}{2}\right) + (\mu-1)\,\mathrm{trace}\left(h^T\right)\right\}\eta(e_1)^2\\ & - \frac{1}{4}\sum_{r=n+1}^{m}\left(\sigma_{11}^r - \sigma_{22}^r - \cdots - \sigma_{nn}^r\right)^2 - \sum_{r=n+1}^{m}\sum_{j=2}^{n}(\sigma_{1j}^r)^2.\end{aligned}$$

Since, we can choose $e_1 = X$ as any unit vector in $T_pM$, therefore the above equation implies (5.5).

Assuming $X = e_1$, from (5.8), the equality case of (5.5) is valid if and only if

$$(5.9)\qquad \sigma_{11}^r = \sigma_{22}^r + \cdots + \sigma_{nn}^r \text{ and } \sigma_{12}^r = \cdots = \sigma_{1n}^r = 0, \quad r \in \{n+1, \ldots, m\}.$$

If $H(p) = 0$, (5.9) implies that $X = e_1$ lies in the relative null space $N_p$. Conversely, if $X = e_1$ lies in the relative null space, then (5.9) is true because $H(p) = 0$ is assumed. Thus **(ii)** is proved.

Now we prove **(iii)**. Assuming the equality case of (5.5) for all unit tangent vectors to $M$ at $p$, in view of (5.9), for each $r \in \{n+1, \dots, m\}$, we have

$$\left\{ \begin{array}{l} 2\sigma_{ii}^r = \sigma_{11}^r + \sigma_{22}^r + \cdots + \sigma_{nn}^r, \quad i \in \{1, ..., n\}, \ r \in \{n+1, \dots, m\}, \\ \sigma_{ij}^r = 0, \quad i \neq j, \ r \in \{n+1, \dots, m\}. \end{array} \right. \tag{5.10}$$

Thus, we have two cases, namely either $n = 2$ or $n \neq 2$. In the first case $p$ is a totally umbilical point, while in the second case $p$ is a totally geodesic point. Since $\xi \in TM$, therefore from Lemma 5.2, each totally umbilical point is a totally geodesic point. Thus in both the cases, $p$ is a totally geodesic point. The proof of converse part is straightforward. □

## 6. $k$-Ricci curvature

Let $L$ be a $k$-plane section of $T_pM$ in an $n$-dimensional Riemannian manifold. Choosing any orthonormal basis $\{e_1, ..., e_k\}$ of $L$, the Ricci curvature $\mathrm{Ric}_L$ of $L$ at $e_1$ is given by

$$\mathrm{Ric}_L(e_1) = K_{12} + K_{13} + \cdots + K_{1k}. \tag{6.1}$$

$\mathrm{Ric}_L(e_1)$ is called a *$k$-Ricci curvature*. The scalar curvature $\tau(L)$ of the $k$-plane section $L$ is given by

$$\tau(L) = \sum_{1 \le i < j \le k} K_{ij}. \tag{6.2}$$

The scalar curvature $\tau(p)$ of $M$ at $p$ is nothing but the scalar curvature of the tangent space of $M$ at $p$. And if $L$ is a 2-plane section, $\tau(L)$ is nothing but the sectional curvature $K$ of $L$. For each integer $k$, $2 \le k \le n$, the Riemannian invariant $\theta_k$ on an $n$-dimensional Riemannian manifold $M$ is defined by [**12**]

$$\theta_k(p) = \left(\frac{1}{k-1}\right) \inf_{L,X} \mathrm{Ric}_L(X), \quad p \in M, \tag{6.3}$$

where $L$ runs over all $k$-plane sections in $T_pM$ and $X$ runs over all unit vectors in $L$. We denote by $L_{i_1 \dots i_k}$ the $k$-plane section spanned by $e_{i_1}, ..., e_{i_k}$. From (6.1) and (6.2), it follows that

$$\tau(L_{i_1 \dots i_k}) = \frac{1}{2} \sum_{i \in \{i_1, \dots, i_k\}} \mathrm{Ric}_{L_{i_1 \dots i_k}}(e_i), \tag{6.4}$$

and

$$\tau(p) = \frac{1}{C_{k-2}^{n-2}} \sum_{1 \le i_1 < \cdots < i_k \le n} \tau(L_{i_1 \dots i_k}). \tag{6.5}$$

Combining (6.3), (6.4) and (6.5), one obtains

$$\tau(p) \ge \frac{n(n-1)}{2} \theta_k(p). \tag{6.6}$$

In view of the equations (3.8) and (6.6), we have the following relationship between the $k$-Ricci curvature and the squared mean curvature for submanifolds in a non-Sasakian $(\kappa, \mu)$-space tangent to the structure vector field.

THEOREM 6.1. *Let $M$ be an $n$-dimensional submanifold in an $m$-dimensional non-Sasakian $(\kappa,\mu)$-space such that $\xi \in TM$. Then, for each integer $k$, $2 \leq k \leq n$, and every point $p \in M$, we have*

$$(6.7) \qquad \theta_k \leq \|H\|^2 + \frac{1}{2n}\{(n-2)(2-\mu)+4\kappa\} + \frac{1}{n(n-1)}\left\{\mathrm{t}_M + 2(\mu+n-2)\operatorname{trace}\left(h^T\right) - \frac{3\mu}{2}\|P\|^2\right\}.$$

## 7. Some applications

An $n$-dimensional submanifold $M$ of an almost contact metric manifold is said to be *invariant* or *anti-invariant* [**28**] according as $\varphi(TM) \subseteq TM$ or $\varphi(TM) \subseteq T^{\perp}M$. A submanifold $M$ in a contact manifold is called an *integral submanifold* [**3**] if every tangent vector of $M$ belongs to the contact distribution defined by $\eta = 0$. In this case we have $d\eta(X,Y) = 0$, $X, Y \in TM$. An integral submanifold $M$ in a contact metric manifold is called a *C-totally real submanifold* [**27**]. Thus $C$-totally real submanifolds are natural examples of anti-invariant submanifolds in a contact metric manifold and satisfy $A_\xi = (\varphi h)^T$.

A submanifold $M$ of an almost contact metric manifold with $\xi \in TM$ is called a *semi-invariant submanifold* [**2**] or a *(contact) CR submanifold* [**28**] if there exists two differentiable distributions $\mathcal{D}$ and $\mathcal{D}^{\perp}$ on $M$ such that **(i)** $TM = \mathcal{D} \oplus \mathcal{D}^{\perp} \oplus \{\xi\}$, **(ii)** the distribution $\mathcal{D}$ is invariant by $\varphi$, i.e., $\varphi(\mathcal{D}) = \mathcal{D}$, and **(iii)** the distribution $\mathcal{D}^{\perp}$ is anti-invariant by $\varphi$, i.e., $\varphi(\mathcal{D}^{\perp}) \subseteq T^{\perp}M$. A *CR*-submanifold becomes invariant or anti-invariant according as $\mathcal{D}^{\perp} = \{0\}$ or $\mathcal{D} = \{0\}$. A *proper CR*-submanifold is neither invariant nor anti-invariant. Every hypersurface $M$ tangent to $\xi$ in an almost contact metric manifold is a proper *CR*-submanifold.

For each non zero vector $X \in T_pM$ such that $X$ is not proportional to $\xi_p$, we denote the angle between $\varphi X$ and $T_pM$ by $\theta(X)$. Then $M$ is said to be *slant* ([**7**],[**20**]) if the angle $\theta(X)$ is constant, i.e., it is independent of the choice of $p \in M$ and $X \in T_pM - \{\xi\}$. The angle $\theta$ of a slant immersion is called the *slant angle* of the immersion. Invariant and anti-invariant immersions are slant immersions with the slant angle $\theta = 0$ and $\theta = \pi/2$ respectively. A *proper* slant immersion is neither invariant nor anti-invariant. For more details see also [**24**].

Let $M$ be an $n$-dimensional invariant submanifold of a contact metric manifold $\tilde{M}$ with the structure $(\varphi,\xi,\eta,\langle\,,\rangle)$, then $\xi \in TM$, $\sigma(X,\xi) = 0$ and $M$ is minimal [**3**]. Choosing an orthonormal basis $\{e_i, \varphi e_i, \xi\}$, $i = 1,\ldots,(n-1)/2$, we get

$$(7.1) \qquad \|P\|^2 = n-1, \ \operatorname{trace}\left(h^T\right) = \operatorname{trace}\left((\varphi h)^T\right) = 0, \ \left\|(\varphi h)^T\right\|^2 = \left\|h^T\right\|^2.$$

Consequently,

$$(7.2) \qquad \mathrm{t}_M = -\frac{1+\kappa-\mu}{1-\kappa}\left\|h^T\right\|^2,$$

$$(7.3) \qquad 2\tau = \frac{1}{2}(n-1)\{2n-(n+1)\mu+4(\kappa-1)\} - \frac{1+\kappa-\mu}{1-\kappa}\left\|h^T\right\|^2 - \|\sigma\|^2.$$

In view of the above equation, we can state the following theorem.

THEOREM 7.1. *Let $M$ be an $n$-dimensional invariant submanifold $M$ in a non-Sasakian $(\kappa,\mu)$-space. Then, we have*

$$\tau \leq \frac{1}{4}(n-1)\{2n-(n+1)\mu+4(\kappa-1)\} - \frac{1+\kappa-\mu}{2(1-\kappa)}\left\|h^T\right\|^2 \tag{7.4}$$

*with equality if and only if $M$ is totally geodesic.*

In view of Theorem 4.3, and equations (7.1) and (7.2) we have the following theorem.

THEOREM 7.2. *Let $M$ be an $n$-dimensional invariant submanifold $M$ in a non-Sasakian $(\kappa,\mu)$-space. For each point $p \in M$ and each plane section $\Pi \subset T_pM$, we have*

$$\begin{aligned} \tau - K(\Pi) \leq\; & \frac{1}{4}(n-1)\{2n-(n+1)\mu+4(\kappa-1)\} + \frac{\mu}{2} - 1 \\ & + \frac{3\mu}{2}\alpha(\Pi) - \{\kappa - 1 + (\mu/2)\}\beta(\Pi) - (\mu-1)\gamma(\Pi) \\ & - \mathfrak{t}_1(\Pi) - \operatorname{trace}(h|_\Pi) - \frac{1+\kappa-\mu}{2(1-\kappa)}\left\|h^T\right\|^2. \end{aligned} \tag{7.5}$$

*The equality in (7.5) holds at $p \in M$ if and only if there exist an orthonormal basis $\{e_1,\dots,e_n\}$ of $T_pM$ and an orthonormal basis $\{e_{n+1},\dots,e_m\}$ of $T_p^\perp M$ such that* **(a)** *$\Pi = Span\{e_1,e_2\}$ and* **(b)** *the shape operators $A_r \equiv A_{e_r}$, $r = n+1,\dots,m$, become of the forms given by (4.3) and (4.4).*

Let $\Pi \subset \{\xi\}_p^\perp$ be a plane section at $p \in M$. We can choose unit vectors $e$ and $Pe$ such that $\Pi = \operatorname{Span}\{e, Pe\}$. Thus, we get

$$\alpha(\Pi) = 1,\ \operatorname{trace}(h|_\Pi) = 0 \text{ and } \det(h|_\Pi) = \det((\varphi h)|_\Pi). \tag{7.6}$$

Thus,

$$\mathfrak{t}_1(\Pi) = \frac{1+\kappa-\mu}{(1-\kappa)}\det(h|_\Pi). \tag{7.7}$$

Hence, using (7.1), (7.6), (7.2) and (7.7) in (4.14), we get

THEOREM 7.3. *Let $M$ be an $n$-dimensional invariant submanifold $M$ in a non-Sasakian $(\kappa,\mu)$-space. Then, for each point $p \in M$ and each plane section $\Pi \subset \{\xi\}_p^\perp$, we have*

$$\begin{aligned} \tau - K(\Pi) \leq\; & \frac{1}{4}(n-1)\{2n-(n+1)\mu+4(\kappa-1)\} + (2\mu-1) \\ & - \frac{1+\kappa-\mu}{(1-\kappa)}\left(\det(h|_\Pi) + \frac{1}{2}\left\|h^T\right\|^2\right). \end{aligned} \tag{7.8}$$

*The equality in (7.8) holds at $p \in M$ if and only if there exist an orthonormal basis $\{e_1,\dots,e_n\}$ of $T_pM$ and an orthonormal basis $\{e_{n+1},\dots,e_m\}$ of $T_p^\perp M$ such that* **(a)** *$e_n = \xi$,* **(b)** *$\Pi = Span\{e_1,e_2\}$ and* **(c)** *the shape operators $A_r \equiv A_{e_r}$, $r = n+1,\dots,m$, become of the forms given by (4.15) and (4.4).*

In view of (7.1) and (5.4), we have

$$\mathfrak{t}_2(X) = \frac{1-(\mu/2)}{1-\kappa}\left\|h^T X\right\|^2 + \frac{\kappa-(\mu/2)}{1-\kappa}\left\|(\varphi h)^T X\right\|^2. \tag{7.9}$$

THEOREM 7.4. *Let $M$ be an $n$-dimensional invariant submanifold in a non-Sasakian $(\kappa,\mu)$-space. Then*
**(i)** *For each unit vector $X \in T_pM$, we have*

$$\begin{aligned}(7.10)\qquad \operatorname{Ric}(X) \leq{}& \kappa + \frac{1}{2}(n-2)(2-\mu) - \frac{3\mu}{2}\|\varphi X\|^2 \\ &+ \left\{(n-2)\left(\kappa - 1 + \frac{\mu}{2}\right)\right\}\eta(X)^2 + (\mu+n-3)\langle h^T X, X\rangle \\ &- \frac{1-(\mu/2)}{1-\kappa}\|h^T X\|^2 - \frac{\kappa-(\mu/2)}{1-\kappa}\|(\varphi h)^T X\|^2,\end{aligned}$$

*with equality case of* (7.10) *if and only if $X$ belongs to the relative null space.*
**(ii)** *For a unit vector $X \in \{\xi\}_p^{\perp}$, we have*

$$\begin{aligned}(7.11)\qquad \operatorname{Ric}(X) \leq{}& \kappa + \frac{1}{2}(n-2)(2-\mu) - \frac{3\mu}{2} + (\mu+n-3)\langle h^T X, X\rangle \\ &- \frac{1-(\mu/2)}{1-\kappa}\|h^T X\|^2 - \frac{\kappa-(\mu/2)}{1-\kappa}\|(\varphi h)^T X\|^2,\end{aligned}$$

*with equality case of* (7.11) *if and only if $X$ belongs to the relative null space.*
**(iii)** *The equality in* (7.10) *holds identically for all unit tangent vectors at $p$ if and only if $p$ is a totally geodesic point.*

In view of Theorem 6.1, we have

THEOREM 7.5. *Let $M$ be an $n$-dimensional invariant submanifold in a non-Sasakian $(\kappa,\mu)$-space. Then, for each integer $k$, $2 \leq k \leq n$, and every point $p \in M$, we have*

$$(7.12)\qquad \theta_k \leq \frac{1}{2n}\left(2n - (n+1)\mu + 4(\kappa-1)\right) - \frac{1+\kappa-\mu}{n(n-1)(1-\kappa)}\|h^T\|^2.$$

For a $\theta$-slant submanifold $M$ of an almost contact metric manifold, we have

$$(7.13)\qquad \langle PX, PY\rangle = \cos^2\theta\,\langle \varphi X, \varphi Y\rangle, \quad \langle FX, FY\rangle = \sin^2\theta\,\langle \varphi X, \varphi Y\rangle$$

for all $X, Y \in TM$. Consequently,

$$(7.14)\qquad \|P\|^2 = (n-1)\cos^2\theta.$$

In particular, if $M$ is an anti-invariant submanifold then $P = 0$. If $M$ is a *CR*-submanifold, then

$$(7.15)\qquad \|P\|^2 = 2l = \dim(\mathcal{D}).$$

Using these equations in the results of previous sections, we can find the corresponding results for *CR*, $\theta$-slant and anti-invariant submanifolds of a non-Sasakian $(\kappa,\mu)$-space tangent to the structure vector field $\xi$.

## References

[1] V. I. Arnold, *Contact geometry: the geometrical method of Gibb's thermodynamics*, Proceedings of the Gibbs Symposium, Yale University, (May 15-17, 1989), American Mathematical Society, 1990, 163-179.
[2] A. Bejancu, *Geometry of CR-submanifolds*, D. Reidel Publ. Co. Boston, 1986.
[3] D.E. Blair, *Riemannian geometry of contact and symplectic manifolds*, Progress in Mathematics, 203. Birkhauser Boston, Inc., Boston, MA, 2002.
[4] D. E. Blair, T. Koufogiorgos and B. J. Papantoniou, *Contact metric manifolds satisfying a nullity condition*, Israel J. Math. **91**(1995), no. 1-3, 189-214.

[5] E. Boeckx, *A class of locally $\varphi$-symmetric contact metric spaces*, Arch. Math. **72**(1997), 466-472.

[6] E. Boeckx, *A full classification of contact metric $(\kappa, \mu)$-spaces*, Illinois J. Math. **44**(2000), no. 1, 212-219.

[7] J. L. Cabrerizo, A. Carriazo, L. M. Fernandez and M. Fernandez, *Slant submanifolds in Sasakian manifolds*, Glasgow Math. J. **42**(2000), no. 1, 125-138.

[8] A. Carriazo, *A contact version of B.-Y. Chen's inequality and its applications to slant immersions*, Kyungpook Math. J. **39**(1999), no. 2, 465-476.

[9] B.-Y. Chen, *Some pinching and classification theorems for minimal submanifolds*, Arch. Math. (Basel) **60**(1993), no. 6, 568-578.

[10] B.-Y. Chen, *A general inequality for submanifolds in complex-space-forms and its applications*, Arch. Math. (Basel) **67**(1996), no. 6, 519-528.

[11] B.-Y. Chen, *Relations between Ricci curvature and shape operator for submanifolds with arbitrary codimensions*, Glasgow Math. J. **41**(1999), 33-41.

[12] B.-Y. Chen, *Riemannian submanifolds*, in Handbook of Differential Geometry, eds. F. Dillen and L. Verstraelen, North Holland, Amsterdam, 2000, 187-418.

[13] S. S. Chern, *Pseudo-groups continus infinis*, colloques Internationaux du C. N. R. S., Strassbourg, 119-136.

[14] F. Defever, I. Mihai and L. Verstraelen, *B.-Y. Chen's inequalities for C-totally real submanifolds of Sasakian space forms*, Boll. Un. Mat. Ital. B(7) **11**(1997), no. 2, 365-374.

[15] H. Geiges, *A brief history of contact geometry and topology*, Expo. Math. **19**(2001), no. 1, 25-53.

[16] J. Gray, *Some global properties of contact structures*, Ann. of Math. **69**(1959), 421-450.

[17] Y. H. Kim and D.-S. Kim, *A basic inequality for submanifolds in Sasakian space forms*, Houston J. Math. **25**(1999), no. 2, 247-157.

[18] T. Koufogiorgos, *Contact Riemannian manifolds with constant $\varphi$-sectional curvature*, Tokyo J. Math. **20**(1997), no. 1, 13-22.

[19] T. Koufogiorgos and C. Tsichlias, *On the existence of a new class of contact metric manifolds*, Canad. Math. Bull. **43**(2000), no. 4, 440-447.

[20] A. Lotta, *Slant submanifolds in contact geometry*, Bull. Math. Soc. Roumanie **39**(1996), 183-186.

[21] S. Sasaki, *On differentiable manifolds with certain structures which are closely related to almost contact structure.* I, Tôhoku Math. J. **12**(1960), 459-476.

[22] S. Sasaki, Y. Hatakeyama, *On differentiable manifolds with certain structures which are closely related to almost contact structure.* II, Tôhoku Math. J. (2) **13**(1961), 281-294.

[23] R. Sharma, *Intrinsic and extrinsic characterizations of contact metric manifolds*, Geometry, analysis and applications (Varanasi, 2000), 119-127, World Sci. Publishing, River Edge, NJ, 2001.

[24] M. M. Tripathi and I. Mihai, *Submanifolds of framed metric manifolds and S-manifolds*, Note Mat. 20 (2000/01), no. 2, 135-164 (2002).

[25] M. M. Tripathi and J.-S. Kim, *C-totally real submanifolds in $(\kappa, \mu)$-contact space forms*, Bull. Austral. Math. Soc. **67**(2003), no. 1, 51-65.

[26] M. M. Tripathi, J.-S. Kim and S.-B. Kim, *A note on Chen's basic equality for submanifolds in a Sasakian space form*, Internat. J. Math. Math. Sci. **2003**(2003), no. 11, 711-716.

[27] S. Yamaguchi, M. Kon and T. Ikawa, *C-totally real submanifolds*, J. Differential Geometry **11**(1976), no. 1, 59-64.

[28] K. Yano and M. Kon, *Structures on manifolds*, Series in Pure Mathematics, **3**. World Scientific. 1984.

DEPARTMENT OF MATHEMATICS AND ASTRONOMY, LUCKNOW UNIVERSITY, LUCKNOW-226 007, INDIA

*E-mail address*: mmt66@satyam.net.in

# Titles in This Series

311 **Clifford J. Earle, William J. Harvey, and Sevín Recillas-Pishmish, Editors,** Complex manifolds and hyperbolic geometry, 2002

310 **Alejandro Adem, Jack Morava, and Yongbin Ruan, Editors,** Orbifolds in mathematics and physics, 2002

309 **Martin Guest, Reiko Miyaoka, and Yoshihiro Ohnita, Editors,** Integrable systems, topology, and physics, 2002

308 **Martin Guest, Reiko Miyaoka, and Yoshihiro Ohnita, Editors,** Differentiable geometry and integrable systems, 2002

307 **Ricardo Weder, Pavel Exner, and Benoit Grébert, Editors,** Mathematical results in quantum mechanics, 2002

306 **Xiaobing Feng and Tim P. Schulze, Editors,** Recent advances in numerical methods for partial differential equations and applications, 2002

305 **Samuel J. Lomonaco, Jr. and Howard E. Brandt, Editors,** Quantum computation and information, 2002

304 **Jorge Alberto Calvo, Kenneth C. Millett, and Eric J. Rawdon, Editors,** Physical knots: Knotting, linking, and folding geometric objects in $\mathbb{R}^3$, 2002

303 **William Cherry and Chung-Chun Yang, Editors,** Value distribution theory and complex dynamics, 2002

302 **Yi Zhang, Editor,** Logic and algebra, 2002

301 **Jerry Bona, Roy Choudhury, and David Kaup, Editors,** The legacy of the inverse scattering transform in applied mathematics, 2002

300 **Sergei Vostokov and Yuri Zarhin, Editors,** Algebraic number theory and algebraic geometry: Papers dedicated to A. N. Parshin on the occasion of his sixtieth birthday, 2002

299 **George Kamberov, Peter Norman, Franz Pedit, and Ulrich Pinkall,** Quaternions, spinors, and surfaces, 2002

298 **Robert Gilman, Alexei G. Myasnikov, and Vladimir Shpilrain, Editors,** Computational and statistical group theory, 2002

297 **Stephen Berman, Paul Fendley, Yi-Zhi Huang, Kailash Misra, and Brian Parshall, Editors,** Recent developments in infinite-dimensional Lie algebras and conformal field theory, 2002

296 **Sean Cleary, Robert Gilman, Alexei G. Myasnikov, and Vladimir Shpilrain, Editors,** Combinatorial and geometric group theory, 2002

295 **Zhangxin Chen and Richard E. Ewing, Editors,** Fluid flow and transport in porous media: Mathematical and numerical treatment, 2002

294 **Robert Coquereaux, Ariel García, and Roberto Trinchero, Editors,** Quantum symmetries in theoretical physics and mathematics, 2002

293 **Donald M. Davis, Jack Morava, Goro Nishida, W. Stephen Wilson, and Nobuaki Yagita, Editors,** Recent progress in homotopy theory, 2002

292 **A. Chenciner, R. Cushman, C. Robinson, and Z. Xia, Editors,** Celestial Mechanics, 2002

291 **Bruce C. Berndt and Ken Ono, Editors,** $q$-series with applications to combinatorics, number theory, and physics, 2001

290 **Michel L. Lapidus and Machiel van Frankenhuysen, Editors,** Dynamical, spectral, and arithmetic zeta functions, 2001